Lecture Notes in Physics

Volume 947

The Lecture Notes in Physics

The series Lecture Notes in Physics (LNP), founded in 1969, reports new developments in physics research and teaching-quickly and informally, but with a high quality and the explicit aim to summarize and communicate current knowledge in an accessible way. Books published in this series are conceived as bridging material between advanced graduate textbooks and the forefront of research and to serve three purposes:

- to be a compact and modern up-to-date source of reference on a well-defined topic
- to serve as an accessible introduction to the field to postgraduate students and nonspecialist researchers from related areas
- to be a source of advanced teaching material for specialized seminars, courses and schools

Both monographs and multi-author volumes will be considered for publication. Edited volumes should, however, consist of a very limited number of contributions only. Proceedings will not be considered for LNP.

Volumes published in LNP are disseminated both in print and in electronic formats, the electronic archive being available at springerlink.com. The series content is indexed, abstracted and referenced by many abstracting and information services, bibliographic networks, subscription agencies, library networks, and consortia.

Proposals should be sent to a member of the Editorial Board, or directly to the managing editor at Springer:

Christian Caron
Springer Heidelberg
Physics Editorial Department I
Tiergartenstrasse 17
69121 Heidelberg/Germany
christian.caron@springer.com

More information about this series at http://www.springer.com/series/5304

Samoil Bilenky

Introduction to the Physics of Massive and Mixed Neutrinos

Second Edition

 Springer

Samoil Bilenky
TRIUMF
Vancouver, BC
Canada

ISSN 0075-8450   ISSN 1616-6361   (electronic)
Lecture Notes in Physics
ISBN 978-3-319-74801-6   ISBN 978-3-319-74802-3   (eBook)
https://doi.org/10.1007/978-3-319-74802-3

Library of Congress Control Number: 2018934708

This Springer imprint is published by the registered company Springer International Publishing AG part
of Springer Nature.
The registered company address is: Gewerbestrasse 11, 6330 Cham, Switzerland

*Dedicated to the memory of the great
neutrino physicist Bruno Pontecorvo*

Preface to the Second Edition

The discovery of neutrino oscillations and the proof that neutrino masses differ from zero and are much smaller than the masses of leptons and quarks is the first (and up to now the only) signature of physics beyond the Standard Model revealed in particle physics experiments. Further investigations of neutrino oscillations have continued in many experiments. The next generation of these investigations, and other experiments now in preparation, will include the search for neutrinoless double β-decay.

After the first edition of this book was published it was found that the mixing angle θ_{13} was different from zero and was relatively large (about $8°$). This finding opened the way to the investigation of such fundamental problems of neutrino mixing as the characteristics of the neutrino mass spectrum (normal or inverted ordering) and charge conjugation parity (CP) violation in the lepton sector.

The important problem of sterile neutrinos will be solved soon; many reactor, accelerator, and source short-baseline experiments investigating evidence in favor of the transition of flavor neutrinos into sterile states, found in Los Alamos Neutrino Detector (LSND) and other experiments, are now ongoing (or in preparation). After the discovery of the Higgs boson at the Large Hadron Collider (LHC) and the establishment of the Standard Model as *the theory of elementary particles in the electroweak region*, belief in the minimal effective Lagrangian (Weinberg) mechanism of the generation of Majorana neutrino masses increased significantly. The search for neutrinoless double β-decay is now a top priority. In the second edition of the book I have presented data on the latest neutrino oscillation and other neutrino data. I have kept the same style of the detailed derivation of major results, but I have also tried to improve and simplify different derivations.

Vancouver, BC, Canada
December 2017

Samoil Bilenky

Preface to the First Edition

For many years the neutrino was considered a massless particle. The theory of a two-component neutrino, which played a crucial role in the creation of the theory of weak interaction, is based on the assumption that the mass of a neutrino is equal to zero. Now we know that neutrinos have nonzero, small masses. In numerous experiments with solar, atmospheric, reactor, and accelerator neutrinos, a new phenomenon, neutrino oscillations, was observed. Neutrino oscillations (periodic transitions between different flavor neutrinos ν_e, ν_μ, ν_τ) are possible only if neutrino mass-squared differences are different from zero and the states of flavor neutrinos are "mixed".

The discovery of neutrino oscillations opened a new era in neutrino physics: an era of investigation of neutrino masses, mixing, magnetic moments, and other neutrino properties. After the establishment of the Standard Model of electroweak interaction at the end of the 1970s, the discovery of neutrino masses was the most important discovery in particle physics. The small neutrino masses cannot be explained by the standard Higgs mechanism of mass generation. For their explanation a new mechanism is needed. Thus, in particle physics, small neutrino mass is the first signature of a new physics beyond the Standard Model.

It took many years of heroic efforts by many physicists to discover neutrino oscillations. After the first period of the discovery and investigation of neutrino oscillations, many challenging problems remained unsolved. One of the most important is the problem of the nature of neutrinos with definite masses. Are they Dirac neutrinos possessing a conserved lepton number which distinguishes neutrinos and antineutrinos, or are they Majorana neutrinos with identical neutrinos and antineutrinos? Many new experiments are now ongoing or in preparation. There is no doubt that exciting results lie ahead.

This book is intended as an introduction to the physics of massive and mixed neutrinos. It is based on numerous lectures that I have given at different universities and schools. I have tried to explain how many of the main results were derived. The details of the derivation can be easily followed by the reader. I hope that this book will be useful for physicists who are working in neutrino physics, for students, for

young physicists who plan to enter this exciting field, and for many scientists who
are interested in the history of neutrino physics and its present status.

Dubna, Russia/Vancouver, BC, Canada   Samoil Bilenky
December 2016

Acknowledgments

I am very happy to express my deep gratitude to my colleagues and collaborators
W. Alberico, J. Bernabeu, A. Bottino, A. Faessler, F. von Feilitzsch, C. Giunti,
A. Grifols, W. Grimus, J. Hosek, C.W. Kim, M. Lindner, M. Mateev, T. Ohlsson,
S. Pascoli, S. Petcov, W. Potzel, F. Simkovic, and T. Schwetz for numerous fruitful
discussions of different aspects of neutrino physics. I am thankful to the theoretical
department of TRIUMF for hospitality.

Contents

Chapter 1
Introduction

The idea of neutrino was put forward by W. Pauli in 1930. This was a dramatic time in physics. After it was established in the Ellis and Wooster experiment (1927) that the average energy of the electrons produced in the β-decay is significantly smaller than the total released energy, only the existence of a neutral particle with a small mass and a large penetration length which is emitted in the β-decay together with the electron, could save the fundamental law of the conservation of energy.

At the time when the neutrino hypothesis was proposed the only known elementary particles were electron and proton. In this sense neutrino (more exactly electron neutrino) is one of the "oldest" elementary particles. However, the existence of the neutrino was established only in the middle of the fifties when neutron, muon, pions, kaons, Λ and other particles were discovered.

We know at present that the 12 fundamental fermions exist in nature: six quarks u, d, c, s, t, b, three charged leptons e, μ, τ and three neutrinos ν_e, ν_μ, ν_τ. They are grouped in the three families, which differ in masses of particles but have universal electroweak interaction with photons and vector $W^\pm$ and Z bosons. In the Lagrangian of the electroweak interaction, neutrinos enter on the same footings as the quarks and charged leptons. In spite of this similarity of the electroweak interaction *neutrinos are special particles*. There are three basic differences between neutrinos and other fundamental fermions.

1. Neutrinos are the only fundamental particles which have equal to zero electric charges. As a result, they have only Charged Current and Neutral Current electroweak interaction. At all available energies cross sections of the interaction of neutrinos with matter is many order of magnitude smaller than the cross section of the electromagnetic interaction of leptons with matter. This is connected with the fact that neutrinos interact with matter via the exchange of the heavy virtual $W^\pm$ and Z^0 bosons whereas charged leptons interact with matter via exchange of virtual γ-quanta.

© Springer International Publishing AG, part of Springer Nature 2018
S. Bilenky, *Introduction to the Physics of Massive and Mixed Neutrinos*,
Lecture Notes in Physics 947, https://doi.org/10.1007/978-3-319-74802-3_1

2. Neutrino masses are many orders of magnitude smaller than the masses of leptons and quarks. It is natural to assume that neutrino masses, unlike quark and lepton masses, are of a non Standard Model origin.
3. Neutrinos are the only fundamental fermions which can be truly neutral Majorana particles. The Majorana nature of neutrinos could be a clue to the solution of the problem of the origin of small neutrino masses.

Because of the smallness of the neutrino cross section, special methods of the detection of neutrino processes must be developed. However, after such methods were developed the observation of neutrino processes allows us to obtain unique information. For example,

1. the measurement of the cross sections of the deep inelastic processes $\nu_\mu(\bar{\nu}_\mu) + N \rightarrow \mu^-(\mu^+) + X$ in the 80s and 90s led to the establishment of the quark structure of the nucleon,
2. the detection of the solar neutrinos allowed us to establish the thermonuclear origin of solar energy and to obtain information about the central invisible part of the sun where the energy is produced,
3. the detection of neutrinos from Supernova SN1987A allowed us to obtain the first information about a mechanism of the gravitational collapse etc.

The measurement of the absolute values of small neutrino masses is a difficult and challenging problem. This problem is still not solved. The observation of neutrino oscillations led to the determination of two neutrino mass-squared differences. From neutrino oscillation data and data of the β-decay experiments on the direct measurement of the neutrino mass it is possible to conclude that

- Neutrino masses are different from zero.
- Neutrino masses are smaller than ~ 2.2 eV, i.e. many order of magnitude smaller than masses of leptons and quarks.

The unified theory of the weak and electromagnetic interactions, the Standard Model, perfectly describes existing experimental data. Discovery of the Higgs boson in the LHC experiments at CERN was impressive confirmation of the Standard Model. However, existence of dark matter and dark energy tell us that a more general, beyond the SM theory must exist. There were many experiments on the search for effects of such a theory. This search continues now in the LHC experiments. *The first evidence for a new, beyond the Standard Model theory was apparently obtained in neutrino experiments in which neutrino oscillations, driven by small neutrino masses, were discovered.*

Discovery of neutrino oscillations signifies not only that neutrino mass-squared differences are different from zero but also that the fields of flavor neutrinos ν_e, ν_μ, ν_τ are unitary combinations ("mixtures") of fields of neutrinos with definite masses. The 3×3 neutrino mixing matrix is characterized by three angles and one *CP* phase.

The phenomenon of the neutrino mixing is similar to the well established quark mixing. However, the quark mixing angles are small and satisfy a hierarchy. The neutrino mixing angles are completely different: two angles are large and one is small. This is also an indication that neutrino and quark mixing have different origin.

The most common explanation of the smallness of the neutrino mass is based on the assumption that the total lepton number is violated at a very large scale (about 10^{15} GeV). If this assumption is correct, neutrinos with definite masses are *Majorana particles*. In this case neutrinos and antineutrinos are identical. The leptons (and quarks) are Dirac particles. This means that leptons and antileptons (quarks and antiquarks) are different particles: they have the same masses but their electric charges differ in sign. Observation of the neutrinoless double β-decay of some even-even nuclei $(A, Z) \rightarrow (A, Z + 2) + e^- + e^-$ would be a proof that neutrinos are Majorana particle.

The first rather long period of the investigation of the problem of neutrino masses, mixing and oscillations is finished. Neutrino oscillations were discovered. Five neutrino oscillation parameters (two mass-squared differences and three mixing angles) are determined with accuracies from about 3% to about 10%. Strong bounds on the half-lives of the neutrinoless double β-decay of different nuclei were obtained. Now a new era of investigation of the problem of neutrino masses, mixing and nature started. The main problems which will be addressed are the following

1. Are neutrinos with definite masses Majorana or Dirac particles?
2. Is the *CP* invariance violated in the lepton sector? What is the value of the *CP* phase?
3. What is the character of the neutrino mass spectrum? Is it normal with smaller mass-squared difference between lighter neutrinos or inverted with smaller mass-squared difference between heavier neutrinos?
4. What are the absolute values of the neutrino masses?
5. Is the number of massive neutrinos equal to the number of the flavor neutrinos (three) or larger than three? In other words are there transitions of flavor neutrinos into sterile neutrino states?
6. ...

Neutrino experiments of the next generation with atmospheric and solar neutrinos and neutrinos from accelerators and reactors have started or are in preparation. New large detectors of atmospheric, solar and supernova neutrinos are under development. Technologies for new neutrino facilities are being developed. *There is no doubt that a new exciting era of neutrino physics is ahead.*

In this book, I intend to give an introduction to the physics of massive and mixed neutrinos. I start with a brief review of the development of the phenomenological $V - A$ current × current theory of the weak interaction starting from Pauli's hypothesis of the neutrino and Fermi's theory of the β-decay. In the next chapter we will consider the Standard Model. The Higgs mechanism of the generation of masses or quarks and leptons is discussed in some details. Then we will consider possible mass terms for neutrinos. We will describe in detail the procedure of the diagonalization of the neutrino mass terms. Next chapter is devoted to the detailed consideration of the general properties of the neutrino mixing matrix. We will consider the standard parametrization of the 3×3 mixing matrix. Then we will present the theory of neutrino oscillations in vacuum. The three-neutrino oscillations are considered in detail. In the next chapter flavor neutrino transitions in matter are discussed. We will derive Wolfenstein equation for neutrino evolution in matter and consider the adiabatic solution of this equation and the resonance MSW effect. The next chapter is dedicated to the neutrinoless double β-decay of even-even nuclei. Basic elements of the theory of the decay are presented. Then we briefly discuss neutrino oscillation experiments and the data obtained. In the next chapter we discuss β-decay experiments on the measurement of the neutrino mass. In the last chapter we will consider neutrino in cosmology.

It is impossible in a book to give a full list of references. Taking into account a limited number of pages available for the book, I give here references mainly to some pioneer neutrino papers, latest experimental papers, relevant reviews and books. I would recommend the web site by C. Giunti and M. Laveder (the Neutrino Unbound, http://www.nu.to.infn.it/) where it is possible to find many references to the neutrino literature (theory and experiment).

In conclusion I would like to name some principal neutrino events.

1930 In a letter addressed to the participants of the nuclear conference in Tuebingen W. Pauli suggested that there exists a new neutral, spin 1/2, weakly interacting particle which is produced together with the electron in the β-decay of nuclei. Pauli called the new particle "neutron". Later E. Fermi and E. Amaldi proposed the name *neutrino* for this particle.

1933–1934 E. Fermi proposed the first theory of the β-decay. Fermi considered the β-decay as four-fermion process in which a neutron is transformed into a proton with the emission of a electron-neutrino pair. He proposed the following Hamiltonian of the β-decay

$$\mathcal{H}_I = G_F \, \bar{p}\gamma^\alpha n \, \bar{e}\gamma_\alpha \nu + \text{h.c.} \tag{1.1}$$

where G_F is the Fermi constant.

1934 Bethe and Peierls estimated the cross section of the interaction of neutrino with nuclei. The estimated value of the cross section was so small that for many years the neutrino was considered as an "undetectable particle".

1946 B. Pontecorvo proposed the first radiochemical method of neutrino detection which was based on the observation of

$$\nu + {}^{37}\text{Cl} \rightarrow e^- + {}^{37}\text{Ar}$$

and other processes. As possible intensive sources of neutrinos Pontecorvo suggested the sun, reactors and radioactive materials which can be produced in reactors.

1947–1948 Pontecorvo, Puppi, Klein, Tiomno and Wheeler advanced the idea of the $\mu - e$ universality of the weak interaction.

1956–1959 In the Reines and Cowen experiments the (anti)neutrino was discovered. In these experiments antineutrinos from the Savannah River reactor (USA) were detected in a large scintillator counter via the observation of the reaction

$$\bar{\nu} + p \to e^+ + n.$$

1957 In the Davis experiment with antineutrinos from a reactor no production of ^{37}Ar in the process

$$\bar{\nu} + {}^{37}Cl \to e^- + {}^{37}Ar$$

was observed. This was the first indication in favor of the existence of the conserved lepton number.

1957 In the Wu et al. experiment with polarized ^{60}Co a large effect of the parity violation in the β-decay was discovered.

1957 Lee and Yang were awarded the Nobel Prize "for their penetrating investigation of the so-called parity laws which has led to important discoveries regarding the elementary particles".

1957 Landau, Lee and Yang and Salam proposed the theory of the massless two-component neutrino. According to this theory the neutrino is the left-handed (or right-handed) particle and the antineutrino is the right-handed (or left-handed) particle.

1958 The helicity of the neutrino was determined in the Goldhaber, Grodzins and Sunyar experiment from the measurement of the circular polarization of γ-quanta in the chain of the reactions

$$e^- + Eu \to \nu + Sm^*$$
$$\downarrow$$
$$Sm + \gamma$$

It was established that the neutrino is the left-handed particle.

1958 Feynman and Gell-Mann, Marshak and Sudarshan proposed the current × current theory of the weak interaction. The Hamiltonian of this theory had the form

$$\mathscr{H}_I = \frac{G_F}{\sqrt{2}} j^\alpha j_\alpha^\dagger. \tag{1.2}$$

Here

$$j_\alpha = 2\left(\bar{p}_L \gamma_\alpha n_L + \bar{\nu}_L \gamma_\alpha e_L + \bar{\nu}_L \gamma_\alpha \mu_L\right) \tag{1.3}$$

is the $\mu - e$ *universal* weak charged current (CC).

1958 Pontecorvo suggested that neutrinos have small masses, the total lepton number is violated and neutrino oscillations similar to $K^0 \leftrightarrows \bar{K}^0$ oscillations could take place. He considered effects of neutrino oscillations in experiments with reactor antineutrinos.

1962 In the Brookhaven neutrino experiment, the first experiment with accelerator high-energy neutrinos, it was established that neutrino which take part in CC weak interaction together with electron and neutrino which take part in CC weak interaction together with muon are *different particles*. They were called electron neutrino ν_e and muon neutrino ν_μ. In order to explain the data of the Brookhaven and other experiments it was necessary to introduce two separately conserved lepton numbers: the electron L_e and the muon L_μ. The weak charged current took the form

$$ j_\alpha = 2 \left(\bar{p}_L \gamma_\alpha n_L + \bar{\nu}_{eL} \gamma_\alpha e_L + \bar{\nu}_{\mu L} \gamma_\alpha \mu_L \right). \tag{1.4} $$

1962 Maki, Nakagawa and Sakata assumed that neutrinos have small masses and the fields of electron and muon neutrinos are connected with the fields of massive neutrinos ν_1 and ν_2 by the mixing relations

$$ \nu_{eL} = \cos\theta \; \nu_{1L} + \sin\theta \; \nu_{2L} $$
$$ \nu_{\mu L} = -\sin\theta \; \nu_{1L} + \cos\theta \; \nu_{2L}, \tag{1.5} $$

where θ is a mixing angle.

1962 Strange particles were included into the weak charged current by N. Cabibbo.

1965 First detection of the atmospheric neutrinos (S. Miyake et al., Kolar gold mine in India, F. Reines et al., South African gold mine).

1967 Glashow (1961), S. Weinberg and A. Salam proposed the unified theory of the weak and electromagnetic (electroweak) interactions (The Standard Model).

1970 In the pioneer experiment by Davis et al. solar neutrinos were detected. In this experiment solar ν_e's were detected by the Pontecorvo radiochemical *Cl–Ar* method via the observation of the reaction

$$ \nu_e + {}^{37}\text{Cl} \rightarrow e^- + {}^{37}\text{Ar}. $$

The threshold of this reaction is 0.81 MeV. Only high-energy solar neutrinos, mainly from the decay ${}^8\text{B} \rightarrow {}^8\text{Be} + e^+ + \nu_e$, were detected in the Davis experiment. The observed rate was two to three times smaller than the rate predicted by the Standard Solar Model. This discrepancy was called *the solar neutrino problem*.

1973 In the experiment with high energy accelerator neutrinos at CERN a new class of the weak interaction, the so called neutral currents (NC), was discovered. In addition to CC deep inelastic processes

$$ \nu_\mu (\bar{\nu}_\mu) + N \rightarrow \mu^- (\mu^+) + X \tag{1.6} $$

new NC processes

$$\nu_\mu(\bar{\nu}_\mu) + N \to \nu_\mu(\bar{\nu}_\mu) + X$$

were observed. The discovery of the neutral currents was the important confirmation of the Standard Model.

80s In CDHS and CHARM experiments on the study of the deep inelastic scattering of neutrinos and antineutrinos on nucleons (CERN) the quark structure of nucleons was established.

1983 In experiments on $p - \bar{p}$ collider at CERN $W^\pm$ and Z^0 bosons were discovered. In 1984 the Nobel Prize was awarded to C. Rubbia and S. van der Meer for this discovery.

1986 Wolfenstein (1978) and Mikheev and Smirnov showed that for solar neutrinos, which were born in the central region of the sun and passed a large amount of matter on the way to the earth, due to the neutrino mixing and coherent scattering of electron neutrinos on electrons resonance matter effects could take place.

1987 Neutrinos from the supernova SN1987A in the Large Magellanic Cloud were detected in the Kamiokande, IMB and Baksan detectors.

1988 Solar neutrinos were detected in the Kamiokande experiment. In this experiment solar neutrinos were detected through the observation of the recoil electrons in the elastic process

$$\nu + e \to \nu + e \tag{1.7}$$

The solar neutrino problem was confirmed.

1988 L. Lederman, M. Schwartz and J. Steinberger were awarded the Nobel Prize for "the discovery of the muon neutrino leading to classification of particles in families".

1991 In the GALLEX and SAGE experiments solar ν_e's were detected by the radiochemical method via the observation of the process

$$\nu_e + {}^{71}Ga \to e^- + {}^{71}Ge. \tag{1.8}$$

Because of the law threshold (0.23 MeV), in the GALLEX and SAGE experiments neutrinos from all reactions of the pp-cycle, including the main reaction $p + p \to d + e^+ + \nu_e$, were detected. The flux of solar neutrinos measured in these experiments was about two times smaller that the flux predicted by the Standard Solar Model.

90s It was proven in experiments on the measurement of the width of the decay $Z \to \nu + \bar{\nu}$ at LEP (CERN) that only three flavor neutrinos (ν_e, ν_μ, ν_τ) exist in Nature.

1995 The Nobel Prize was awarded to F. Reines "for the detection of the neutrino".

1998 In the Super-Kamiokande experiment a large azimuth angle asymmetry of high-energy atmospheric muon neutrino events was observed. This was the first model-independent evidence for neutrino oscillations driven by a neutrino mass-squared difference $\Delta m_{23}^2 \simeq 2.5 \times 10^{-3}$ eV^2.

2000 In the experiment DONUT at Fermilab the first direct evidence of the existence of the third neutrino ν_τ was obtained.

2002 In the solar neutrino experiment SNO solar neutrinos were detected through the observation of the CC reaction

$$\nu_e + d \rightarrow e^- + p + p \tag{1.9}$$

and the NC reaction

$$\nu + d \rightarrow \nu + n + p. \tag{1.10}$$

This experiment solved the solar neutrino problem in a model-independent way: it was proven that solar ν_e's on the way from the central part of the sun to the earth are transformed into other types of neutrinos.

2002 R. Davis and M. Koshiba were awarded the Nobel Prize for "pioneering contributions to astrophysics, in particular for the detection of cosmic neutrinos".

2002 In the reactor experiment KamLAND, $\bar{\nu}_e$'s from 57 reactors in Japan were detected through the observation of the reaction

$$\bar{\nu}_e + p \rightarrow e^+ + n \tag{1.11}$$

The average distance between reactors and detector was about 180 km. In this experiment a model-independent evidence for neutrino oscillations driven by a solar neutrino mass-squared difference $\Delta m_{12}^2 \simeq 8 \times 10^{-5}$ eV^2 was obtained.

2004 In the long-baseline accelerator neutrino experiment K2K the evidence for neutrino oscillations obtained in the atmospheric neutrino experiment Super-Kamiokande was confirmed. In this experiment, neutrinos from the accelerator at KEK were detected by the Super-Kamiokande detector at a distance of about 250 km.

2006 In the long-baseline neutrino experiment MINOS, the Super-Kamiokande atmospheric neutrino evidence for neutrino oscillations was also confirmed. In the MINOS experiment, neutrinos from the accelerator at Fermilab were detected by the detector in the Soudan mine at a distance of 735 km.

2007 A new solar neutrino experiment BOREXino started. In this experiment monochromatic ^7Be solar neutrinos with the energy 0.86 MeV were detected in real time.

2010 The first off-axis long baseline accelerator neutrino experiment T2K in Japan with near (280 m) and far (295 km) detectors started. First indications in favor of non zero value of the oscillation parameter $\sin^2 2\theta_{13}$ was obtained.

2012 The small parameter $\sin^2 2\theta_{13}$ was measured in the Daya Bay, RENO and Double Chooz reactor neutrino experiments. This measurements open a way for the determination of the neutrino mass spectrum and CP angle δ.

2015 The Nobel prize was awarded to T. Kajita and A. McDonald "for the discovery of neutrino oscillations, which shows that neutrinos have mass".

2016 First data of the new off-axis long baseline accelerator neutrino experiment NOvA at Fermilab were obtained. There are two detectors in the experiment: near detector (1 km from the target) and far detector (810 km from the target).

Chapter 2
Weak Interaction Before the Standard Model

All existing weak interaction data are in a perfect agreement with the Standard Model. Before this theory was created, there was a long phenomenological period of the development of the theory of the weak interaction. In this introductory chapter we will briefly consider this period.

2.1 Pauli Hypothesis of Neutrino

The only weak process which was known in the 20s and 30s was the β-decay of nuclei. In 1914 Chadwick discovered that the energy spectrum of electrons from β-decay is continuous. If β-decay is a process of the transition of a nucleus (A,Z) into a nucleus (A,Z+1) and the electron (as it was believed at that time), from conservation of energy and momentum follows that the electron must have a fixed kinetic energy approximately equal to $Q \simeq m_{A,Z} - m_{A,Z+1} - m_e$ where $m_{A,Z}$ ($m_{A,Z+1}$) is the mass of the initial (final) nucleus and m_e is the mass of the electron.

For many years continuous β spectra were interpreted as the result of the loss of energy of electrons in the target. However, in 1927 Ellis and Wooster performed a crucial calorimetric β-decay experiment. They measured the total energy released in a RaE (^{210}Bi) source which was put inside of a calorimeter. For the β-decay of ^{210}Bi the total energy release is $Q = 1.05$ MeV. In the Ellis and Wooster experiment it was found that the average energy per one β-decay is equal to (344 ± 34) keV which is in an agreement with the average energy of the electrons (390 keV). Thus, it was proved that the energy carried by the β-decay electron was smaller than the total released energy.

There were two possibilities to explain this experimental data

1. To assume that in β-decay together with the electron a neutral penetrating particle, which is not detected in experiments, is produced. The total released

© Springer International Publishing AG, part of Springer Nature 2018   11
S. Bilenky, *Introduction to the Physics of Massive and Mixed Neutrinos*,
Lecture Notes in Physics 947, https://doi.org/10.1007/978-3-319-74802-3_2

energy is shared between the electron and the new particle. As a result, electrons produced in β-decay, will have a continuous spectrum.
2. To assume that in β-decay the energy is not conserved.

The idea of new particle was proposed by W. Pauli. The second point of view was advocated by N. Bohr.

Pauli wrote about his idea in a letter to H. Geiger and L. Meitner who participated in the nuclear conference at Tübingen (December 4, 1930). Pauli asked them to inform the participants of the conference on his proposal.

Pauli called the new particle "neutron". He assumed that the "neutron" has spin 1/2, small mass ("the mass of the neutrons should be of the same order of magnitude as the electron mass and in any event not larger than 0.01 of the proton mass") and large penetration length. Pauli assumed that the "neutron"[1] is emitted together with the electron in the β-decay of nuclei.

Below there is Pauli's letter translated into English.

Dear Radioactive Ladies and Gentlemen,

As the bearer of these lines, to whom I graciously ask you to listen, will explain to you in more detail, how because of the "wrong" statistics of the N and Li$_6$ nuclei and the continuous beta spectrum, I have hit upon a desperate remedy to save the "exchange theorem" of statistics and the law of conservation of energy. Namely, the possibility that there could exist in the nuclei electrically neutral particles, that I wish to call neutrons, which have spin 1/2 and obey the exclusion principle and which further differ from light quanta in that they do not travel with the velocity of light. The mass of the neutrons should be of the same order of magnitude as the electron mass and in any event not larger than 0.01 proton masses. The continuous beta spectrum would then become understandable by the assumption that in beta decay a neutron is emitted in addition to the electron such that the sum of the energies of the neutron and the electron is constant.

I agree that my remedy could seem incredible because one should have seen those neutrons very earlier if they really exist. But only the one who dare can win and the difficult situation, due to the continuous structure of the beta spectrum, is lighted by a remark of my honored predecessor, Mr. Debye, who told me recently in Brussels: "Oh, It's well better not to think to this at all, like new taxes". From now on, every solution to the issue must be discussed. Thus, dear radioactive people, look and judge. Unfortunately, I cannot appear in Tübingen personally since I am indispensable here in Zürich because of a ball on the night of 6/7 December. With my best regards to you, and also to Mr. Back.

Your humble servant W. Pauli

At the time when Pauli proposed the idea of the existence of the "neutron", nuclei were considered as bound states of protons and electrons. As it is seen from Pauli's letter he assumed that his new particle "exists in the nuclei". This assumption allowed him to solve the problem of the spin of some nuclei. Let us consider the

---

[1] Later E. Fermi and E. Amaldi proposed to call the Pauli particle neutrino (from Italian, *neutral, small*).

nucleus $^7\mathrm{N}_{14}$. According to the proton-electron model this nucleus is a bound state of 14 protons and 7 electrons. Because spins of protons and electrons are equal to 1/2 the spin of $^7\mathrm{N}_{14}$ must be half-integer. However, from the analysis of the spectrum of $^7\mathrm{N}_{14}$ molecules it was found that nucleus $^7\mathrm{N}_{14}$ satisfies Bose-Einstein statistics and, according to the spin-statistic theorem, the spin of this nucleus must be integer. An odd number of "neutrons" in $^7\mathrm{N}_{14}$ would make its spin integer.

In 1932 neutron, a heavy particle with a mass approximately equal to the mass of the proton, was discovered by J. Chadwick. Soon after this discovery Heisenberg, Majorana and Ivanenko suggested that nuclei are bound states of protons and neutrons. This hypothesis (which, as we know today, is the correct one) could successfully describe all nuclear data.

The problem of the spin of $^{14}\mathrm{N}_7$ and other nuclei disappeared.[2] But what about β-decay of nuclei and continuous β-spectrum? In the framework of proton-neutron structure of nuclei these problems were solved by E. Fermi in 1933–1934 on the basis of the Pauli's hypothesis of neutrino.

2.2 Fermi Theory of β-Decay

Fermi was the first who understood that *electron-neutrino pair was produced in the transition*[3]

$$n \to p + e^- + \bar{\nu}. \tag{2.1}$$

and that the Quantum Field Theory provides adequate apparatus for the description of such processes.

Fermi assumed that the Hamiltonian of the process (2.1) is analogous to the Hamiltonian of the electromagnetic transition

$$p \to p + \gamma. \tag{2.2}$$

The simplest electromagnetic Hamiltonian which induces this transition has the form of the scalar product of the electromagnetic current $j_\alpha^{EM} = \bar{p}(x)\gamma_\alpha p(x)$ and electromagnetic field $A^\alpha(x)$

$$\mathcal{H}^{\mathrm{EM}}(x) = e\,\bar{p}(x)\gamma_\alpha p(x)\,A^\alpha(x). \tag{2.3}$$

Here e is the electric charge of the proton, $p(x)$ is the proton field ($\bar{p}(x) = p^\dagger(x)\gamma^0$ is the conjugated field) and γ_α ($\alpha = 0, 1, 2, 3$) are the Dirac matrices.

---

[2]Nucleus $^{14}\mathrm{N}_7$ is the bound state of seven protons and seven neutrons and has an integer spin.
[3]We know today that in the β-decay together with the electron an antineutrino $\bar{\nu}$ is produced. Later we will explain the difference between neutrino and antineutrino.

By analogy, Fermi suggested that the Hamiltonian of the decay (2.1) was the scalar product of the vector $\bar{p}(x)\gamma_\alpha n(x)$ which provided the $n \rightarrow p$ transition and the vector $\bar{e}(x)\gamma_\alpha \nu(x)$ which provided the emission of the electron-antineutrino pair:

$$\mathcal{H}^\beta(x) = G_F\, \bar{p}(x)\gamma_\alpha n(x)\, \bar{e}(x)\gamma_\alpha \nu(x) + \text{h.c.} \qquad (2.4)$$

Here G_F is a constant (which is called the Fermi constant).

Let us stress an important difference between the Hamiltonians (2.3) and (2.4). The Hamiltonian (2.3) describes the interaction of two fermions and a boson while the Hamiltonian (2.4) describes the interaction of four fermions. As a consequence of that, the Fermi constant G_F and the electromagnetic charge e have *different dimensions*. In the system of the units $\hbar = c = 1$, which we are using, e is a dimensionless quantity whereas the Fermi constant G_F has the dimension $[M]^{-2}$. We will return to a discussion of this point later.

The largest contributions to the probability of the β-decay come from transitions in which electron and antineutrino are produced in states with orbital momenta equal to zero (S-states). Such transitions are called allowed. For allowed transitions it follows from the Fermi Hamiltonian (2.4) that spins and parities of the initial and final nuclei must be equal (Fermi selection rules):

$$\Delta J = 0, \quad \pi_i = \pi_f. \qquad (2.5)$$

Here $\Delta J = J_f - J_i$ (J_i, π_i and J_f, π_f are spins and parities of initial and final nucleus).

From the conservation of the total angular momentum it follows that in the case of the allowed transitions which satisfies the Fermi selection rule electron and antineutrino are produced in a state with the total spin S equal to zero (singlet state). If electron and (anti)neutrino are produced in the triplet state ($S = 1$) in this case we have

$$\Delta J = \pm 1, 0 \quad \pi_i = \pi_f \quad (0 \rightarrow 0 \text{ is forbidden}). \qquad (2.6)$$

These selection rules are called the Gamov-Teller selection rules.

In experiments there were observed β-decays of nuclei which satisfy the Fermi as well as the Gamov-Teller selection rules. This means that *the total Hamiltonian of the β-decay in addition to the Fermi Hamiltonian (2.4) must include additional term(s)*.

2.3 General Four-Fermion Hamiltonian of β-Decay

The Fermi Hamiltonian had the form of the scalar product of vector × vector. The most general Hamiltonian of the Fermi type, in which only fields but not their derivatives enter, has the form of the sum of the products of scalar × scalar, vector × vector, tensor × tensor, axial × axial and pseudoscalar × pseudoscalar:

$$\mathcal{H}_I^{\beta}(x) = \sum_{i=S,V,T,A,P} G_i\, \bar{p}(x) O^i n(x)\, \bar{e}(x) O_i \nu(x) + \text{h.c.} \tag{2.7}$$

Here

$$O^i \to 1\ (S),\ \gamma^{\alpha}\ (V),\ \sigma^{\alpha\beta}\ (T),\ \gamma^{\alpha}\gamma_5\ (A),\ \gamma_5\ (P). \tag{2.8}$$

and G_i are coupling constants, which have dimensions $[M]^{-2}$.

The Hamiltonian (2.7) could describe all β-decay data. Transitions, which satisfy the Fermi selection rules, are due to V and S terms and transitions which satisfy the Gamov-Teller selection rules, are due to A and T terms.

In the Fermi Hamiltonian (2.4) only one fundamental constant G_F entered. The Hamiltonian (2.7) was characterized by five (!) interaction constants. Analogy and economy which were the basis of the Fermi theory were lost.

There was a general belief that there are "dominant" terms in the interaction (2.7). Such terms were searched for many years via analysis of the data of different β-decay experiments. This search did not lead, however, to a definite result: some experiments were in favor of V and A terms, other were in favor of S and T terms. Up to 1957, when the violation of parity in the β-decay (and other weak processes) was discovered, the situation with the Hamiltonian of the β-decay remained uncertain.

2.4   Violation of Parity in β-Decay

For many years physicists believed that the conservation of parity (the invariance under space inversion) is a general law of nature. The discovery of violation of parity in the β-decay and other weak processes was a great surprise. In the beginning it looked that this discovery made the theory of the β-decay (and other weak processes) more complicated. In reality, as we will see later, this discovery lead to the creation of a simple theory of the weak interaction which allowed to describe all existed data.

The violation of parity in the weak interaction was one of the most important discoveries in the physics of the twentieth century. In 1957 Lee and Yang were awarded the Nobel Prize "for their penetrating investigation of the so-called parity laws which has led to important discoveries regarding the elementary particles".

The investigation of the decays of strange particles at the beginning of the 50s created the so-called $\theta - \tau$ problem.[4] As one of the possible solutions of the $\theta - \tau$

---

[4]A strange particle which decayed into π^+ and π^0 was called θ^+ and a strange particle which decayed into π^+, π^- and π^+ was called τ^+. From experimental data it followed that the *masses*

problem Lee and Yang proposed the hypothesis of the non-conservation of parity (1956). They analyzed existed experimental data and came to the conclusion that there was an evidence that parity is conserved in the strong and electromagnetic interactions, but there were no data which proved that parity was conserved in the β-decay and other weak decays. They concluded: "...as for weak interactions parity conservation is so far only extrapolated hypothesis unsupported by experimental evidence". Lee and Yang proposed different experiments which would allow to test the parity conservation in weak decays.

The first experiments in which large violation of parity was discovered was performed by Wu et al. at the beginning of 1957.

In this experiment the β-decay of polarized ^{60}Co was investigated. Let us consider the emission of the electron with momentum $\mathbf{p}$ in the β-decay of a nucleus with polarization $\mathbf{P}$. From the invariance under rotations (conservation of the total momentum) it follows that the decay probability can depend only on the scalar products $\mathbf{p} \cdot \mathbf{p}$ and $\mathbf{P} \cdot \mathbf{p}$. Taking into account that the decay probability depends linearly on the polarization of a nucleus we obtain the following general expression for the probability of the emission of the electron with momentum $\mathbf{p}$ by a nucleus with polarization $\mathbf{P}$

$$w_{\mathbf{P}}(\mathbf{p}) = w_0 \left(1 + \alpha \mathbf{P} \cdot \mathbf{k}\right) = w_0 \left(1 + \alpha P \cos\theta\right). \qquad (2.9)$$

Here $\mathbf{k} = \frac{\mathbf{p}}{p}$ is a unit vector in the direction of the electron momentum, θ is the angle between the vectors $\mathbf{P}$ and $\mathbf{p}$, and w_0 and α are functions of p^2.

Under the inversion of a coordinate system momentum $\mathbf{p}$ and polarization $\mathbf{P}$ are transformed *differently*. Namely, momentum is transformed as a vector

$$p_i' = -p_i \qquad (2.10)$$

while polarization is transformed as a pseudovector

$$P_i' = +P_i. \qquad (2.11)$$

Here p_i (P_i) are components of a vector of momentum (pseudovector of polarization) in some right-handed system and p_i' (P_i') are components of the same momentum (same polarization) in the inverted (left-handed) system.

From (2.10) and (2.11) it follows that under the inversion the scalar product $\mathbf{P} \cdot \mathbf{p}$ is transformed as a pseudoscalar

$$\mathbf{P}' \cdot \mathbf{p}' = -\mathbf{P} \cdot \mathbf{p} \qquad (2.12)$$

---

and lifetimes of θ^+ and τ^+ are the same. The study of the Dalitz plot of the decay of τ^+ showed that the total angular momentum of π^+, π^-, π^+ was equal to zero and, consequently, the parity of τ^+ was equal to -1. If τ^+ and θ^+ were the same particle in this case the spin of θ^+ must be equal to zero. The parity of the two pions produced in S-state was equal to $+1$ and we were confronted with the following problem: the same particle decayed into states with different parities.

while $\mathbf{p} \cdot \mathbf{p}$ is transformed as a scalar

$$\mathbf{p}' \cdot \mathbf{p}' = +\mathbf{p} \cdot \mathbf{p}. \tag{2.13}$$

If the invariance under the inversion holds (parity is conserved), in this case the decay probability in a right-handed system and in an inverted left-handed system is the same

$$w_{\mathbf{P}'}(\mathbf{p}') = w_{\mathbf{P}}(\mathbf{p}). \tag{2.14}$$

From (2.9)–(2.11) and (2.14) we conclude that in the case of the conservation of parity $\alpha = 0$ and the probability of the emission of the electron by the polarized nucleus does not depend on the angle θ.

In the Wu et al. experiment it was found that $\alpha \simeq -0.7$ (i.e. electrons are emitted mainly in the direction opposite to the polarization of the nucleus). Thus, it was discovered that parity in the β-decay is not conserved.

Let us discuss the Hamiltonian of the β-decay. The Hamiltonian (2.7) is a scalar. It conserves the parity. In order to take into account the results of the Wu et al. and other experiments we must assume that the *Hamiltonian of the β-decay is the sum of a scalar and a pseudoscalar*. Such a Hamiltonian can be built if we add to five scalars which enter into the Hamiltonian (2.7) additional five pseudoscalars which are formed from products of the scalar $\bar{p}(x)n(x)$ and pseudoscalar $\bar{e}(x)\gamma_5 v(x)$, vector $\bar{p}(x)\gamma^\alpha n(x)$ and pseudovector $\bar{e}(x)\gamma_\alpha \gamma_5 v(x)$, etc. The most general Hamiltonian of the β-decay takes the form

$$\mathcal{H}_I^\beta(x) = \sum_{i=S,V,T,A,P} \bar{p}(x)O_i n(x)\, \bar{e}(x)O^i(G_i - G_i'\gamma_5)v(x) + \text{h.c.}, \tag{2.15}$$

where the constants G_i characterize the scalar part of the Hamiltonian, the constants G_i' characterize the pseudoscalar part and the matrices O^i are given by (2.8).

The Hamiltonian (2.15) is characterized by ten fundamental interaction constants. From the Wu et al. experiment it followed that scalar and pseudoscalar terms of the Hamiltonian must be of the same order. This means that the constants $|G_i|$ and $|G_i'|$ (at least some of them) must be of the same order.

In 1957–1958 enormous progress in the development of the theory of the weak interaction was reached. *Two fundamental steps were done which brought us to the modern effective Hamiltonian of the β-decay and other weak processes.*

2.5 Two-Component Neutrino Theory

The first step was the theory of the two-component neutrino.

Soon after the discovery of the parity violation Landau, Lee and Yang and Salam came to an idea of *a possible connection of the violation of parity observed in the β-decay and other weak processes with neutrinos.*

The neutrino field $v(x)$ satisfies the Dirac equation

$$(i\gamma^\alpha \partial_\alpha - m)v(x) = 0 \tag{2.16}$$

where m is the neutrino mass.

Let us present the field $v(x)$ in the form

$$v(x) = v_L(x) + v_R(x). \tag{2.17}$$

where

$$v_{L,R}(x) = (\frac{1 \mp \gamma_5}{2})v(x) \tag{2.18}$$

are left-handed and right-handed components of the field $v(x)$.

From (2.16) and (2.17) we obtain two coupled equations for $v_L(x)$ and $v_R(x)$

$$i\gamma^\alpha \partial_\alpha v_L(x) - mv_R(x) = 0 \quad i\gamma^\alpha \partial_\alpha v_R(x) - mv_L(x) = 0. \tag{2.19}$$

Let us assume that $m = 0$. In this case we obtain Weil equations for $v_L(x)$ and $v_R(x)$

$$i\gamma^\alpha \partial_\alpha v_{L,R}(x) = 0. \tag{2.20}$$

Thus, for $m = 0$, the neutrino field can be $v_L(x)$ (or $v_R(x)$). Such a theory can be valid only if parity is violated. In fact, under the inversion of coordinates the field $v(x)$ is transformed as follows:

$$v'(x') = \eta\gamma^0 v(x). \tag{2.21}$$

Here $x' = (x^0 - \mathbf{x})$ and η is a phase factor. From (2.21) it follows that under the inversion a left-handed (right-handed) component of the field is transformed into a (right-handed) (left-handed) component:

$$v'_{L(R)}(x') = \eta\gamma^0 v_{R(L)}(x). \tag{2.22}$$

Thus, Eq. (2.20) is not invariant under the inversion.

A method of the measurement of the neutrino mass was proposed by Fermi and Perrin in 1934. In order to determine the neutrino mass they proposed to perform a precise measurement of the high-energy part of β-spectrum in which the neutrino energy is small. At the time of the discovery of the violation of parity from β-decay experiments for the neutrino mass it was found the bound much smaller that the mass of the electron ($m < 200\,\text{eV}$).

Landau, Lee and Yang and Salam assumed that the neutrino mass was equal to zero and that the neutrino field was $v_L(x)$ (or $v_R(x)$). By the reasons, which will be clear later, this theory was called the two-component neutrino theory. There were two major consequences of the two-component theory.

1. Parity is strongly violated in the β-decay and in other processes in which neutrino participate. The most general Hamiltonian of the β-decay in the case of parity violation is given by expression (2.15). In the case of the two-component theory we have

$$G_i' = G_i \quad \text{(if neutrino field is } \nu_L(x)\text{)} \tag{2.23}$$

and

$$G_i' = -G_i \quad \text{(if neutrino field is } \nu_R(x)\text{)} \tag{2.24}$$

and the most general Hamiltonian of the β-decay take the form

$$\mathcal{H}_I^\beta(x) = \sum_{i=S,V,T,A,P} G_i \, \bar{p}(x) O_i n(x) \, \bar{e}(x) O^i (1 \mp \gamma_5) \nu(x) + \text{h.c.} \tag{2.25}$$

From this expression it followed that effects of violation of parity in the β-decay would be large (maximal).

2. The neutrino helicity (projection of the spin on the direction of momentum) is equal to $-1(+1)$ in the case if the neutrino field is $\nu_L(x)$ ($\nu_R(x)$).

   In fact, from the Dirac equation for the massless neutrino we have

$$\not{p} \, u^r(p) = 0, \tag{2.26}$$

where $\not{p} = \gamma_\alpha p^\alpha$. The spinor $u^r(p)$ describes a particle with helicity equal to r ($r = \pm 1$). We have

$$\boldsymbol{\Sigma} \cdot \mathbf{k} \, u^r(p) = r \, u^r(p), \tag{2.27}$$

where $\boldsymbol{\Sigma}$ is the operator of the spin and $\mathbf{k}$ is the unit vector in the direction of the momentum $\mathbf{p}$. For the operator of the spin we have

$$\boldsymbol{\Sigma} = \gamma_5 \boldsymbol{\alpha} = \gamma_5 \gamma^0 \boldsymbol{\gamma}. \tag{2.28}$$

From (2.26) and (2.28) we have

$$\boldsymbol{\Sigma} \cdot \mathbf{k} \, u^r(p) = \gamma_5 \, u^r(p). \tag{2.29}$$

Thus, for a massless particle operator γ_5 is the operator of the helicity. From (2.27) we find

$$\gamma_5 \, u^r(p) = r u^r(p). \tag{2.30}$$

Similarly, for the spinor $u^r(-p)$ which describes the state with negative energy $-p^0$ and momentum $-\mathbf{p}$ we have

$$\gamma_5 \, u^r(-p) = -r \, u^r(-p). \tag{2.31}$$

From (2.30) we find that $\frac{1-\gamma_5}{2}$ is the projection operator:

$$\frac{1-\gamma_5}{2} \, u^{-1}(p) = u^{-1}(p), \qquad \frac{1-\gamma_5}{2} \, u^1(p) = 0. \tag{2.32}$$

From (2.31) we have

$$\frac{1-\gamma_5}{2} \, u^1(-p) = u^1(-p), \qquad \frac{1-\gamma_5}{2} \, u^{-1}(-p) = 0. \tag{2.33}$$

From these relations for the left-handed neutrino field we find

$$\nu_L(x) = \int N_p \left(u^{-1}(p) \, c_{-1}(p) \, e^{-ipx} + u^1(-p) \, d_1^\dagger(p) \, e^{ipx} \right) d^3 p. \tag{2.34}$$

Analogously, for the right-handed neutrino field we have

$$\nu_R(x) = \int N_p \left(u^1(p) \, c_1(p) \, e^{-ipx} + u^{-1}(-p) \, d_{-1}^\dagger(p) \, e^{ipx} \right) d^3 p. \tag{2.35}$$

The neutrino helicity was measured in 1958 in a spectacular Goldhaber, Grodzins and Sunyar experiment. In this experiment the helicity of the neutrino was determined from the measurement of the circular polarization of γ-quanta in the chain of reactions

$$e^- + \text{Eu} \to \nu + \text{Sm}^*$$
$$\downarrow$$
$$\text{Sm} + \gamma \tag{2.36}$$

The authors concluded "...our result is compatible with 100% negative helicity of neutrino emitted in orbital electron capture".

The Goldhaber et al. experiment confirmed the theory of the two-component neutrino. It was established that the neutrino is the left-handed particle and the neutrino field is $\nu_L(x)$.[5]

---

[5]Let us stress that the experiment by Goldhaber et al. does not exclude that the neutrino has a small mass. In fact, if in the Hamiltonian of the β-decay enters $\nu_L(x)$ and the neutrino mass is not equal to zero in this case the longitudinal polarization of the neutrino for $m \ll E$ is equal to $P_{\parallel} \simeq -1 + \frac{m^2}{2E^2} \simeq -1$.

2.6 μ-e Universal Charged Current. Current × Current Theory

The next decisive step in the construction of the Hamiltonian of the β-decay and other weak processes was done by Feynman and Gell-Mann, Marshak and Sudarshan in 1957–1958. Generalizing the theory of the two-component neutrino, Feynman and Gell-Mann, Marshak and Sudarshan assumed that *in the Hamiltonian of the weak interaction enter left-handed components of all fermion fields*. In this case the most general four-fermion Hamiltonian of the β-decay has the form

$$\mathcal{H}_I^\beta = \sum_{i=S,V,T,A,P} G_i \, \bar{p}_L O_i n_L \, \bar{e}_L O^i v_L + \text{h.c.,} \tag{2.37}$$

where O_i are Dirac matrices (see (2.8)).

We have

$$\bar{e}_L O_i v_L = \bar{e} \frac{1+\gamma_5}{2} O_i \frac{1-\gamma_5}{2} v. \tag{2.38}$$

It is obvious that

$$\frac{1+\gamma_5}{2} \left(1;\ \sigma_{\alpha\beta};\ \gamma_5\right) \frac{1-\gamma_5}{2} = 0. \tag{2.39}$$

Therefore, S, T and P terms in the Hamiltonian (2.37) are equal to zero. Moreover A and V terms are connected by the relation:

$$\frac{1+\gamma_5}{2} \gamma_\alpha \gamma_5 \frac{1-\gamma_5}{2} = -\frac{1+\gamma_5}{2} \gamma_\alpha \frac{1-\gamma_5}{2}. \tag{2.40}$$

Thus, if we assume that only left-handed components of the fields enter into the four-fermion Hamiltonian, we come to the unique expression for the Hamiltonian of the β-decay

$$\mathcal{H}_I^\beta = \frac{G_F}{\sqrt{2}} 4\, \bar{p}_L \gamma_\alpha n_L \, \bar{e}_L \gamma^\alpha v_L + \text{h.c.}$$

$$= \frac{G_F}{\sqrt{2}} \bar{p} \gamma_\alpha (1-\gamma_5) n\, \bar{e} \gamma^\alpha (1-\gamma_5) v + \text{h.c.} \tag{2.41}$$

The Hamiltonian (2.41) is the simplest possible four-fermion Hamiltonian of the β-decay which takes into account large violation of parity. Like the Fermi Hamiltonian (2.4), it is characterized by only one interaction constant.[6]

---

[6]In order to keep the numerical value of the Fermi constant the coefficient $\frac{1}{\sqrt{2}}$ was introduced in (2.41). It is interesting that the title of the Feynman and Gell-Mann paper is "Theory of the Fermi interaction".

The theory proposed by Feynman and Gell-Mann, Marshak and Sudarshan was a very successful one: the Hamiltonian (2.41) allowed to describe all existing β-decay data. We know today that (2.41) is the correct effective Hamiltonian of the β-decay, of the process $\bar{v} + p \to n + e^+$, and other connected processes.[7]

Until now we have considered only the Hamiltonian of the β-decay. At the time when parity violation was discovered the following processes in which a muon-neutrino pair was involved were also known

$$\mu^- + (A, Z) \to v + (A, Z - 1) \quad (\mu\text{-capture}) \tag{2.42}$$

$$\mu^+ \to e^+ + v + \bar{v} \quad (\mu\text{-decay}). \tag{2.43}$$

In 1947 B. Pontecorvo suggested the existence of a $\mu - e$ *universal weak interaction*, which is characterized by the Fermi constant G_F. He compared the probability of the μ-capture (2.42) with the probability of the K-capture

$$e^- + (A, Z) \to v + (A, Z - 1) \tag{2.44}$$

and found that the constant of the interaction of the muon-neutrino pair with nucleons is of the same order as the Fermi constant G_F. The idea of a $\mu - e$ universal weak interaction was also proposed by Puppi, Klein and Tiomno and Wheeler. In order to build a $\mu - e$ universal theory of the weak interaction, Feynman and Gell-Mann introduced the notion of *the charged weak current*

$$j^\alpha = 2 \left(\bar{p}_L \gamma^\alpha n_L + \bar{v}_L \gamma^\alpha e_L + \bar{v}_L \gamma^\alpha \mu_L \right) \tag{2.45}$$

and assumed that the Hamiltonian of the weak interaction has the current $\times$ current form

$$\mathcal{H}_I = \frac{G_F}{\sqrt{2}} j^\alpha j^+_\alpha, \tag{2.46}$$

where G_F is the Fermi constant.

Two remarks are in order.

1. The hadron part of the current has the form

$$j^\alpha = v^\alpha - a^\alpha,$$

---

[7]What about numerous experiments from which it followed that S and T terms are the dominant terms of the Hamiltonian of the β-decay? In the Feynman and Gell-Mann paper it was written "These theoretical arguments seem to the authors to be strong enough to suggest that the disagreement with [6]He recoil experiment and with some other less accurate experiments indicates that these experiments are wrong". In fact, later experiments did not confirm results of all experiments which indicated in favor of the dominance of S and T terms.

where $v^\alpha = \bar{p}\gamma^\alpha n$ and $a^\alpha = \bar{p}\gamma^\alpha \gamma_5 n$ are the vector and axial currents.[8] Notice that Fermi transitions of nuclei are due to the vector current and Gamov-Teller transitions are due to the axial current.

2. The current j^α provides transitions $n \rightarrow p$, $e^- \rightarrow v$, etc. in which $\Delta Q = Q_f - Q_i = 1$ ($Q_i(Q_f)$ is the initial (final) charge). By this reason the current j^α is called the charged current (CC).

There are two types of terms in the Hamiltonian (2.46): nondiagonal and diagonal. The nondiagonal terms are given by

$$\mathscr{H}_I^{\text{nondiag}} = \frac{G_F}{\sqrt{2}} 4\, \{[(\bar{p}_L\gamma^\alpha n_L)(\bar{e}_L\gamma_\alpha v_L) + \text{h.c.}]$$
$$+ [(\bar{p}_L\gamma^\alpha n_L)(\bar{\mu}_L\gamma_\alpha v_L) + \text{h.c.}]$$
$$+ [(\bar{e}_L\gamma^\alpha v_L)(\bar{v}_L\gamma_\alpha \mu_L) + \text{h.c.}]\} \qquad (2.47)$$

The first term of this expression is the Hamiltonian of β-decay of the neutron (2.1), of the process $\bar{v} + p \rightarrow e^+ + n$ and other connected processes. The second term of (2.47) is the Hamiltonian of the process $\mu^- + p \rightarrow v + n$ and other connected processes. Finally the third term of (2.47) is the Hamiltonian of the μ-decay (2.43) and other processes.

The diagonal terms of the Hamiltonian (2.46) are given by

$$\mathscr{H}^{\text{diag}} = \frac{G_F}{\sqrt{2}} 4[(\bar{v}_L\gamma^\alpha e_L)(\bar{e}_L\gamma_\alpha v_L) + (\bar{v}_L\gamma^\alpha \mu_L)(\bar{\mu}_L\gamma_\alpha v_{\mu L}) + (\bar{p}_L\gamma^\alpha n_L)(\bar{n}_L\gamma_\alpha p_L)]$$
$$(2.48)$$

The first term of the (2.48) is the Hamiltonian of the processes of elastic scattering of neutrino and antineutrino on an electron

$$v + e \rightarrow v + e \qquad (2.49)$$

and

$$\bar{v} + e \rightarrow \bar{v} + e, \qquad (2.50)$$

of the process $e^+ + e^- \rightarrow \bar{v} + v$ and other processes. Such processes were not known in the 50s. Their existence was predicted by the current × current theory.

The cross sections of the processes (2.49) and (2.50) are very small. The observation of such processes was a challenge. After many years of efforts, the cross section of the process (2.50) was measured by F. Reines et al. in an experiment with antineutrinos from a reactor. At that time the Standard Model already existed. According to the Standard Model, to the matrix elements of the processes (2.49) and

---

[8]This is the reason why the Feynman and Gell-Mann, Marshak and Sudarshan theory is called the $V - A$ theory.

(2.50) contributes not only Hamiltonian (2.48) but also additional (neutral current) Hamiltonian. The result of the experiment by F. Reines et al. was in agreement with the Standard Model.

2.7 Theory with Vector W Boson

In the Feynman and Gell-Mann paper it was mentioned that the current × current Hamiltonian of the weak interaction (2.46) could originate from the exchange of a heavy intermediate charged vector meson.[9] We will discuss here briefly this hypothesis. Let us assume that there exists a charged vector $W^{\pm}$ boson and that the Lagrangian of the weak interaction has the form of a scalar product of the current j_{α} given by Eq. (2.45) and the vector field W^{α}

$$\mathscr{L}_I = -\frac{g}{2\sqrt{2}}\, j_{\alpha}\, W^{\alpha} + \text{h.c.},\tag{2.51}$$

where g is a dimensionless interaction constant.

If the Lagrangian of the weak interaction has the form (2.51), the β-decay of the neutron proceeds in the following three steps: (1) neutron produces the virtual W^--boson and is transferred into proton; (2) the virtual W^--boson propagates; (3) the virtual W^--boson decays into a electron-antineutrino pair (see Fig. 2.1).

In the Feynman diagram, the propagator of the W-boson contains a factor $\frac{-1}{Q^2-m_W^2}$, where $Q = p_n - p_p$ is the momentum transfer and m_W is the mass of the W-boson. If the W-boson is a heavy particle (with a mass much larger than the mass of the proton), in this case Q^2 in the W-propagator can be safely neglected and the matrix element of the β-decay of the neutron can be obtained from the Hamiltonian (2.46) in which the Fermi constant is given by the relation

$$\frac{G_F}{\sqrt{2}} = \frac{g^2}{8\, m_W^2}.\tag{2.52}$$

In a similar way it can be shown that in the region of relatively small energies, the matrix elements of all weak processes with the virtual W-boson can be obtained from the current × current Hamiltonian (2.46) in which the Fermi constant is given by relation (2.52).[10]

---

[9]"We have adopted the point of view that the weak interactions all arise from the interaction of a current J_{α} with itself, possibly via an intermediate charged vector meson of high mass" (Feynman and Gell-Mann).

[10]From the point of view of the theory with the W-boson, the current × current Hamiltonian with the Fermi constant (2.52) is the effective Hamiltonian of the weak interaction.

Fig. 2.1 Feynman diagram
of the process
$n \to p + e^- + \bar{\nu}$ in the
theory with the $W^\pm$-boson

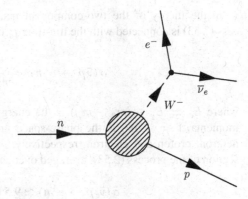

Thus, the theory with the vector $W^\pm$-boson could explain the current × current
structure of the weak interaction Hamiltonian and the fact that the Fermi constant
has the dimension $[M]^{-2}$.

We know today that the intermediate charged $W^\pm$-boson exists. The $W^\pm$-boson
is one of the heaviest particles: its mass is equal to $m_W \simeq 80.4 \, \text{GeV}$.

The first idea of the charged vector boson, mediator of the weak interaction, was
suggested by O. Klein in 1938, soon after the Fermi β-decay theory had appeared.
Fermi built the first Hamiltonian of the β-decay by analogy with electrodynamics.
O. Klein noticed that the analogy would be more complete if weak interaction was
originated from an interaction which had the form of a product of a current and a
vector field (like the electromagnetic interaction).

2.8 First Observation of Neutrinos: Lepton Number Conservation

The proof of the existence of neutrino was obtained in the mid-fifties in the
experiment by F. Reines and C.L. Cowan. In this experiment (anti)neutrinos from
the Savannah River reactor were detected through the observation of the process

$$\bar{\nu} + p \to e^+ + n. \tag{2.53}$$

Antineutrinos are produced in a reactor β-decays of neutron-rich nuclei, products of
the fission of uranium and plutonium. The energies of antineutrinos from a reactor
are less than $\sim 10 \, \text{MeV}$. About 2.3×10^{20} antineutrinos per second were emitted by
the Savannah River reactor. The flux of $\bar{\nu}$'s in the Reines and Cowan experiment
was about $10^{13} \, \text{cm}^{-2} \, \text{s}^{-1}$.

In the theory of the two-component neutrino, the cross section of the process (2.53) is connected with the life-time τ_n of the neutron by the relation

$$\sigma(\bar{\nu}p \to e^+n) = \frac{2\pi^2}{m_e^5 f \tau_n} p_e E_e, \tag{2.54}$$

where $E_e \simeq E_{\bar{\nu}} - (m_n - m_p)$ is the energy of the positron, p_e is the positron momenta, $f = 1.686$ is the phase-space factor, m_n, m_p, m_e are masses of the neutron, proton and electron, respectively. From (2.54) it follows that the cross section of the process (2.53), averaged over antineutrino spectrum, is equal to

$$\bar{\sigma}(\bar{\nu}_e p \to e^+n) \simeq 9.5 \times 10^{-44} \text{ cm}^2. \tag{2.55}$$

A liquid scintillator (1.4×10^3 l) loaded with $CdCl_2$ was used as a target in the experiment. Positron, produced in the process (2.53), slowed down in the scintillator and annihilated with electron, producing two γ- quanta with energies $\simeq 0.51$ MeV and opposite momenta. A neutron, produced in the process was captured by Cd within about 5 µs, producing γ-quantum. The γ-quanta were detected by 110 photomultipliers. Thus, the signature of the $\bar{\nu}$-event in the Reines and Cowan experiment was two γ-quanta from the $e^+ - e^-$-annihilation in coincidence with a delayed γ-quantum from the neutron capture by cadmium. For the cross section of the process (2.53) the value

$$\sigma_\nu = (11 \pm 2.6) \, 10^{-44} \text{ cm}^2 \tag{2.56}$$

was obtained in the experiment. This value was in agreement with the predicted value (2.55).

The particle which is produced in the β-decay together with electron is called antineutrino. It is a direct consequence of the quantum field theory that antineutrino can produce a positron in the inverse β-decay (2.53) and other similar processes. Can antineutrinos produce electrons in processes of interaction with nucleons? The answer to this question was obtained from an experiment which was performed in 1956 by Davis et al. with antineutrinos from the Savannah River reactor. In this experiment ^{37}Ar from the process

$$\bar{\nu} + {}^{37}Cl \to e^- + {}^{37}Ar \tag{2.57}$$

was searched for. The process (2.57) was not observed in the experiment. It was shown that the ^{37}Ar production rate was about five times smaller than the rate expected if antineutrinos could produce electrons via the weak interaction.

Thus, it was established that antineutrinos from a reactor can produce positrons (the Reines-Cowan experiment) but can not produce electrons. In order to explain this fact we assume that exist *conserving lepton charge number L*, the same for $\bar{\nu}$ and e^+. Let us put $L(\bar{\nu}) = L(e^+) = -1$. According to the quantum field theory the lepton charges of the corresponding antiparticles are opposite: $L(\nu) = L(e^-) = 1$.

We also assume that the lepton numbers of proton, neutron and other hadrons are equal to zero. Conservation of the lepton number could explain the negative result of the Davis experiment. According to the law of conservation of the lepton number a neutrino is produced together with e^+ in the β^+-decay

$$(A, Z) \rightarrow (A, Z - 1) + e^+ + \nu$$

2.9   Discovery of Muon Neutrino: Electron and Muon Lepton Numbers

When the universal $V - A$ theory of weak interaction was formulated by Feynman and Gell-Mann, Marshak and Sudarshan they considered only one type of neutrinos. There existed, however, an idea, expressed by different physicists, that neutrinos which take part in the weak interaction together with an electron and a muon could be different. Let us call neutrinos which participate in weak processes together with electrons and neutrinos which participate in weak processes together with muons, correspondingly, the electron and muon neutrinos (ν_e and ν_μ). The charged current of the current × current theory takes in this case the form

$$j^\alpha = 2 \left(\bar{p}_L \gamma^\alpha n_L + \bar{\nu}_{eL} \gamma^\alpha e_L + \bar{\nu}_{\mu L} \gamma^\alpha \mu_L \right) \tag{2.58}$$

Are ν_e and ν_μ the same or different particles? The answer to this fundamental question was obtained in the famous Brookhaven accelerator neutrino experiment.

The first indication that ν_e and ν_μ are different particles was obtained from an analysis of the $\mu^+ \rightarrow e^+ \gamma$ data. The probability of the decay $\mu^+ \rightarrow e^+ \gamma$ was calculated by Feinberg in the nonrenormalizable theory with W-boson. It was found that if ν_e and ν_μ are identical particles for the ratio R of the probability of the decay $\mu^+ \rightarrow e^+ \gamma$ to the probability of the decay $\mu^+ \rightarrow e^+ \nu \bar{\nu}$ is given by

$$R \simeq \frac{\alpha}{24\pi} \simeq 10^{-4} \tag{2.59}$$

The decay $\mu^+ \rightarrow e^+ \gamma$ was not observed in experiment. At the time of the Brookhaven experiment, for the upper bound of the ratio R was found the value

$$R < 10^{-8}, \tag{2.60}$$

which is much smaller than (2.59).

A direct proof of the existence of the second (muon) type of neutrino was obtained by L.M. Lederman, M. Schwartz, J. Steinberger et al. in the first experiment with accelerator neutrinos in 1962. The idea of the experiment was proposed by B. Pontecorvo in 1959.

A beam of π^+'s in the Brookhaven experiment was obtained by the bombardment of Be target by protons with an average energy of about 15 GeV. In the decay channel (about 21 m long) practically all π^+'s decay. After the channel there was shielding (13.5 m of iron), in which charged particles were absorbed. After the shielding there was the neutrino detector (aluminium spark chamber, 10 tons) in which the production of charged leptons was observed.

The dominant decay channel of the π^+-meson is

$$\pi^+ \to \mu^+ + \nu_\mu. \tag{2.61}$$

According to the universal $V - A$ theory, the ratio R of the width of the decay

$$\pi^+ \to e^+ + \nu_e \tag{2.62}$$

to the width of the decay (2.61) is equal to

$$R = \frac{m_e^2}{m_\mu^2} \frac{(1 - \frac{m_e^2}{m_\pi^2})^2}{(1 - \frac{m_\mu^2}{m_\pi^2})^2} \simeq 1.2 \times 10^{-4}. \tag{2.63}$$

Thus, the decay $\pi^+ \to e^+\nu_e$ is strongly suppressed with respect to the decay $\pi^+ \to \mu^+ + \nu_\mu$.[11] From (2.63) follows that the neutrino beam in the Brookhaven experiment was practically a pure ν_μ beam (with a small about 1% admixture of ν_e from decays of muons and kaons).

Let us assume that $\nu_\mu \equiv \nu_e = \nu$. In this case neutrinos produced in the decay $\pi^+ \to \mu^+ + \nu$ can produce muons and electrons in the reactions

$$\nu + N \to \mu^- + X \text{ and } \nu + N \to e^- + X. \tag{2.64}$$

Due to the $\mu - e$ universality of the weak interaction one could expect to observe in the detector (practically) equal number of muons and electrons.

If ν_μ and ν_e are different particles in this case neutrinos emitted in the decay $\pi^+ \to \mu^+ + \nu_\mu$ can produce only muons in the reaction

$$\nu_\mu + N \to \mu^- + X. \tag{2.65}$$

---

[11]The reason for this suppression can be easily understood. Indeed, let us consider the decay (2.62) in the rest frame of the pion. The helicity of the neutrino is equal to -1. If we neglect the mass of the e^+, the helicity of the positron will be equal to $+1$ (the helicity of the positron will be the same in this case as the helicity of the antineutrino). Thus, the projection of the total angular momentum on the neutrino momentum will be equal to -1. The spin of the pion is equal to zero and consequently the process (2.62) in the limit $m_e \to 0$ is forbidden. These arguments explain the appearance of the small factor $(\frac{m_e}{m_\mu})^2$ in (2.63).

Table 2.1 Lepton numbers
of particles

Lepton number	$\nu_e \ e^-$	$\nu_\mu \ \mu^-$	Hadrons, γ
L_e	1	0	0
L_μ	0	1	0

In the Brookhaven experiment 29 muon events were detected. The observed six electron candidates could be explained by the background. The measured cross section was in agreement with the $V - A$ theory. Thus, it was proved that ν_μ and ν_e are different particles.[12]

The results of the Brookhaven and other experiments suggested that the total electron L_e and muon L_μ lepton numbers are conserved:

$$\sum_i L_e^{(i)} = \text{const}; \quad \sum_i L_\mu^{(i)} = \text{const} \tag{2.66}$$

The lepton numbers of particles are given in Table 2.1. The lepton numbers of antiparticles are opposite to the lepton numbers of the corresponding particles.

For many years all experimental data were in an agreement with (2.66). At present it is established that (2.66) is an approximate phenomenological rule. It is violated in neutrino oscillations due to small neutrino masses and neutrino mixing. Later we will discuss neutrino oscillations in details.

2.10 Strange Particles. Quarks. Cabibbo Current

The current $\times$ current Hamiltonian (2.46) with CC current (2.45) is the effective Hamiltonian of such processes in which leptons, neutrinos and nonstrange hadrons are participating. The strange particles were discovered in cosmic rays in the fifties. Decays of strange particles were studied in details in accelerator experiments. From the investigation of the semi-leptonic decays

$$K^+ \to \mu^+ + \nu_\mu, \quad \Lambda \to n + e^- + \bar{\nu}_e,$$
$$\Sigma^- \to n + e^- + \bar{\nu}_e, \quad \Xi^- \to \Lambda + \mu^- + \bar{\nu}_\mu$$

and others the following *three phenomenological rules* were established.

I. The strangeness S in the decays of strange particles is changed by one

$$|\Delta S| = 1.$$

---

[12]In 1963 in the CERN with the invention of the magnetic horn the intensity and purity of neutrino beams were greatly improved. In the more precise 45 tons spark-chamber experiment and in the large bubble chamber experiment the Brookhaven result was confirmed.

For example, in accordance with this rule the decay

$$\Xi^- \to n + e^- + \bar{\nu}_e \tag{2.67}$$

is forbidden. From existing experimental data for the ratio of the width of the decay (2.67) to the total decay width of Ξ^- it was found the following upper bound

$$R(\Xi^- \to ne^-\bar{\nu}_e) < 3.2 \times 10^{-3}. \tag{2.68}$$

II. In the decays of the strange particles the rule

$$\Delta Q = \Delta S$$

is satisfied. Here $\Delta Q = Q_f - Q_i$ and $\Delta S = S_f - S_i$, where S_i and S_f are the initial and final total strangeness of the hadrons and Q_i and Q_f are the initial and final total electric charges of hadrons (in the unit of the proton charge).

From this rule it followed that the decay

$$\Sigma^+ \to n + e^+ + \nu_e \tag{2.69}$$

is forbidden. We will present the results of modern experiments. From these experiments the following upper bound was found

$$R(\Sigma^+ \to ne^+\nu_e) < 5 \times 10^{-6}. \tag{2.70}$$

III. The decays of strange particles are suppressed with respect to the decays of non strange particles.

In 1964 independently Gell-Mann and Zweig proposed the idea of the existence of three quarks u, d, s, constituents of strange and nonstrange hadrons. The quantum numbers of the quarks are presented in Table 2.2 Let us build from the quark fields the charged currents which changes the charge by one. If we accept the Feynman-Gell-Mann, Marshak-Sudarshan prescription (into the weak current enter left-handed components of the fermion fields) there are only two possibilities to build such currents from the fields of u, d and s quarks:

$$\bar{u}_L \gamma_\alpha d_L \quad \text{and} \quad \bar{u}_L \gamma_\alpha s_L. \tag{2.71}$$

Table 2.2 Quantum numbers of quarks (Q is the charge, S is the strangeness, B is the baryon number)

Quark	Q	S	B
u	2/3	0	1/3
d	−1/3	0	1/3
s	−1/3	−1	1/3

The first current changes the charge by one and does not change the strangeness ($\Delta Q = 1$, $\Delta S = 0$). The second current changes the charge and the strangeness by one ($\Delta Q = 1$, $\Delta S = 1$). The matrix elements of these currents *automatically satisfy rules I. and II.* Let us stress that this was one of the first arguments in favor of quark structure of the hadron current.

The weak interaction of the strange particles was included into the current $\times$ current theory by N. Cabibbo in 1962.

He introduced an angle θ_C and suggested that the hadronic charged current had the form

$$j_\alpha^{\text{Cabibbo}} = \cos\theta_C \; j_\alpha^{1+i2} + \sin\theta_C \; j_\alpha^{4+i5}. \tag{2.72}$$

Here

$$j_\alpha^{1+i2} = j_\alpha^1 + ij_\alpha^2, \quad j_\alpha^{4+i5} = j_\alpha^4 + ij_\alpha^5 \tag{2.73}$$

where j_α^i are components of the $SU(3)$ octet current.

The first term of (2.72) does not change the strangeness and the second term changes the strangeness by one. Cabibbo showed that the current (2.72) could describe existed at that time experimental data. From the analysis of the data on the investigation of the decays of strange particles he found that $\sin\theta_C \simeq 0.2$.

In terms of quark fields the Cabibbo current had the form

$$j_\alpha^{\text{Cabibbo}}(x) = 2\,(\cos\theta_C \; \bar{u}_L(x)\gamma_\alpha \, d_L(x) + \sin\theta_C \; \bar{u}_L(x)\gamma_\alpha \, s_L(x)). \tag{2.74}$$

The total weak charged current takes the form

$$j_\alpha^{\text{CC}}(x) = 2\,(\bar{\nu}_{eL}(x)\gamma_\alpha \, e_L(x) + \bar{\nu}_{\mu L}(x)\gamma_\alpha \, \mu_L(x) + \bar{u}_L(x)\gamma_\alpha \, d_L^{\text{mix}}(x)), \tag{2.75}$$

where

$$d_L^{\text{mix}}(x) = \cos\theta_C \, d_L(x) + \sin\theta_C \, s_L(x). \tag{2.76}$$

2.11 Charmed Quark

It was shown in 1970 by Glashow, Illiopulos and Maiani (GIM) that the charged current (2.74) induces a neutral current which does not change electric charge ($\Delta Q = 0$) and change the strangeness by one ($|\Delta S| = 1$). As a result, the predicted width of the decay

$$K^+ \to \pi^+ + \nu + \bar{\nu}. \tag{2.77}$$

was many orders of magnitude larger than *the upper bound* of the width obtained in experiments.

In order to solve this problem Glashow, Illiopulos and Maiani assumed that there exists a fourth "charmed" quark c with charge 2/3 and that there is an additional term in the weak current into which enters the field of the new quark c_L and the combination of d_L and s_L fields orthogonal to the Cabibbo combination (2.76). The weak currents took the form

$$j_\alpha^{CC}(x) = 2\,(\bar{\nu}_{eL}(x)\gamma_\alpha\, e_L(x) + \bar{\nu}_{\mu L}(x)\gamma_\alpha\, \mu_L(x) + \bar{u}_L(x)\gamma_\alpha\, d_L^{mix}(x) + \bar{c}_L(x)\gamma_\alpha\, s_L^{mix}(x)),$$

$$(2.78)$$

where

$$d_L^{mix}(x) = \cos\theta_C d_L(x) + \sin\theta_C s_L(x)$$
$$s_L^{mix}(x) = -\sin\theta_C d_L(x) + \cos\theta_C s_L(x)\,. \qquad (2.79)$$

It can be shown that in the theory with the charged current (2.78) neutral current which changes the strangeness does not appear.

The relations (2.79) mean that *the fields of d and s quarks enter into the charged current in the mixed form*. The phenomenon of mixing was perfectly confirmed by experiment.

The existence of the c-quark means the existence of a new family of "charmed" particles. This prediction of the theory was perfectly confirmed by experiment. In 1974 the J/Ψ particles, bound states of $c-\bar{c}$, were discovered. In 1976 the $D^{\pm,0}$ mesons, bound states of charmed and nonstrange quarks, were discovered etc. All data on the investigation of the weak decays and neutrino reactions were in agreement with the current $\times$ current theory with the current given by (2.78).

In 1975 the third charged lepton τ was discovered in experiments at $e^+ - e^-$ colliders. In the framework of the Standard Model, which we will consider in the next chapter, the existence of the third charged lepton requires the existence of the corresponding third type of neutrino (ν_τ) and an additional pair of quarks: the t (top) quark with electric charge 2/3 and the b (bottom) quark with electric charge $-1/3$. All these predictions of the SM were perfectly confirmed by numerous experiments.

Taking into account the existence of τ and ν_τ and t and b quarks and assuming the universality, for the charged current we have the following expression

$$j_\alpha^{CC}(x) = 2\,(\bar{\nu}_{eL}(x)\gamma_\alpha\, e_L(x) + \bar{\nu}_{\mu L}(x)\gamma_\alpha\, \mu_L(x) + \bar{\nu}_{\tau L}(x)\gamma_\alpha\, \tau_L(x)$$
$$+\bar{u}_L(x)\gamma_\alpha\, d_L^{mix}(x) + \bar{c}_L(x)\gamma_\alpha\, s_L^{mix}(x) + \bar{t}_L(x)\gamma_\alpha\, b_L^{mix}(x)).$$

$$(2.80)$$

Here

$$d_L^{\text{mix}}(x) = \sum_{q=u,s,b} V_{uq}\, q_L(x),$$

$$s_L^{\text{mix}}(x) = \sum_{q=u,s,b} V_{cq}\, q_L(x),$$

$$b_L^{\text{mix}}(x) = \sum_{q=u,s,b} V_{tq}\, q_L(x). \tag{2.81}$$

are mixed quark fields and V is an unitary 3×3 mixing matrix. We know today that the Lagrangian of the CC weak interaction has the form

$$\mathscr{L}_I^{CC}(x) = -\frac{g}{2\sqrt{2}}\, j_\alpha^{CC}(x)\, W^\alpha(x) + \text{h.c.} \tag{2.82}$$

where $W^\alpha(x)$ is the field of the vector $W^\pm$-bosons.

2.12 Concluding Remarks

The theory of the weak interaction started with the famous Fermi paper "An attempt of a theory of beta radiation". The Fermi theory was based on

- The Pauli neutrino hypothesis.
- The proton-neutron structure of nuclei.
- The assumption that an electron-neutrino pair is produced in the process of a neutron into a proton transition.
- The assumption that in analogy with electromagnetic interaction the weak interaction is the vector one.

Later in accordance with experimental data this last assumption was generalized and other terms (scalar, tensor, axial and pseudoscalar) were included into the Hamiltonian.

The discovery of the parity violation in the β-decay and other weak processes played a revolutionary role in the development of the theory of the weak interaction. Soon after this discovery the two-component theory of massless neutrino was proposed. According to this theory in the Hamiltonian of the weak interaction the left-handed (or right-handed) component of the neutrino field enters. In less than 1 year this theory was confirmed by experiment. It was proved that neutrino is a left-handed particle.

The next fundamental step was the universal current × current $V - A$ theory of the weak interaction which was based on the assumption that only left-handed components of the fields enter into charged current.

The electron neutrino was discovered in the fifties in the first reactor neutrino experiment. A few years later in the first accelerator neutrino experiment the muon neutrino was discovered.

After the hypothesis of quarks was proposed, the weak charged current started to be considered as quark and lepton current. One of the fundamental ideas which was put forward in the process of the phenomenological development of the theory was the idea of the quark mixing. At the very early stage of the development of the theory the idea of the existence of the charged heavy vector intermediate $W^{\pm}$ boson was proposed.

It was a long (about 40 years) extremely important period of the development of the physics of the weak interaction with a lot of bright, courageous ideas.[13] The theory which was finally proposed allowed to describe data of a large number of experiments. The unified theory of the weak and electromagnetic interactions, the Standard Model, could not appear without the phenomenological $V - A$ theory.

---

[13] And also many wrong ideas which we did not discussed here.

Chapter 3
The Standard Model of the Electroweak Interaction

3.1 Introduction

We will consider here the Glashow-Weinberg-Salam theory of the weak and electromagnetic interactions, which usually is called the Standard Model (SM). This theory is one of the greatest achievements of particle physics of the twentieth century. The SM predicted the existence of new particles (charmed, bottom, top), a new class of the weak interaction (Neutral currents), $W^{\pm}$ and Z^0 vector bosons and masses of these particles, the existence of the third type of neutrino (ν_τ), the existence of the scalar Higgs boson etc. All predictions of the Standard Model are in perfect agreement with existing experimental data. In 2012 in the ATLAS and CMS experiments at LHC (CERN) a scalar particle with the mass $\simeq 125\,\text{GeV}$ was discovered. All existing data (production cross section, decay rates) are compatible with the assumption that the discovered particle is the predicted by the Standard Model Higgs boson. The current $\times$ current theory of the weak interaction, which we considered in the previous chapter, was a very successful theory. In the lowest order of the perturbation theory this theory allowed to describe all experimental data existed at the 60s. However, the current $\times$ current theory and also the theory with the $W^{\pm}$ vector boson were unrenormalizable theories: the infinities of the higher orders of the perturbation theory could not be excluded in these theories by the renormalization of the masses and other physical constants.

This was the main reason why, in spite of big phenomenological success, these theories for many years were not considered as satisfactory ones. The Standard Model was born in the end of the 60s in an attempt to build a renormalizable theory of the weak interaction. The only renormalizable physical theory, that was known at that time, was quantum electrodynamics. The renormalizable theory of the weak interaction was build in the framework of *the unification of the weak and electromagnetic (electroweak) interactions*. This theory was proposed by Glashow, Weinberg and Salam. It was proved by t'Hooft and Veltman that the Standard Model is a renormalizable theory.

© Springer International Publishing AG, part of Springer Nature 2018
S. Bilenky, *Introduction to the Physics of Massive and Mixed Neutrinos*,
Lecture Notes in Physics 947, https://doi.org/10.1007/978-3-319-74802-3_3

The Standard Model is built first as a gauge invariant theory of massless lepton, neutrino, quark and gauge boson fields. Then we assume that there exist also a scalar Higgs field, which interact with gauge boson, quark and lepton fields and possess (due to the symmetry) degenerate vacuum and nonzero vacuum expectation values. In such a system the symmetry is spontaneously broken. Spontaneous symmetry breaking generates mass terms of all fields except the electromagnetic field and apparently neutrino fields.

In the first two introductory sections of this chapter we will consider two major ingredients of the Standard Model: $SU(2)$ local gauge invariance and spontaneous symmetry breaking.

3.2 $SU(2)$ Yang-Mills Local Gauge Invariance

Let us assume that

$$\psi(x) = \begin{pmatrix} \psi^{(+1)}(x) \\ \psi^{(-1)}(x) \end{pmatrix}, \tag{3.1}$$

is the doublet of a $SU(2)$ group ($\psi^{(\pm 1)}(x)$ are spin 1/2 fields). If the masses of $\psi^{(\pm 1)}(x)$ fields are equal to m, the free Lagrangian of the field $\psi(x)$ is given by the expression

$$\mathscr{L}_0(x) = \bar{\psi}(x)\,(i\,\gamma^\alpha \partial_\alpha + m)\,\psi(x). \tag{3.2}$$

The Lagrangian (3.2) is invariant under *the global phase $SU(2)$ transformation*

$$\psi'(x) = U\,\psi(x), \quad \bar{\psi}'(x) = \bar{\psi}(x)\,U^+. \tag{3.3}$$

Here

$$U = e^{i\frac{1}{2}\tau\cdot\Lambda}, \tag{3.4}$$

where Λ_i are *arbitrary constants* ($\tau\cdot\Lambda = \sum_{i=1}^3 \tau_i\,\Lambda_i$, τ_i are the Pauli matrices).

Let us stress that the Lagrangian $\mathscr{L}_0(x)$ is invariant under the transformation (3.3) because (for the constant Λ_i) the derivative $\partial_\alpha\,\psi(x)$ is transformed in the same way as the field $\psi(x)$.

From the general Noether's theorem it follows that the invariance under the transformation (3.3) leads to the conservation of the isovector current

$$\partial_\alpha j_i^\alpha(x) = 0, \quad j_i^\alpha(x) = \bar{\psi}(x)\,\gamma^\alpha\,\frac{1}{2}\,\tau_i\,\psi(x). \tag{3.5}$$

From (3.5) it follows that the total isotopic spin $T_i = \int j_i^0(x)d^3x$ is conserved. The operator $U = e^{i\frac{1}{2}\boldsymbol{\tau}\cdot\boldsymbol{\Lambda}}$ is the operator of a rotation in a three-dimensional isotopic space around the direction of the vector $\boldsymbol{\Lambda}$ by the angle $|\boldsymbol{\Lambda}|$. Thus the global *SU*(2) invariance is the invariance under the rotations which are the same in all space-time points. However, because different space-time points are independent, the requirement of the invariance under *the local gauge SU*(2) *transformation*

$$\psi'(x) = U(x)\,\psi(x), \quad U(x) = e^{i\frac{1}{2}\,\boldsymbol{\tau}\cdot\boldsymbol{\Lambda}(x)}, \tag{3.6}$$

where $\Lambda_i(x)$ are *arbitrary functions of* x, is the more natural one.

For the derivative $\partial_\alpha\,\psi(x)$ we have

$$\partial_\alpha\,\psi(x) = U^+(x)\,U(x)\,\partial_\alpha\,U^+(x)\psi'(x)$$
$$= U^+(x)\,(\partial_\alpha + U(x)\,\partial_\alpha\,U^+(x))\,\psi'(x). \tag{3.7}$$

Because of the second term in this expression the free Lagrangian (3.2) is not invariant under the transformation (3.6). In fact, let us consider an infinitesimal *SU*(2) transformation

$$U(x) \simeq 1 + i\,\frac{1}{2}\,\boldsymbol{\tau}\cdot\boldsymbol{\Lambda}(x), \quad U^+(x) \simeq 1 - i\,\frac{1}{2}\,\boldsymbol{\tau}\cdot\boldsymbol{\Lambda}(x), \tag{3.8}$$

where parameters $\Lambda_i(x)$ are small and in all expansions over $\Lambda_i(x)$ we will keep only linear terms.

From (3.7) and (3.8) we find

$$\partial_\alpha\,\psi(x) = U^+(x)\,(\partial_\alpha - i\,\frac{1}{2}\,\boldsymbol{\tau}\cdot\partial_\alpha\,\Lambda(x))\,\psi'(x). \tag{3.9}$$

From (3.6) follows that

$$\psi(x) = U^+(x)\,\psi'(x). \tag{3.10}$$

Comparing (3.9) and (3.10) we conclude that the field $\psi(x)$ and the derivative $\partial_\alpha\psi(x)$ are *transformed differently*. This is the reason why the free Lagrangian (3.2) is not invariant under local gauge transformations (3.6).

In order to build a theory which is invariant under local gauge transformations we need to assume that exist vector-isovector bosons and the field of these bosons $A_\alpha^i(x)$ ($i = 1, 2, 3$) is transformed in such a way that the term $\frac{1}{2}\,\boldsymbol{\tau}\cdot\mathbf{A}_\alpha(x)$ is "absorbed".

Let us consider the covariant derivative

$$D_\alpha\,\psi(x) = (\partial_\alpha + i\,g\,\frac{1}{2}\,\boldsymbol{\tau}\cdot\mathbf{A}_\alpha(x))\,\psi(x), \tag{3.11}$$

where g is a dimensionless constant. We have

$$D_\alpha\,\psi(x) = U^+(x)\,U(x)\,D_\alpha\,U^+(x)\,\psi'(x). \tag{3.12}$$

We will consider now the term $U(x)\,D_\alpha\,U^+(x)$. Using (3.8) we find

$$U(x)\,D_\alpha\,U^+(x) = \partial_\alpha - i\,\frac{1}{2}\,\boldsymbol{\tau}\cdot\partial_\alpha\,\Lambda(x) + i\,g\,U(x)\,\frac{1}{2}\,\boldsymbol{\tau}\cdot\mathbf{A}_\alpha(x)\,U^+(x). \tag{3.13}$$

For the last term of (3.13) we have

$$U(x)\,\frac{1}{2}\,\boldsymbol{\tau}\cdot\mathbf{A}_\alpha(x)\,U^+(x) = \frac{1}{2}\,\boldsymbol{\tau}\cdot\mathbf{A}_\alpha(x) + i\,[\frac{1}{2}\,\boldsymbol{\tau}\cdot\Lambda(x)\,,\,\frac{1}{2}\,\boldsymbol{\tau}\cdot\mathbf{A}_\alpha(x)]$$

$$= \frac{1}{2}\,\boldsymbol{\tau}\cdot\mathbf{A}_\alpha(x) - \frac{1}{2}\,\boldsymbol{\tau}\cdot(\boldsymbol{\Lambda}\times\mathbf{A}_\alpha(x)), \tag{3.14}$$

where we take into account that $[\frac{1}{2}\tau_i,\,\frac{1}{2}\tau_k] = i\,e_{ikl}\,\frac{1}{2}\tau_l$. From (3.13) and (3.14) we obtain

$$U(x)\,D_\alpha\,U^+(x) = \partial_\alpha + i\,g\,\frac{1}{2}\,\boldsymbol{\tau}\cdot\mathbf{A}'_\alpha(x) = D'_\alpha. \tag{3.15}$$

Here

$$\mathbf{A}'_\alpha(x) = \mathbf{A}_\alpha(x) - \frac{1}{g}\,\partial_\alpha\,\Lambda(x) - \Lambda(x)\times\mathbf{A}_\alpha(x). \tag{3.16}$$

Thus, from (3.12) and (3.15) we have

$$D_\alpha\,\psi(x) = U^+(x)\,D'_\alpha\,\psi'(x) \tag{3.17}$$

Comparing (3.10) and (3.17) we conclude that under the local gauge $SU(2)$ transformations, which include the phase transformation (3.6) of the spinor field $\psi(x)$ and the gauge transformation (3.16) of the vector field $\mathbf{A}_\alpha(x)$,[1] the covariant derivative $D_\alpha\,\psi(x)$ and the field $\psi(x)$ are *transformed in the same way*.

   Thus, if in the free Lagrangian (3.2) we will make the change

$$\partial_\alpha\,\psi(x) \to D_\alpha\,\psi(x) \tag{3.18}$$

we obtain the Lagrangian

$$\mathscr{L}_1(x) = \bar{\psi}(x)\,(i\,\gamma^\alpha\,D_\alpha + m)\,\psi(x) \tag{3.19}$$

which is invariant under the local gauge transformations (3.6) and (3.16).

---

[1] The field $\mathbf{A}_\alpha(x)$ is called *the gauge field*.

The Lagrangian $\mathcal{L}_1(x)$ is the sum of the free Lagrangian of the field $\psi(x)$ and the Lagrangian of the interaction of the field $\psi(x)$ and the vector field $\mathbf{A}_\alpha(x)$. The total Lagrangian must include also the free Lagrangian of the field $\mathbf{A}_\alpha(x)$, which is invariant under the gauge transformation (3.16). In order to build the free Lagrangian of the field $\mathbf{A}_\alpha(x)$ let us consider the commutator $[D_\alpha, D_\beta]$. We have

$$[D_\alpha, D_\beta] = i\, g\, \frac{1}{2}\, \boldsymbol{\tau} \cdot \mathbf{F}_{\alpha\beta}(x), \tag{3.20}$$

where

$$\mathbf{F}_{\alpha\beta}(x) = \partial_\alpha \mathbf{A}_\beta(x) - \partial_\beta \mathbf{A}_\alpha(x) - g\, \mathbf{A}_\alpha(x) \times \mathbf{A}_\beta(x) \tag{3.21}$$

is the stress tensor. From (3.15) we find the following relation

$$U\,[D_\alpha, D_\beta]\,U^+ = [D'_\alpha, D'_\beta]. \tag{3.22}$$

Further, from (3.20) and (3.22) we have

$$U(x)\frac{1}{2}\,\boldsymbol{\tau} \cdot \mathbf{F}_{\alpha\beta}(x)\,U^+(x) = \frac{1}{2}\,\boldsymbol{\tau} \cdot \mathbf{F}'_{\alpha\beta}(x), \tag{3.23}$$

where

$$\mathbf{F}'_{\alpha\beta}(x) = \partial_\alpha \mathbf{A}'_\beta(x) - \partial_\beta \mathbf{A}'_\alpha(x) - g\, \mathbf{A}'_\alpha(x) \times \mathbf{A}'_\beta(x). \tag{3.24}$$

Finally, from (3.8) from (3.23) we find that the stress tensor $\mathbf{F}_{\alpha\beta}(x)$ is transformed as an isotopic vector:

$$\mathbf{F}'_{\alpha\beta}(x) = \mathbf{F}_{\alpha\beta}(x) - \Lambda(x) \times \mathbf{F}_{\alpha\beta}(x). \tag{3.25}$$

Thus, the scalar product $\mathbf{F}_{\alpha\beta} \cdot \mathbf{F}^{\alpha\beta}$ is invariant.

The free Lagrangian of the vector field $\mathbf{A}_\alpha(x)$, which is invariant under the transformation (3.16), can be chosen in the form

$$\mathcal{L}'_0(x) = -\frac{1}{4}\,\mathbf{F}_{\alpha\beta}(x) \cdot \mathbf{F}^{\alpha\beta}(x). \tag{3.26}$$

The total Lagrangian of the spinor field $\psi(x)$ and the vector field $\mathbf{A}_\alpha(x)$, which is invariant under the transformations (3.6) and (3.16), is given by the following expression

$$\mathcal{L}(x) = \bar{\psi}(x)\left(i\,\gamma^\alpha(\partial_\alpha + i\,g\,\frac{1}{2}\,\boldsymbol{\tau} \cdot \mathbf{A}_\alpha(x)) + m\right)\psi(x) - \frac{1}{4}\,\mathbf{F}_{\alpha\beta}(x)\mathbf{F}^{\alpha\beta}(x). \tag{3.27}$$

Thus, the requirement of the local gauge $SU(2)$ invariance can be satisfied if exists the vector field $\mathbf{A}_\alpha(x)$ and the Lagrangian of the interaction has the form of the product of the isotopic vector current

$$\mathbf{j}_\alpha(x) = \bar{\psi}(x)\,\gamma_\alpha\,\frac{1}{2}\,\boldsymbol{\tau}\,\psi(x) \tag{3.28}$$

and the vector field:

$$\mathscr{L}_I(x) = -g\,\mathbf{j}_\alpha(x)\cdot\mathbf{A}^\alpha(x) = -g\sum_{i=1}^{3} j_\alpha^i(x)\,A^{\alpha i}(x), \tag{3.29}$$

The Lagrangian (3.29) can be written in the form

$$\mathscr{L}_I(x) = \left(-\frac{g}{2\sqrt{2}}\,j_\alpha(x)\,W^\alpha(x) + \text{h.c}\right) - g\,j_\alpha^3(x)\,A^{\alpha 3}(x). \tag{3.30}$$

Here

$$j_\alpha(x) = 2\,j_\alpha^{1+i2}(x), \quad W_\alpha(x) = \frac{1}{\sqrt{2}}\,A_\alpha^{1-i2}(x), \tag{3.31}$$

where

$$j_\alpha^{1\pm i2} = j_\alpha^1 \pm i j_\alpha^2, \quad A_\alpha^{1\pm i2} = A_\alpha^1 \pm i A_\alpha^2. \tag{3.32}$$

For the current $j_\alpha(x)$ we have

$$j_\alpha(x) = 2\,\bar{\psi}(x)\,\gamma_\alpha\,\frac{1}{2}\,(\tau_1 + i\tau_2)\,\psi(x) = 2\,\bar{\psi}^{(+1)}(x)\,\gamma_\alpha\,\psi^{(-1)}(x), \tag{3.33}$$

where $\psi^{(+1)}(x)$ and $\psi^{(-1)}(x)$ are components of the isotopic doublet with the third projections of the isotopic spin I_3 equal to $\frac{1}{2}$ and $-\frac{1}{2}$, correspondingly.

According to the Gell-Mann and Nishijima relation we have

$$Q = I_3 + \frac{1}{2}\,Y. \tag{3.34}$$

Here Q is the electric charge in the units $(-e)$ (e is the electric charge of the electron) and Y is the hypercharge ($Y = Q^{(1)} + Q^{(-1)}$). From (3.33) and (3.34) follows that the current $j_\alpha(x)$ changes the charges of particles by one ($\Delta Q = 1$). Thus, due to the conservation of the total electric charge the field $W^\alpha(x)$ is the field of the vector particles with electric charges equal to ± 1.

For the current $j_\alpha^3(x)$ we have

$$j_\alpha^3(x) = \bar{\psi}(x) \, \gamma_\alpha \, \frac{1}{2} \, \tau_3 \, \psi(x) = \frac{1}{2} (\bar{\psi}^{(+1)}(x) \, \gamma_\alpha \, \psi^{(+1)}(x) - \bar{\psi}^{(-1)}(x) \, \gamma_\alpha \, \psi^{(-1)}(x)).$$

$$(3.35)$$

The current $j_\alpha^3(x)$ does not change the electric charges of particles and hence $A_\alpha^3(x)$ is the field of neutral, vector particles.

Thus, we have built the $SU(2)$ local gauge invariant Yang-Mills theory with gauge fields which include charged as well as neutral vector fields. Such a theory will be used as a basis for the theory of the weak and electromagnetic interactions.

In conclusion we make the following remarks.

1. After the change

$$\partial_\alpha \, \psi(x) \rightarrow (\partial_\alpha + i \, g \, \frac{1}{2} \, \boldsymbol{\tau} \cdot \mathbf{A}_\alpha(x)) \, \psi(x) \qquad (3.36)$$

in the free Lagrangian of the spinor field $\psi(x)$ we came to the interaction Lagrangian (3.29), which has the form of the product of the isovector current $\mathbf{j}_\alpha(x)$ and the isovector field $\mathbf{A}^\alpha(x)$. The constant g is the interaction constant g.

It is necessary, however, to stress that the requirements of the local gauge invariance do not fix the form of the interaction Lagrangian. For example, to the interaction Lagrangian (3.29) we can add a tensor term

$$\mathscr{L}_I^T = \mu \, \bar{\psi}(x) \, \sigma_{\alpha\beta} \, \frac{1}{2} \boldsymbol{\tau} \, \psi(x) \, \mathbf{F}_{\alpha\beta}, \qquad (3.37)$$

which is invariant under the transformations (3.6) and (3.16). Let us stress, however, that in order to "absorb" the term $\partial_\alpha \Lambda(x)$ in (3.9) and to ensure the local gauge $SU(2)$ invariance we need to perform the change (3.36) which induces the interaction (3.29).[2] Thus, the interaction (3.29) is *the minimal gauge invariant interaction* of the spinor and vector gauge fields.

2. The mass term of the vector field $\frac{1}{2} m_A^2 \, \mathbf{A}_\alpha(x) \, \mathbf{A}^\alpha(x)$ is not invariant under the transformation (3.16). Thus, the local gauge invariance requires that $W_\alpha(x)$ and $A_\alpha^3(x)$ are fields of massless particles. This is the reason why in a realistic theory the local $SU(2)$ gauge symmetry is violated. In the next section we will discuss the mechanism of the spontaneous symmetry breaking.

---

[2]Let us stress that this is minimum what we have to do to ensure the local gauge $SU(2)$ invariance.

3.3 Spontaneous Symmetry Breaking. Brout-Englert-Higgs Mechanism

Let us consider the complex scalar field $\phi(x)$ and assume that the Lagrangian of the field is given by the expression

$$\mathscr{L} = \partial_\alpha \phi^\dagger \, \partial^\alpha \phi - V(\phi^\dagger \phi). \tag{3.38}$$

Here

$$V(\phi^\dagger \phi) = -\mu^2 \, \phi^\dagger \phi + \lambda \, (\phi^\dagger \phi)^2, \tag{3.39}$$

where μ^2 and λ are positive constants.

The first term of the Lagrangian (3.38) is the kinetic term. The second term $V(\phi^\dagger \phi)$ is the so called potential. Let us notice that the quadratic term $-\mu^2 \, \phi^\dagger \phi$ is not a mass term (it differs from a mass term by the sign).

The Lagrangian (3.38) is invariant under the global transformation

$$\phi'(x) = e^{i\Lambda} \, \phi(x), \tag{3.40}$$

where Λ is a constant arbitrary phase.

Let us find the minimum of the Hamiltonian

$$\mathscr{H} = \partial_0 \phi^\dagger \, \partial_0 \phi + \sum_{i=1}^{3} \partial_i \phi^\dagger \, \partial_i \phi + V(\phi^\dagger \phi). \tag{3.41}$$

Equation (3.39) can be written in the form

$$V(\phi^\dagger \phi) = \lambda \, (\phi^\dagger \phi - \frac{\mu^2}{2\lambda})^2 - \frac{\mu^4}{4\lambda}. \tag{3.42}$$

From this expression it is evident that the Hamiltonian reaches its minimum at the constant value of the field which satisfies the condition

$$\phi_0^\dagger \phi_0 = \frac{\mu^2}{2\lambda}. \tag{3.43}$$

Thus, the Hamiltonian reaches the minimum at

$$\phi = \phi_0 = \frac{v}{\sqrt{2}} e^{i\alpha}, \tag{3.44}$$

where α is a real, constant phase and

$$v = \sqrt{\frac{\mu^2}{\lambda}}. \tag{3.45}$$

Thus, in the case of the complex scalar field with potential (3.39):

- The Hamiltonian is minimal at *different from zero constant (vacuum) values of the field*.
- The minimum of the Hamiltonian is reached at an infinite number of vacuum fields given by (3.44). Obviously that this freedom is due to the global invariance of the theory.

The system possesses *one (any) vacuum field*. That means that in the case of the interaction (3.39) the vacuum state violates global invariance. Such a violation is called spontaneous.[3] We can choose

$$\phi_0 = \frac{v}{\sqrt{2}}. \tag{3.47}$$

Let us introduce real fields $\chi_1(x)$ and $\chi_2(x)$ which are connected with the complex field $\phi(x)$ by the relation

$$\phi(x) = \frac{v}{\sqrt{2}} + \frac{\chi_1 + i\chi_2}{\sqrt{2}}. \tag{3.48}$$

The fields $\chi_{1,2}(x)$ are determined in such a way that their vacuum values are equal to zero.

Notice that in the Quantum Field Theory instead of (3.44) we have

$$\langle 0|\phi(x)|0\rangle = \frac{v}{\sqrt{2}} e^{i\alpha}, \tag{3.49}$$

---

[3] A typical example of the spontaneous symmetry braking is the ferromagnetism. As a function of the magnetization $\mathbf{M}(t)$ in the homogeneous case the energy density is given by the expression

$$E = (\partial_t \mathbf{M})^2 + \alpha_1(\mathbf{M}\mathbf{M}) + \alpha_2(\mathbf{M}\mathbf{M})^2, \tag{3.46}$$

which follows from many-body theory which takes into account collective effects. Here $\alpha_2 > 0$ and $\alpha_1 = \beta(T - T_c)$, where $\beta > 0$ and T_c is the Curie temperature. It is obvious that the energy is invariant under the rotations. At $T < T_c$ the energy reaches minima at $(\mathbf{M}\mathbf{M})_0 = -\frac{\alpha_1}{2\alpha_2}$. From this equation it follows that $(\mathbf{M})_0 = \sqrt{-\frac{\alpha_1}{2\alpha_2}} \mathbf{e}$, where $\mathbf{e}$ is any unit vector. This degeneracy is due to the rotational symmetry of the Hamiltonian. At $T < T_c$ the magnetization of a homogenous ferromagnetic is a constant vector $(\mathbf{M})_0 = \sqrt{-\frac{\alpha_1}{2\alpha_2}} \mathbf{e}_0$. Thus, existence of the ferromagnetism violates the rotational symmetry of the initial Hamiltonian which follows from the symmetry of the basic Hamiltonian.

where $\langle 0|\phi(x)|0\rangle$ is the vacuum expectation value (vev) of the field $\phi(x)$. This relation means that

- vacuum states $|0\rangle$ are not empty states,
- due to global symmetry the vacuum states are degenerate.

Because a physical system possess one (any) vacuum state we can choose

$$\langle 0|\phi(x)|0\rangle = \frac{v}{\sqrt{2}}. \tag{3.50}$$

Let us introduce two quantum hermitian fields $\chi_1(x)$ and $\chi_2(x)$ in the following way:

$$\phi(x) = \frac{v + \chi_1(x) + i\chi_2(x)}{\sqrt{2}}. \tag{3.51}$$

It is evident that

$$\langle 0|\chi_{1,2}(x)|0\rangle = 0. \tag{3.52}$$

Thus, in the vacuum state $|0\rangle$, we have chosen, there are no particles, quanta of the fields $\chi_1(x)$ and $\chi_2(x)$.

Let us return back to the classical theory of the field $\phi(x)$. From (3.38) and (3.48) we obtain the following expression for the Lagrangian of the system[4]

$$\mathcal{L} = \frac{1}{2}\partial^\alpha \chi_1 \partial_\alpha \chi_1 + \frac{1}{2}\partial^\alpha \chi_2 \partial_\alpha \chi_2 - \frac{\lambda}{4}[\chi_1(\chi_1 + 2v) + \chi_2^2]^2. \tag{3.53}$$

The quadratic over χ_1 term

$$\lambda v^2 \chi_1^2 = \frac{1}{2}2\mu^2 \tag{3.54}$$

is the mass term of the field $\chi_1(x)$. There is no mass term of the field $\chi_2(x)$. Thus, the Lagrangian (3.53) is the Lagrangian of two interacting real scalar fields: the field $\chi_1(x)$ with mass $\sqrt{2}\mu$ and massless field $\chi_2(x)$.

According to *the general Goldstone theorem* spontaneous breaking of a continuous symmetry generates massless (Goldstone) particles. In our example imaginary part of the field $\phi(x)$, the field $\chi_2(x)$, is the goldstone field. Its appearance is connected with violation of the global invariance.

---

[4]The constant term $-\frac{\mu^4}{4\lambda}$ can be omitted.

We can come to the same conclusion if we parameterize the field $\phi(x)$ in the form

$$\phi(x) = \frac{v + h(x)}{\sqrt{2}} e^{i\frac{\theta(x)}{v}}, \tag{3.55}$$

where $h(x)$ and $\theta(x)$ are real fields. Their vacuum values are equal to zero. For the Lagrangian of the system we find the following expression

$$\mathscr{L} = \frac{1}{2}\partial^\alpha h\, \partial_\alpha h + \frac{1}{2}\partial^\alpha \theta\, \partial_\alpha \theta + (\frac{h}{v} + \frac{h^2}{2v^2})\partial^\alpha \theta\, \partial_\alpha \theta - \frac{\lambda}{4}[h(h + 2v)]^2. \tag{3.56}$$

The Lagrangian (3.56) is the Lagrangian of two interacting scalar real fields: the field $h(x)$ with mass $\sqrt{2}\mu$ and massless Goldstone field $\theta(x)$.

The appearance of Goldstone bosons is a problem for theories with spontaneous symmetry breaking: massless scalar bosons were not observed in experiments. In local gauge invariant theories, discussed in the previous section, vector gauge bosons are massless and this is another problem for a realistic theory. Brout, Englert and Higgs showed that in the theory based on the local gauge invariance and spontaneous violation of the symmetry Goldstone scalar bosons do not appear and gauge bosons are massive.

It is obvious that the global invariance with constant phase Λ do not allow us to remove the Goldstone massless field $\theta(x)$. Let us consider the scalar complex field $\phi(x)$ and vector gauge field $A_\alpha(x)$ with the following Lagrangian

$$\mathscr{L} = [(\partial^\alpha + ig A^\alpha)\phi]^\dagger (\partial_\alpha + ig A_\alpha)\phi - V(\phi^\dagger \phi) - \frac{1}{4}F^{\alpha\beta}F_{\alpha\beta}. \tag{3.57}$$

Here

$$F_{\alpha\beta} = \partial_\alpha A_\beta - \partial_\beta A_\alpha, \tag{3.58}$$

g is a real dimensionless constant and the potential $V(\phi^\dagger \phi)$ is given by (3.39).

The Lagrangian (3.57) is invariant under the local gauge transformations

$$\phi'(x) = e^{i\Lambda(x)}\phi(x), \quad A'_\alpha(x) = A_\alpha(x) - \frac{1}{g}\partial_\alpha \Lambda(x), \tag{3.59}$$

where $\Lambda(x)$ is an arbitrary function of x.

The Hamiltonian of the system has the minimum at the constant value of the scalar field which satisfies the condition

$$\phi_0^\dagger \phi_0 = \frac{v^2}{2}, \tag{3.60}$$

where the constant v is given by (3.45). Thus, we have

$$\phi_0 = \frac{v}{\sqrt{2}} e^{i\,\alpha}, \tag{3.61}$$

where α is an arbitrary phase. If we choose

$$\phi_0 = \frac{v}{\sqrt{2}} \tag{3.62}$$

we will spontaneously violate the symmetry.

The complex field $\phi(x)$ can be presented in the form

$$\phi(x) = \frac{v + h(x)}{\sqrt{2}} e^{i\frac{\theta(x)}{v}}, \tag{3.63}$$

where $h(x)$ and $\theta(x)$ are real functions. Their vacuum values are equal to zero.

After spontaneous symmetry breaking the Lagrangian (3.57) continue to be invariant under the transformation (3.59). We can choose the gauge function $\Lambda(x)$ in such a way that

$$\phi(x) = \frac{v + h(x)}{\sqrt{2}}. \tag{3.64}$$

Such gauge is called the unitary one.

From (3.57) and (3.64) for the Lagrangian of the real scalar field $\chi(x)$ and the real vector field $A_\alpha(x)$ we obtain the following expression

$$\mathscr{L} = \frac{1}{2}\partial^\alpha h\,\partial_\alpha h + \frac{g^2}{2} A^\alpha A_\alpha (v+h)^2 - \frac{\lambda}{4}[h(h+2v)]^2 - \frac{1}{4}F^{\alpha\beta}F_{\alpha\beta}. \tag{3.65}$$

The Lagrangian (3.65) is the Lagrangian of the interacting real scalar field $h(x)$ with mass $m_h = \sqrt{2}\mu$ and the real vector field $A_\alpha(x)$ with mass $m_A = gv$. There is no massless Goldstone field in the system.

This mechanism of the generation of the mass of the vector field is called *the Brout-Englert-Higgs mechanism*. The field $\phi(x)$ is called Higgs field.

Thus, from the local gauge invariant theory of the interacting complex massless scalar field and the real massless vector gauge field, after spontaneous breaking of symmetry we came to the theory of a massive neutral vector field and a massive scalar Higgs field. The massless vector field is characterized by two degrees of freedom (two projections of the spin) while the massive vector field is characterized by three degrees of freedom (three projections of the spin). Hence, as a result of the spontaneous breaking of the symmetry, the Goldstone degree of freedom of the complex scalar field became an additional degree of freedom of the vector field (2+2 degrees of freedom became 1+3 degrees of freedom).

3.4 The Standard Model for Leptons and Quarks

In this section we will consider the unified theory of the weak and electromagnetic (electroweak) interactions (The Standard Model). The Standard Model is based on the following principles:

1. The local gauge $SU_L(2) \times U_Y(1)$ invariance of the Lagrangian of massless fields.
2. The unification of the weak and electromagnetic interactions.
3. The Brout-Englert-Higgs mechanism of the generation of masses of particles.

The Standard Model is the theory of spin 1/2 quarks, charged leptons, neutrinos, spin 1 gauge vector bosons and spin 0 Higgs bosons. The Lagrangian of the theory is built in such a way to include the phenomenological Lagrangian of the $V - A$ charged current interaction, which describes the β-decay of nuclei, μ-decay, π-decay, decay of strange particles, neutrino processes and many other processes.

We have seen in the first chapter that into the charged current enter left-handed components of the lepton and quark fields. Let us assume that[5]

$$\psi_{eL}^{lep} = \begin{pmatrix} \nu_{eL}' \\ e_L' \end{pmatrix}, \ \psi_{\mu L}^{lep} = \begin{pmatrix} \nu_{\mu L}' \\ \mu_L' \end{pmatrix}, \ \psi_{\tau L}^{lep} = \begin{pmatrix} \nu_{\tau L}' \\ \tau_L' \end{pmatrix} \qquad (3.66)$$

and

$$\psi_{1L}^{q} = \begin{pmatrix} u_L' \\ d_L' \end{pmatrix} \ \psi_{2L}^{q} = \begin{pmatrix} c_L' \\ s_L' \end{pmatrix} \ \psi_{3L}^{q} = \begin{pmatrix} t_L' \\ b_L' \end{pmatrix} \qquad (3.67)$$

are doublets of the $SU_L(2)$ group.

In the framework of the Quantum Field Theory it is natural to suggest that the $SU_L(2)$ symmetry is a local one. The local $SU_L(2)$ symmetry requires existence of a gauge vector field $\mathbf{A}_\alpha(x)$. It will be ensured if in the free Lagrangian

$$\mathscr{L}(x) = \sum_{l=e,\mu,\tau} \bar{\psi}_{lL}^{lep}(x) i\gamma^\alpha \partial_\alpha \psi_{lL}^{lep}(x) + \sum_{i=1}^{3} \bar{\psi}_{iL}^{q}(x) i\gamma^\alpha \partial_\alpha \psi_{iL}^{q}(x) + \dots \qquad (3.68)$$

we will make the following change

$$\partial_\alpha \psi_{lL}^{lep}(x) \to (\partial_\alpha + ig\frac{1}{2}\boldsymbol{\tau}\mathbf{A}_\alpha(x))\psi_{lL}^{lep}(x), \quad \partial_\alpha \psi_{iL}^{q}(x) \to (\partial_\alpha + ig\frac{1}{2}\boldsymbol{\tau}\mathbf{A}_\alpha(x))\psi_{iL}^{q}(x).$$

$$(3.69)$$

---

[5]The meaning of primes will be clear later.

We come to the following Lagrangian of the interaction of leptons and quarks with gauge vector bosons

$$\mathscr{L}_{\mathscr{I}} = -g\mathbf{j}_\alpha(x)\mathbf{A}^\alpha(x). \tag{3.70}$$

Here

$$\mathbf{j}_\alpha(x) = \mathbf{j}_\alpha^{\text{lep}}(x) + \mathbf{j}_\alpha^q(x), \tag{3.71}$$

where

$$\mathbf{j}_\alpha^{\text{lep}}(x) = \sum_{l=e,\mu,\tau} \bar\psi_{lL}^{\text{lep}}(x)\frac{1}{2}\boldsymbol{\tau}\gamma_\alpha\psi_{lL}^{\text{lep}}(x) \tag{3.72}$$

and

$$\mathbf{j}_\alpha^q(x) = \sum_{i=1}^{3} \bar\psi_{iL}^q(x)\frac{1}{2}\boldsymbol{\tau}\gamma_\alpha\psi_{iL}^q(x) \tag{3.73}$$

are lepton and quark isovector currents.

Let us stress the following

1. $SU_L(2)$ is the nonabelian group. The interaction constant enters into the stress tensor (see (3.21)). This means that the interaction constants for different lepton and quark multiplets must be the same (*universality*).
2. The interaction (3.70) is *the minimal interaction* compatible with requirements of the local $SU_L(2)$ gauge invariance.

We can rewrite the Lagrangian (3.70) in the form

$$\mathscr{L}_I = (-\frac{g}{2\sqrt{2}}\, j_\alpha^{CC}\, W^\alpha + h.c.) - g\, j_\alpha^3 A^{3\alpha}. \tag{3.74}$$

Here

$$j_\alpha^{CC} = 2\, j_\alpha^{1+i2} = 2\,(\bar u'_L\,\gamma_\alpha\,d'_L + \bar c'_L\,\gamma_\alpha\,s'_L + \bar t'_L\,\gamma_\alpha\,b'_L) + 2\sum_{l=e,\mu,\tau}\bar\nu'_{lL}\,\gamma_\alpha\,l'_L \tag{3.75}$$

is the charged current of quarks and leptons,

$$W_\alpha = \frac{1}{\sqrt{2}}\, A_\alpha^{1-i2} = \frac{1}{\sqrt{2}}\,(A_\alpha^1 - i\, A_\alpha^2) \tag{3.76}$$

is the field of the charged $W^{\pm}$-bosons and the current j_{α}^3 is given by the following expression

$$j_{\alpha}^3 = \frac{1}{2} \sum_{q=u,c,t} \bar{q}_L' \gamma_{\alpha} q_L' - \frac{1}{2} \sum_{q=d,s,b} \bar{q}_L' \gamma_{\alpha} q_L' + \frac{1}{2} \sum_{l=e,\mu,\tau} \bar{\nu}_{lL}' \gamma_{\alpha} \nu_{lL}' - \frac{1}{2} \sum_{l=e,\mu,\tau} \bar{l}_L' \gamma_{\alpha} l_L'.$$

$$(3.77)$$

The first term of the expression (3.74) is the CC Lagrangian. The last term of this expression is the Lagrangian of the interaction of quarks and leptons with A_{α}^3, the field of neutral vector particles. It is obvious that this term can not be identified with the Lagrangian of the electromagnetic interaction: in the current j_{α}^3 enter only left-handed components of the fields while in the electromagnetic current enter left-handed and right-handed components of quark and charged lepton fields. Thus, the CC weak interaction which violates parity and the electromagnetic interaction which conserve parity can not be unified on the basis of the local $SU_L(2)$ group.

In order to build the unified theory of the weak and electromagnetic interactions it is necessary to enlarge the symmetry group. A new interaction Lagrangian must include the CC Lagrangian and the Lagrangian of the electromagnetic interaction. The minimal group on the basis of which the weak and electromagnetic interactions can be unified is the direct product $SU_L(2) \times U_Y(1)$, where $U_Y(1)$ is the group of the hypercharge.

The local $SU_L \times U_Y(1)$ gauge invariance will be ensured if in the free Lagrangian the left-handed lepton and quark doublets we make the following change

$$\partial_{\alpha} \psi_{lL}^{\text{lep}}(x) \rightarrow (\partial_{\alpha} + i g \frac{1}{2} \boldsymbol{\tau} \mathbf{A}_{\alpha}(x) + i g' \frac{1}{2} Y_L^{\text{lep}} B_{\alpha}(x)) \psi_{lL}^{\text{lep}}(x) \qquad (3.78)$$

and

$$\partial_{\alpha} \psi_{iL}^q(x) \rightarrow (\partial_{\alpha} + i g \frac{1}{2} \boldsymbol{\tau} \mathbf{A}_{\alpha}(x) + i g' \frac{1}{2} Y_L^q B_{\alpha}(x)) \psi_{iL}^q(x). \qquad (3.79)$$

In the free Lagrangian of the right-handed lepton and quark singlets we will perform the change

$$\partial_{\alpha} l_R'(x) \rightarrow (\partial_{\alpha} + i g' \frac{1}{2} Y_R^{\text{lep}} B_{\alpha}(x)) l_R'(x), \qquad (3.80)$$

$$\partial_{\alpha} q_R'(x) \rightarrow (\partial_{\alpha} + i g' \frac{1}{2} Y_R^{\text{up}} B_{\alpha}(x)) q_R'(x), \quad q = u, c, t \qquad (3.81)$$

and

$$\partial_{\alpha} q_R'(x) \rightarrow (\partial_{\alpha} + i g' \frac{1}{2} Y_R^{\text{down}} B_{\alpha}(x)) q_R'(x), \quad q = d, s, b. \qquad (3.82)$$

Here $B_\alpha(x)$ is $U_Y(1)$ gauge vector field and the interaction constants are written in the form $g'\frac{1}{2}Y_L^{\text{lep}}$, $g'\frac{1}{2}Y_L^q$, etc. Because $U_Y(1)$ is an abelian group, Y_L^{lep}, Y_L^q etc. are arbitrary constants. We will choose them in accordance with the Gell-Mann-Nishijima relation

$$Q = I_3 + \frac{1}{2}Y, \tag{3.83}$$

where Q is the electric charge. For the left-handed doublets and right-handed singlets we have, correspondingly,

$$Y_L = Q^{(1)} + Q^{(-1)}, \quad Y_R = 2Q. \tag{3.84}$$

From these relations we find

$$Y_L^{\text{lep}} = -1, \ Y_L^q = \frac{1}{3}, \ Y_R^{\text{lep}} = -2, \ Y_R^{up} = \frac{4}{3}, \ Y_R^{down} = -\frac{2}{3}. \tag{3.85}$$

From (3.78)–(3.82) for the Lagrangian of the minimal interaction of the quark, lepton and vector fields we will find the following expression

$$\mathscr{L}_I = -g\,\mathbf{j}_\alpha\,\mathbf{A}^\alpha - g'\,\frac{1}{2}\,j_\alpha^Y\,B^\alpha, \tag{3.86}$$

where the isovector current $\mathbf{j}_\alpha$ is given by (3.71) and $\frac{1}{2}j_\alpha^Y(x)$ is the total hypercurrent

$$\frac{1}{2}j_\alpha^Y(x) = \frac{1}{2}j_\alpha^{Y\text{lep}}(x) + \frac{1}{2}j_\alpha^{Yq}(x). \tag{3.87}$$

Here

$$\frac{1}{2}j_\alpha^{Y\text{lep}} = \sum_{l=e,\mu,\tau} \bar\psi_{lL}^{\text{lep}}\frac{1}{2}Y_L^{\text{lep}}\gamma_\alpha\psi_{lL}^{\text{lep}} + \sum_{l=e,\mu,\tau} \bar l_R'\frac{1}{2}Y_R^{\text{lep}}\gamma_\alpha l_R' \tag{3.88}$$

is the lepton hypercurrent and[6]

$$\frac{1}{2}j_\alpha^{Yq} = \sum_{i=1}^{3} \bar\psi_{iL}^q\frac{1}{2}Y_L^q\gamma_\alpha\psi_{iL}^q + \sum_{q=u,c,t} \bar q_R'\frac{1}{2}Y_R^{up}\gamma_\alpha q_R' + \sum_{q=d,s,b} \bar q_R'\frac{1}{2}Y_R^{down}\gamma_\alpha q_R' \tag{3.89}$$

is the quark hypercurrent.

---

[6]Let us notice that hypercharges of the right-handed neutrino singlets ν_{lR}' are equal to zero. Thus neutrino singlets ν_{lR}' can not contribute to the hypercurrent. As we will see later, the right-handed neutrino fields can enter only into neutrino mass terms.

The electric charges of left-handed and right-handed components of the lepton fields are the same. Thus for the leptonic hypercurrent we have

$$\frac{1}{2} j_\alpha^{Ylep} = (-1) \sum_{l=e,\mu,\tau} \bar{l}'_L \gamma_\alpha l'_L + (-1) \sum_{l=e,\mu,\tau} \bar{l}'_R \gamma_\alpha l'_R - j_\alpha^{3lep} = j_\alpha^{EMlep} - j_\alpha^{3lep},$$

(3.90)

where

$$j_\alpha^{EMlep} = (-1) \sum_{l=e,\mu,\tau} \bar{l}' \gamma_\alpha l'$$

(3.91)

is the electromagnetic current of the charged leptons.

Analogously for the quarks we have

$$\frac{1}{2} j_\alpha^{Yq} = j_\alpha^{EMq} - j_\alpha^{3q},$$

(3.92)

where

$$j_\alpha^{EMq} = \frac{2}{3} \sum_{q=u,c,t} \bar{q}' \gamma_\alpha q' - \frac{1}{3} \sum_{q=d,s,b} \bar{q}' \gamma_\alpha q'$$

(3.93)

is the quark electromagnetic current.

For the total hypercurrent of leptons and quarks we obviously find

$$\frac{1}{2} j_\alpha^{Y} = j_\alpha^{EM} - j_\alpha^{3}.$$

(3.94)

The total $SU_L \times U_Y(1)$ invariant Lagrangian of the minimal interaction of quarks and leptons with gauge vector bosons takes the form

$$\mathcal{L}_I = -g \mathbf{j}_\alpha \mathbf{A}^\alpha - g'(j_\alpha^{EM} - j_\alpha^{3}) B^\alpha.$$

(3.95)

From the requirement of the local gauge $SU_L \times U_Y(1)$ invariance follows that fields of gauge vector bosons, quarks and leptons are massless. The Standard Model is based of the Brout-Englert- Higgs mechanism of the generation of masses of leptons, quarks and intermediate vector bosons. This mechanism assumes existence of a scalar Higgs field with nonzero vacuum expectation values and degeneracy of vacuum states. The degeneracy is a result of a gauge symmetry of the Lagrangian. Because a physical system occupies one vacuum state a gauge symmetry in the case of the Higgs field is spontaneously broken.

In order for the theory to be invariant under $SU_L(2) \times U_Y(1)$ transformations the Higgs fields must have definite $SU_L(2) \times U_Y(1)$ transformation properties. Let us assume that the Higgs field $H(x)$ is the $SU_L(2)$ doublet

$$H(x) = \begin{pmatrix} H_+(x) \\ H_0(x) \end{pmatrix}, \tag{3.96}$$

where $H_+(x)$ is the scalar complex field of particles with electric charges equal to ± 1 and $H_0(x)$ is the complex field of particles with electric charge equal to zero. From the Gell-Mann-Nishijima relation (3.83) follows that the hypercharge of the field $H(x)$ is equal to one ($Y_H = 1 + 0 = 1$).

For the free Lagrangian of the Higgs doublet we have

$$\mathscr{L}_0 = \partial_\alpha H^\dagger \partial^\alpha H - V(H^\dagger H). \tag{3.97}$$

Here the potential $V(H^\dagger H)$ is given by the expression

$$V(H^\dagger H) = -\mu^2 H^\dagger H + \lambda (H^\dagger H)^2 = \lambda \left(H^\dagger H - \frac{\mu^2}{2\lambda} \right)^2 - \frac{\mu^4}{4\lambda}, \tag{3.98}$$

where μ^2 and λ are positive constants.

In order to ensure $SU_L(2) \times U_Y(1)$ invariance of the theory in (3.97) we must change the derivative $\partial_\alpha H$ by the covariant derivative:

$$\partial_\alpha H \to (\partial_\alpha + i\, g\, \frac{1}{2}\, \boldsymbol{\tau} \cdot \mathbf{A}_\alpha + i\, g'\, \frac{1}{2}\, B_\alpha)\, H, \tag{3.99}$$

where $\mathbf{A}_\alpha$ and B_α are $SU_L(2)$ and $U_Y(1)$ gauge fields. The Lagrangian takes the form

$$\mathscr{L} = \left((\partial_\alpha + i\, g\, \frac{1}{2}\, \boldsymbol{\tau} \cdot \mathbf{A}_\alpha + i\, g'\, \frac{1}{2}\, B_\alpha)\, H \right)^\dagger \left((\partial^\alpha + i\, g\, \frac{1}{2}\, \boldsymbol{\tau} \cdot \mathbf{A}^\alpha + i\, g'\, \frac{1}{2}\, B^\alpha)\, H \right)$$

$$-V(H^\dagger H). \tag{3.100}$$

The potential (3.98) reaches a minimum at such constant values of the Higgs field which satisfy the relation

$$(H^\dagger H)_0 = \frac{v^2}{2}, \tag{3.101}$$

where

$$v^2 = \frac{\mu^2}{\lambda}. \tag{3.102}$$

Because of the conservation of the electric charge the vacuum expectation value of the charged field H_+ is equal to zero. From (3.101) follows that we can choose H_0

in the form

$$H_0 = \begin{pmatrix} 0 \\ \frac{v}{\sqrt{2}} \end{pmatrix}. \tag{3.103}$$

The complex scalar doublet $H(x)$ can be presented as follows

$$H(x) = e^{i \frac{1}{2} \frac{\boldsymbol{\tau} \cdot \boldsymbol{\theta}(x)}{v}} \begin{pmatrix} 0 \\ \frac{v+h(x)}{\sqrt{2}} \end{pmatrix}, \tag{3.104}$$

where $\theta_i(x)$ ($i = 1, 2, 3$) and $h(x)$ are real functions. The parametrization (3.104) was chosen in such a way that the vacuum values of the functions $\theta_i(x)$ and $H(x)$ are equal to zero.

After the spontaneous symmetry breaking the Lagrangian of the system, we are considering, remains invariant under the $SU(2) \times U(1)$ local gauge transformations. Let us choose the gauge in such a way that

$$H(x) = \begin{pmatrix} 0 \\ \frac{v+h(x)}{\sqrt{2}} \end{pmatrix}. \tag{3.105}$$

With this choice of the gauge (which is called the unitary gauge) we find the following expression for the Lagrangian (3.100)

$$\mathscr{L} = \frac{1}{2} \partial_\alpha h \, \partial^\alpha h + H^\dagger (\frac{g}{2} \boldsymbol{\tau} \cdot \mathbf{A}_\alpha + \frac{g'}{2} B_\alpha)(\frac{g}{2} \boldsymbol{\tau} \cdot \mathbf{A}^\alpha + \frac{g'}{2} B^\alpha) H - V. \tag{3.106}$$

Let us consider the different terms of this expression. Taking into account that

$$\tau_i \, \tau_k = \delta_{ik} + i \, e_{ikl} \, \tau_l \tag{3.107}$$

we have

$$\boldsymbol{\tau} \cdot \mathbf{A}_\alpha \, \boldsymbol{\tau} \cdot \mathbf{A}^\alpha = \mathbf{A}_\alpha \, \mathbf{A}^\alpha = 2 \, W_\alpha^\dagger \, W^\alpha + A_\alpha^3 \, A^{3\alpha}, \tag{3.108}$$

where W^α is the field of the charged $W^\pm$ bosons given by Eq. (3.76).
Further, from (3.105) we find

$$H^\dagger \, \boldsymbol{\tau} \cdot \mathbf{A}_\alpha \, H = -\frac{1}{2} (v + h)^2 \, A_\alpha^3. \tag{3.109}$$

From (3.98), (3.105), (3.108) and (3.109) we obtain the following expression for the Lagrangian (3.106)

$$\mathscr{L} = \frac{1}{2} \partial_\alpha h \, \partial^\alpha h + \frac{g^2}{4} (v+h)^2 \, W_\alpha^\dagger \, W^\alpha + \frac{(g^2 + g'^2)}{8} (v+h)^2 \, Z_\alpha \, Z^\alpha - \frac{\lambda}{4} (2vh + h^2)^2, \tag{3.110}$$

where

$$Z_\alpha = \frac{g}{\sqrt{g^2 + g'^2}} A_\alpha^3 - \frac{g'}{\sqrt{g^2 + g'^2}} B_\alpha. \tag{3.111}$$

The field A_α given by the relation

$$A_\alpha = \frac{g'}{\sqrt{g^2 + g'^2}} A_\alpha^3 + \frac{g}{\sqrt{g^2 + g'^2}} B_\alpha. \tag{3.112}$$

is orthogonal to Z_α.

The Lagrangian (3.110) includes the mass terms of the vector W^α and Z^α fields and the mass term of the scalar field h:

$$\mathscr{L}^m = m_W^2\, W_\alpha^\dagger\, W^\alpha + \frac{1}{2}\, m_Z^2\, Z_\alpha\, Z^\alpha - \frac{1}{2}\, m_h^2\, h^2. \tag{3.113}$$

Here

$$m_W^2 = \frac{1}{4}\, g^2\, v^2, \quad m_Z^2 = \frac{1}{4}\, (g^2 + g'^2)\, v^2, \quad m_h^2 = 2\,\lambda\, v^2 = 2\,\mu^2. \tag{3.114}$$

Thus, after spontaneous symmetry breaking $W^\alpha(x)$ became the field of the charged vector $W^\pm$ bosons with the mass $m_W = \frac{1}{2} g\, v$, $Z^\alpha(x)$ became the field of neutral vector Z^0 bosons with the mass $m_Z = \frac{1}{2}\sqrt{g^2 + g'^2}\, v$. The field $A_\alpha(x)$ remained massless.

Three Goldston degrees of freedom of the Higgs doublet provided the masses of the $W^\pm$ and Z^0 bosons. The fourth degree of freedom, the neutral scalar field $h(x)$, is the field of the scalar Higgs bosons with the mass $m_h = \sqrt{2}\mu$. Higgs boson was discovered at LHC in CERN in 2012.

Let us introduce weak (Weinberg) angle θ_W by the relation

$$g' = g\, \tan\theta_W. \tag{3.115}$$

Taking into account that

$$\frac{g}{\sqrt{g^2 + g'^2}} = \cos\theta_W, \quad \frac{g'}{\sqrt{g^2 + g'^2}} = \sin\theta_W. \tag{3.116}$$

From (3.111) and (3.112) we find

$$A_\alpha^3 = \cos\theta_W Z_\alpha + \sin\theta_W A_\alpha, \quad B_\alpha = -\sin\theta_W Z_\alpha + \cos\theta_W A_\alpha. \tag{3.117}$$

Let us now consider the interaction Lagrangian (3.95). It can be written in the form

$$\mathscr{L}_I = \left(-\frac{g}{2\sqrt{2}} j_\alpha^{CC} W^\alpha + \text{h.c}\right) + \mathscr{L}_I^0, \tag{3.118}$$

where

$$j_\alpha^{CC} = 2j_\alpha^{1+i2} = 2(j_\alpha^1 + ij_\alpha^2) \tag{3.119}$$

and

$$\mathscr{L}_I^0 = -g\, j_\alpha^3 A^{3\alpha} - g'\, (j_\alpha^{EM} - j_\alpha^3) B^\alpha. \tag{3.120}$$

The first term of (3.118) is the Lagrangian of the CC interaction of the quarks and leptons with $W^\pm$ bosons and the second term is the Lagrangian of the interaction of the quarks and leptons with neutral vector bosons. Taking into account (3.117) we have

$$\mathscr{L}^0 = -g\sin\theta_W j_\alpha^{EM} A^\alpha - \frac{g}{2\cos\theta_W} j_\alpha^{NC} Z^\alpha. \tag{3.121}$$

Here

$$j_\alpha^{NC} = 2(j_\alpha^3 - \sin^2\theta_W j_\alpha^{EM}) \tag{3.122}$$

The first term of the Lagrangian (3.121) is a product of the electromagnetic current and the massless vector field $A^\alpha(x)$. It can be identified with the Lagrangian of the electromagnetic interaction

$$\mathscr{L}^{EM} = -e j_\alpha^{EM} A^\alpha \tag{3.123}$$

if the constants g and $\sin\theta_W$ satisfy the following unification constraint

$$g\sin\theta_W = e. \tag{3.124}$$

Here $(-e)$ is the charge of the electron. The massless vector field A^α is the electromagnetic field. After the spontaneous breaking of the $SU_L(2) \times U_Y(1)$ symmetry the Lagrangian of the system is invariant under the transformations of the local $U_{EM}(1)$ group.

The second term of (3.121) is the Lagrangian of the interaction of quarks and massive neutral vector Z^0 bosons. The weak current (3.122), which does not change the electric charges of quarks and leptons is called the neutral current (NC). Before the SM appeared, only CC interaction was unknown. The unification of the CC weak and electromagnetic interactions on the basis of the local gauge $SU_L(2) \times U_Y(1)$ group allowed *to predict the existence of the massive neutral vector Z^0-boson and* *a new type of the weak interaction (NC).* Neutral current processes were discovered in 1973 at CERN in experiments on the bubble chamber Gargamelle.

We will now turn to the consideration of the generation of the masses of quarks and leptons. Let us start with quarks. The mass term of the quark field $q(x)$ has the form

$$\mathscr{L}_m = -m_q \bar{q}\, q = -m_q \bar{q}_L\, q_R + \text{h.c.,} \qquad (3.125)$$

where m_q is the mass of the q-quark ($q = d, u, s \ldots$) In the SM left-handed fields are components of $SU(2)$ doublets and right-handed fields are $SU(2)$ singlets. Thus, the quark mass term is not invariant under the $SU(2) \times U(1)$ transformations.

Masses and mixing of the quarks are generated in the SM via the mechanism of the spontaneous symmetry breaking. Let us assume that in the total Lagrangian of the Standard Model there is the following Lagrangian of the Yukawa interaction of the quark and Higgs fields

$$\mathscr{L}_Y^{\text{down}} = -\sqrt{2} \sum_{a,q} \bar{\psi}_{iL}\, Y_{iq}^{\text{down}}\, q_R'\, H + \text{h.c.,} \qquad (3.126)$$

where Y_{iq}^{down} is a complex 3×3 matrix. Because ψ_{iL} and H are the $SU(2)$ doublets and q_R' are singlets, it is obvious that the Lagrangian $\mathscr{L}_Y^{\text{down}}$ is the $SU_L(2)$ scalar. Let us also require $U_Y(1)$ invariance, i.e. the conservation of the hypercharge. The hypercharges of the quark and Higgs doublets are equal to 1/3 and 1, correspondingly. The Lagrangian (3.126) conserves the hypercharge if $2e_{q_R'} + 1 = 1/3$. Thus, $e_{q_R'} = -1/3$, i.e. the right-handed fields q_R' in (3.126) are the "down" fields d_R', s_R', b_R'.

From (3.103) and (3.126) after the spontaneous symmetry breaking we find

$$\mathscr{L}_Y^{\text{down}} = -\overline{D}_L'\, Y^{down}\, D_R'\, (v + h) + \text{h.c.} \qquad (3.127)$$

Here

$$D_{L,R}' = \begin{pmatrix} d_{L,R}' \\ s_{L,R}' \\ b_{L,R}' \end{pmatrix}. \qquad (3.128)$$

The first term of (3.127) is the mass term of the down quarks and the second term is the Lagrangian of the interaction of the down quarks and the Higgs bosons.

In order to generate the mass term of up quarks we will use the conjugated Higgs doublet

$$\tilde{H} = i\, \tau_2 H^*. \qquad (3.129)$$

The hypercharge of the doublet $\tilde{H}$ is equal to -1. From (3.103) and (3.129) we have

$$\tilde{H} = \begin{pmatrix} \frac{v+h}{\sqrt{2}} \\ 0 \end{pmatrix}. \tag{3.130}$$

We will assume that in addition to (3.126) the following Lagrangian of the Yukawa interaction of quarks and Higgs bosons enters in the total Lagrangian

$$\mathscr{L}_Y^{\text{up}} = -\sqrt{2} \sum_{i,q} \overline{\psi}_{iL}^q \, Y_{iq}^{\text{up}} \, q_R' \, \tilde{H} + \text{h.c.}, \tag{3.131}$$

where Y^{up} is a complex 3×3 matrix. From the conservation of the hypercharge we have $2e_{q_R'} + (-1) = 1/3$. Thus, $e_{q_R'} = 2/3$ and the index q in (3.131) runs over u, c, t.

After the spontaneous symmetry breaking we find from (3.130) and (3.131)

$$\mathscr{L}_Y^{\text{up}} = -\overline{U}_L' \, Y^{\text{up}} \, U_R' \, (v+h) + \text{h.c.}, \tag{3.132}$$

where

$$U_{L,R}' = \begin{pmatrix} u_{L,R}' \\ c_{L,R}' \\ t_{L,R}' \end{pmatrix}. \tag{3.133}$$

The first term of (3.132) is the mass term of the up quarks and the second term is the Lagrangian of the interaction of the up quarks and the scalar Higgs bosons.

Let us now bring the mass terms of up and down quarks to the diagonal form. The complex matrices Y^{up} and Y^{down} can be diagonalized by the biunitary transformations (see Appendix B). We have

$$Y^{\text{up}} = V_L^{\text{up}} \, y^{\text{up}} \, V_R^{\text{up}\dagger} \qquad Y^{\text{down}} = V_L^{\text{down}} \, y^{\text{down}} \, V_R^{\text{down}\dagger}. \tag{3.134}$$

Here $V_{L,R}^{\text{up}}$ and $V_{L,R}^{\text{down}}$ are unitary 3×3 matrices and y^{up} and y^{down} are diagonal matrices with positive diagonal elements.

From (3.127), (3.132) and (3.134) we find the following expressions

$$\mathscr{L}_Y^{\text{up}} = -\overline{U} \, m^{\text{up}} \, U \, (1 + \frac{h}{v}) = -\sum_{q=u,c,t} m_q \, \overline{q} \, q \, (1 + \frac{h}{v}) \tag{3.135}$$

and

$$\mathscr{L}_Y^{\text{down}} = -\overline{D} \, m^{\text{down}} \, D \, (1 + \frac{h}{v}) = -\sum_{q=d,s,b} m_q \, \overline{q} \, q \, (1 + \frac{h}{v}). \tag{3.136}$$

Here

$$U = U_L + U_R = \begin{pmatrix} u \\ c \\ t \end{pmatrix}, \quad D = D_L + D_R = \begin{pmatrix} d \\ s \\ b \end{pmatrix}, \qquad (3.137)$$

and

$$m^{\text{up}} = \begin{pmatrix} m_u & 0 & 0 \\ 0 & m_c & 0 \\ 0 & 0 & m_t \end{pmatrix}, \quad m^{\text{down}} = \begin{pmatrix} m_d & 0 & 0 \\ 0 & m_s & 0 \\ 0 & 0 & m_b \end{pmatrix}. \qquad (3.138)$$

Matrices $U_{L,R}$ and $D_{L,R}$ in (3.137) are given by the relation

$$U_{L,R} = V_{L,R}^{\text{up}\dagger} U'_{L,R}, \quad D_{L,R} = V_{L,R}^{\text{down}\dagger} D'_{L,R}. \qquad (3.139)$$

For the quark masses we have

$$m_q = y_q v, \quad q = d, u, s, c, b, t \qquad (3.140)$$

Thus masses of quarks, generated by the Brout-Englert-Higgs mechanism, are proportional to the vacuum expectation value v. The dimensionless Yukawa constants y_q are not constrained by the symmetry of the Standard Model. They are parameters of the model. It follows from (3.135) that $q(x)$ is the field of q-quark with the mass m_q.

Let us consider now the charged current of the quarks. From (3.119), (3.128) and (3.133) we find

$$j_\alpha^{CCq}(x) = 2\, \bar{U}'_L(x)\, \gamma_\alpha\, D'_L(x) \qquad (3.141)$$

We will write down the charged current in terms of the fields of quarks with definite masses. Taking into account (3.139) and (3.141) we find

$$j_\alpha^{CCq}(x) = 2\, \bar{U}_L(x)\, \gamma_\alpha\, V\, D_L(x), \qquad (3.142)$$

where the matrix V is given by the relation

$$V = (V_L^{\text{up}})^\dagger\, V_L^{\text{down}} \qquad (3.143)$$

From (3.137) and (3.142) follows that the CC can be presented in the following form

$$j_\alpha^{CCq}(x) = 2\, [\bar{u}_L(x)\, \gamma_\alpha\, d_L^{\text{mix}}(x) + \bar{c}_L(x)\, \gamma_\alpha\, s_L^{\text{mix}}(x) + \bar{t}_L(x)\, \gamma_\alpha\, b_L^{\text{mix}}(x)]. \qquad (3.144)$$

Here

$$d_L^{\text{mix}}(x) = \sum_{q=d,s,b} V_{uq}\, q_L(x)$$

$$s_L^{\text{mix}}(x) = \sum_{q=d,s,b} V_{cq}\, q_L(x)$$

$$b_L^{\text{mix}}(x) = \sum_{q=d,s,b} V_{tq}\, q_L(x). \tag{3.145}$$

From (3.143) follows that V is a unitary matrix

$$V^\dagger V = 1. \tag{3.146}$$

We came to an important conclusion: due to the spontaneous breaking of the electroweak symmetry *the left-handed components of fields of the down quarks enter into the CC of the SM in mixed form*. The unitary 3×3 mixing matrix V is called the Cabibbo-Kobayashi-Maskawa (CKM) mixing matrix.

Let us stress that the mixing of quarks is due to the fact that the unitary matrices V_L^{up} and V_L^{down}, which connect left-handed primed and physical fields of up and down quarks, are *different*. It follows from (3.144) and (3.145) that because of the mixing the charged current changes the flavor of quarks ($d \rightarrow u$, $s \rightarrow u$, $c \rightarrow s$ etc.).

Let us now express the electromagnetic current through the fields of physical quarks. From (3.123) we have

$$j_\alpha^{EMq} = \frac{2}{3}\,(\bar{U}'_L\,\gamma_\alpha\,U'_L + \bar{U}'_R\,\gamma_\alpha\,U'_R) - \frac{1}{3}\,(\bar{D}'_L\,\gamma_\alpha\,D'_L + \bar{D}'_R\,\gamma_\alpha\,D'_R). \tag{3.147}$$

Taking into account the unitarity of the matrices $V_{L,R}^{up}$ and $V_{L,R}^{down}$ we find

$$j_\alpha^{EMq} = \frac{2}{3}\,\bar{U}\,\gamma_\alpha\,U - \frac{1}{3}\,D\,\gamma_\alpha\,D. \tag{3.148}$$

From (3.137) and (3.148) we have

$$j_\alpha^{EMq}(x) = \sum_{q=d,u,s,\dots} e_q\,\bar{q}(x)\,\gamma_\alpha\,q(x), \tag{3.149}$$

where $e_{u,c,t} = \frac{2}{3}$ and $e_{d,s,b} = -\frac{1}{3}$. Thus, we come to the standard expression for the electromagnetic current of quarks, which is diagonal in quark flavors.

Finally let us consider the quark neutral current of the Standard Model. From (3.122) we find

$$j_\alpha^{NCq} = \bar{U}'_L\,\gamma_\alpha\,U'_L - D'_L\,\gamma_\alpha\,D'_L - 2\,\sin^2\theta_W\,j_\alpha^{EM} \tag{3.150}$$

In order to come to the fields of the physical quarks we will use the relations (3.139). Taking into account that V_L^{up} and V_L^{down} are unitary matrices we find

$$
\begin{aligned}
j_\alpha^{\text{NCq}} &= \bar{U}_L \, \gamma_\alpha \, U_L - D_L \, \gamma_\alpha \, D_L - 2 \sin^2 \theta_W \, j_\alpha^{\text{EM}} \\
&= \sum_{q=u,c,t} \bar{q}_L \, \gamma_\alpha \, q_L - \sum_{q=d,s,b} \bar{q}_L \, \gamma_\alpha \, q_L - 2 \sin^2 \theta_W \, j_\alpha^{\text{EM}}
\end{aligned} \quad (3.151)
$$

From this expression we conclude that the *neutral current of the SM is diagonal in quark flavors.*

We will now consider the generation of masses of leptons. The $SU_L(2) \times U_Y(1)$ invariant Lagrangian of the Yukawa interaction of lepton and Higgs fields has the form

$$
\mathscr{L}_Y^{\text{lep}} = -\sqrt{2} \sum_{l_1,l_2} \overline{\psi}_{l_1 L}^{\text{lep}} \, Y_{l_1 l_2}^{\text{lep}} \, l_{2R}' \, H + \text{h.c.}, \quad (3.152)
$$

where Y^{lep} is a 3×3 complex matrix and H is the doublet of Higgs fields. If we choose for the field $H(x)$ the expression (3.105) the symmetry will be spontaneously broken and for the Lagrangian $\mathscr{L}_Y^{\text{lep}}$ we find the following expression

$$
\mathscr{L}_Y^{\text{lep}} = -\bar{L}_L' \, Y^{\text{lep}} \, L_R' \, (v + h) + \text{h.c.} \quad (3.153)
$$

Here

$$
L_{L,R}' = \begin{pmatrix} e_{L,R}' \\ \mu_{L,R}' \\ \tau_{L,R}' \end{pmatrix}. \quad (3.154)
$$

Let us now diagonalize the matrix Y^{lep}. We have

$$
Y^{\text{lep}} = U_L \, y^{\text{lep}} \, U_R^\dagger. \quad (3.155)
$$

where $U_{L,R}$ are unitary matrices and y^{lep} is a diagonal matrix with positive elements. From (3.153) and (3.155) we find

$$
\mathscr{L}_Y^{\text{lep}} = -\bar{L}_L \, y^{\text{lep}} \, L_R \, (v + h) + \text{h.c.}, \quad (3.156)
$$

where

$$
L_L = U_L^\dagger \, L_L', \quad L_R = U_R^\dagger \, L_R'. \quad (3.157)
$$

From (3.156) we obtain the following expression for the Lagrangian (3.153)

$$L_Y^{lep}(x) = -\bar{L}(x)\,m^{lep}\,L(x)(1 + \frac{h(x)}{v}) = -\sum_{l=e,\mu,\tau} m_l\,\bar{l}(x)\,l(x)\,(1 + \frac{h(x)}{v}).$$

(3.158)

Here

$$L = L_L + L_R = \begin{pmatrix} e \\ \mu \\ \tau \end{pmatrix}, \quad m^{lep} = \begin{pmatrix} m_e & 0 & 0 \\ 0 & m_\mu & 0 \\ 0 & 0 & m_\tau \end{pmatrix}.$$

(3.159)

and

$$m_l = y_l\,v, \quad l = e, \mu, \tau.$$

(3.160)

The first term of the Lagrangian (3.158) is the standard mass term of the charged leptons. The field $l(x)$ is the field of the charged leptons $l^\pm$ with the mass m_l ($l = e, \mu, \tau$). The second term of (3.158) is the Lagrangian of the interaction of the lepton and Higgs fields.

It is a common belief that neutrino masses are generated by a new beyond the Standard Model mechanism. In the next chapter we will discuss in details the problem of neutrino masses.

Let us consider now the lepton charged current. It can be written in the following matrix form

$$j_\alpha^{CClep}(x) = 2\,\bar{v}_L'(x)\,\gamma_\alpha\,L_L'(x),$$

(3.161)

where

$$v_L' = \begin{pmatrix} v_{eL}' \\ v_{\mu L}' \\ v_{\tau L}' \end{pmatrix}.$$

(3.162)

Taking into account (3.157), we have for the leptonic charged current

$$j_\alpha^{CClep}(x) = 2\,\bar{v}_L(x)\,\gamma_\alpha\,L_L(x) = 2\sum_{l=e,\mu,\tau} \bar{v}_{lL}(x)\,\gamma_\alpha\,l_L(x).$$

(3.163)

Here $l(x)$ is the field of lepton l with the mass m_l and

$$v_L = U_L^\dagger\,v_L' = \begin{pmatrix} v_{eL} \\ v_{\mu L} \\ v_{\tau L} \end{pmatrix}.$$

(3.164)

The field ν_{lL} ($l = e, \mu, \tau$) is called *the flavor neutrino field*. Let us consider now the electromagnetic current of the charged leptons. Taking into account the unitarity of the matrices $U_{L,R}$ we find from (3.91)

$$j_\alpha^{\text{EMlep}}(x) = -\bar{L}'_L(x)\, \gamma_\alpha\, L'_L(x) - \bar{L}'_R(x)\, \gamma_\alpha\, L'_R(x) = -\bar{L}(x)\, \gamma_\alpha\, L(x). \qquad (3.165)$$

From (3.165) and (3.159) we obtain the following standard expression for the electromagnetic current of the leptons

$$j_\alpha^{\text{EMlep}}(x) = \sum_{l=e,\mu,\tau} (-1)\, \bar{l}(x)\, \gamma_\alpha\, l(x). \qquad (3.166)$$

For the lepton neutral current we have from (3.122)

$$j_\alpha^{\text{NClep}}(x) = \bar{\nu}'_L(x)\, \gamma_\alpha\, \nu'_L(x) - \bar{L}'_L(x)\, \gamma_\alpha\, L'_L(x) - 2\sin^2\theta_W\, j_\alpha^{\text{EM}}(x). \qquad (3.167)$$

In terms of the flavor neutrino fields and fields of physical leptons from (3.167) we find the following expression for the neutrino and charged leptons neutral current

$$j_\alpha^{\text{NClep}}(x) = \sum_{l=e,\mu,\tau} \bar{\nu}_{lL}(x)\, \gamma_\alpha\, \nu_{lL}(x) - \sum_{l=e,\mu,\tau} \bar{l}_L(x)\, \gamma_\alpha\, l_L(x) - 2\sin^2\theta_W\, j_\alpha^{\text{EM}}(x), \qquad (3.168)$$

where $j_\alpha^{EM}(x)$ is given by (3.166).

Let us notice that if in the Standard Model neutrino masses are equal to zero, the total SM Lagrangian is invariant under the global transformations

$$\nu'_{lL}(x) = e^{i\,\Lambda_l}\, \nu_{lL}(x), \quad l'(x) = e^{i\,\Lambda_l}\, l(x), \quad q'(x) = q(x) \qquad (3.169)$$

where Λ_l ($l = e, \mu, \tau$) are arbitrary constant phases.

From the invariance under the transformations (3.169) follows that the total electron L_e, muon L_μ and tau L_τ lepton numbers are conserved :

$$\sum_i L_e^i = \text{const} \quad \sum_i L_\mu^i = \text{const} \quad \sum_i L_\tau^i = \text{const.} \qquad (3.170)$$

The flavor lepton numbers of the particles are presented in Table 3.1. The lepton numbers of the antiparticles are opposite to the lepton numbers of the corresponding particles. The conservation of the total flavor lepton numbers means that in the CC

Table 3.1 Lepton numbers of the particles

	ν_e, e^-	ν_μ, μ^-	ν_τ, τ^-	Quarks,W,Z,γ
L_e	1	0	0	0
L_μ	0	1	0	0
L_τ	0	0	1	0

decays together with a μ^+ a muon neutrino ν_μ is produced, in the process of the CC interaction of an electron antineutrino $\bar{\nu}_e$ with a nucleon a e^+ is produced etc. We know at present that the law of the conservation of electron, muon and tau lepton numbers is an approximate one. As we will discuss in details later, due to small neutrino masses and neutrino mixing the conservation law (3.170) is strongly violated in neutrino oscillations.

Let us summarize some results considered before. The SM Lagrangian of the interaction of quarks and leptons with $W^\pm$ and Z^0 bosons and γ-quanta is given by the sum of the CC Lagrangian, the NC Lagrangian and the electromagnetic Lagrangian:

$$\mathscr{L}_I = \left(-\frac{g}{2\sqrt{2}} j_\alpha^{CC} W^\alpha + \text{h.c} \right) - \frac{g}{2\cos\theta_W} j_\alpha^{NC} Z^\alpha - e j_\alpha^{EM} A^\alpha. \tag{3.171}$$

The Fermi constant G_F, which characterizes the effective four-fermion weak interaction induced by the exchange of the virtual W-boson at $Q^2 \ll m_W^2$, is connected with the constant g and mass of the W-boson by the relation

$$\frac{G_F}{\sqrt{2}} = \frac{g^2}{8\,m_W^2}. \tag{3.172}$$

The value of the Fermi constant is well known from investigation of the μ-decay and other low-energy CC weak processes. We have

$$G_F = 1.1663787(6) \times 10^{-5}\,\text{GeV}^{-2} \tag{3.173}$$

The mass of the W-boson is given by the relation (see (3.114))

$$m_W = \frac{1}{2} g\, v. \tag{3.174}$$

From (3.172) and (3.174) we have

$$\frac{g^2}{8m_W^2} = \frac{1}{2v^2} = \frac{G_F}{\sqrt{2}}. \tag{3.175}$$

Thus the constant v, the vacuum expectation value of the Higgs field, is determined by the Fermi constant:

$$v = (\sqrt{2}\, G_F)^{-1/2} \simeq 246\,\text{GeV}. \tag{3.176}$$

From (3.114) and (3.115) we obtain the following relation between the masses of the W and Z bosons

$$\frac{m_W}{m_Z} = \cos\theta_W. \tag{3.177}$$

Further, taking into account the unification condition (3.124), we find the following expressions for the masses of the W and Z bosons

$$m_W = \left(\frac{\pi \alpha}{\sqrt{2} G_F}\right)^{1/2} \frac{1}{\sin \theta_W}, \quad m_Z = \left(\frac{\pi \alpha}{\sqrt{2} G_F}\right)^{1/2} \frac{1}{\sin \theta_W \cos \theta_W}, \quad (3.178)$$

where $\alpha = \frac{e^2}{4\pi}$ is the fine structure constant. We have

$$\left(\frac{\pi \alpha}{\sqrt{2} G_F}\right)^{1/2} = 37.28039(1) \cdot \text{GeV} \quad (3.179)$$

The parameter $\sin^2 \theta_W$ characterizes the neutral current. The value of this parameter was determined from the data of numerous experiments on the investigation of NC processes. From the existing data it was found that

$$\sin^2 \theta_W = 0.23126(5) \quad (3.180)$$

If we take into account radiative corrections, the relation (3.178) for the mass of the W boson is modified. We have in this case

$$m_W = \left(\frac{\pi \alpha}{\sqrt{2} G_F}\right)^{1/2} \frac{1}{\sin \theta_W (1 - \Delta r)}, \quad (3.181)$$

where the term Δr is due to the radiative corrections. For this term the value $\Delta r = 0.03639 \pm 0.00036 \pm 0.00011$ was obtained. From existing data for the masses of the $W^{\pm}$ and Z^0 bosons the following values were found

$$m_Z = (91.1876 \pm 0.0021) \cdot \text{GeV} \quad m_W = (80.385 \pm 0.015) \cdot \text{GeV}. \quad (3.182)$$

These values are in a perfect agreement with the prediction of the Standard Model.

3.5 Concluding Remarks

The unified theory of the weak and electromagnetic interactions (The Standard Model) is based on the following basic principles

1. Local $SU_L(2) \times U_Y(1)$ invariance of the Lagrangian of massless quark, lepton and vector fields with left-handed doublets and right-handed singlets.
2. Brout-Englert-Higgs mechanism (with Higgs doublet) of the generations of the masses of $W^{\pm}$ and Z^0 bosons, the quarks and the charged leptons.
3. The unification of the weak and electromagnetic interactions.

The SM Lagrangian contains as a low-energy limit the effective classical current ×
current Lagrangian of the CC weak interaction.

In order the Standard Model will be a renormalizable theory it is necessary that
the sum of the electric charges of the particles, the fields of which are components
of the doublets, is equal to zero:

$$3\left(\frac{2}{3} + \left(-\frac{1}{3}\right)\right) N_f^{\text{quarks}} + (0 + (-1)) N_f^{\text{leptons}} = 0. \tag{3.183}$$

Here N_f^{quarks} and N_f^{leptons} are the numbers of the quark and lepton families. We took
into account in (3.183) that there exist three colored quarks of each type. Thus, we
have

$$N_f^{\text{quarks}} = N_f^{\text{leptons}}. \tag{3.184}$$

After the τ-lepton was discovered (1975), the Standard Model allowed to predict
the existence of the third neutrino ν_τ, $SU(2)$ partner of τ, and the third family of
quarks (b and t). The Standard Model predicted a new class of the weak interaction,
Neutral Currents, existence of the vector $W^\pm$ and Z^0 bosons and the masses of these
particles. It predicted existence of the neutral, scalar Higgs boson. All predictions
of the Standard Model were confirmed by experiments.

On the basis of this agreement of the Standard Model with experiment we can
conclude that in the framework of the basic principles on which the SM is based
Nature chooses *the simplest possibilities* ($SU(2)$ is the simplest nonabelian group,
the SM interaction Lagrangian is the simplest minimal one, the Higgs doublet is the
minimal possibility etc). Massless two-component neutrinos in the Standard Model
is the simplest possibility. Future experiments will allow to test this last possibility.

Chapter 4
Neutrino Mass Terms

4.1 Introduction

The neutrino mass term is the central object of the theory of massive and mixed neutrinos. It determines neutrino masses, neutrino mixture, neutrino nature (Dirac or Majorana) and the number of massive and sterile neutrinos.

We have seen in the previous chapter how the mass term of the charged leptons is generated by the standard Brout-Englert-Higgs mechanism. Here we will consider all phenomenologically possible neutrino mass terms. Our discussion will be general, based only on Lorentz invariance. We will only use the fact that a mass term of any spin-1/2 field is a sum of Lorentz-invariant products of left-handed and right-handed components of the field.

It was established by the LEP experiments at CERN that three flavor neutrinos ν_e, ν_μ, ν_τ exist in nature. The flavor neutrinos take part in CC and NC weak processes via the standard charged current and neutral current interactions

$$\mathscr{L}_I^{CC} = -\frac{g}{2\sqrt{2}} j_\alpha^{CC} W^\alpha + \text{h.c.}, \quad \mathscr{L}_I^{NC} = -\frac{g}{2\cos\theta_W} j_\alpha^{NC} Z^\alpha, \tag{4.1}$$

where

$$j_\alpha^{CC}(x) = 2 \sum_{l=e,\mu,\tau} \bar{\nu}_{lL}(x)\, \gamma_\alpha\, l_L(x), \quad j_\alpha^{NC} = \sum_{l=e,\mu,\tau} \bar{\nu}_{lL}(x)\, \gamma_\alpha\, \nu_{lL}(x). \tag{4.2}$$

The flavor fields $\nu_{lL}(x)$ ($l = e, \mu, \tau$) must enter into the neutrino mass term. The structure of mass term depends on

- other fields (if any) which enter into the mass term,
- the conservation of the total lepton number $L = L_e + L_\mu + L_\tau$.

© Springer International Publishing AG, part of Springer Nature 2018
S. Bilenky, *Introduction to the Physics of Massive and Mixed Neutrinos*,
Lecture Notes in Physics 947, https://doi.org/10.1007/978-3-319-74802-3_4

4.2 Dirac Mass Term

The mass term of the charged leptons is given by the expression

$$\mathcal{L}^{\text{lep}}(x) = - \sum_{l=e,\mu,\tau} m_l \, \bar{l}(x) \, l(x) = - \sum_{l=e,\mu,\tau} m_l \, \bar{l}_L(x) \, l_R(x) + \text{h.c.} \qquad (4.3)$$

Because of the conservation of the total electric charge the SM Lagrangian is invariant under the global transformation

$$l'_L(x) = e^{-i\Lambda} l_L(x), \quad l'_R(x) = e^{-i\Lambda} l_R(x), \quad W'_\alpha(x) = e^{i\Lambda} W_\alpha(x), \quad v'_{lL} = v_{lL} \ \ etc. \qquad (4.4)$$

where Λ is an arbitrary constant. The mass term (4.3) is *the Dirac mass term* and $l(x)$ is the Dirac field of the leptons l^- and antileptons l^+.

Let us now consider neutrinos. By analogy with leptons we will assume that in addition to the flavor left-handed fields $v_{lL}(x)$ three right-handed neutrino fields $v_{lR}(x)$ enter into the Lagrangian. In this case the most general neutrino mass term have the form

$$\mathcal{L}^{\text{D}}(x) = - \sum_{l',l} \bar{v}_{l'L}(x) \, M^{\text{D}}_{l'l} \, v_{lR}(x) + \text{h.c.} \qquad (4.5)$$

Here the indices l and l' takes values e, μ, τ and M^{D} is a 3×3 complex, nondiagonal matrix.

Let us present the mass term (4.5) in the diagonal form. The complex matrix M^{D} can be diagonalized by the biunitary transformation (see Appendix B).

$$M^{\text{D}} = U^\dagger \, m \, V. \qquad (4.6)$$

Here U and V are unitary matrices and $m_{ik} = m_i \delta_{ik}, m_i > 0$.
From (4.5) and (4.6) we find

$$\mathcal{L}^{\text{D}}(x) = - \sum_{i=1}^{3} m_i \, \bar{v}_i(x) v_i(x), \qquad (4.7)$$

Thus, $v_i(x) = v_{iL}(x) + v_{iR}(x)$ is the field of neutrino with mass m_i. Flavor fields $v_{lL}(x)$ are connected with left-handed components of the fields of neutrinos with definite masses by the relations

$$v_{lL}(x) = \sum_{i=1}^{3} U_{li} \, v_{iL}(x) \quad (l = e, \mu, \tau) \qquad (4.8)$$

From (4.2) and (4.8) we conclude that fields of massive neutrinos enter into the charged current in the mixed form (like quark fields). The unitary 3×3 mixing matrix U is called Pontecorvo-Maki-Nakagawa-Sakata (PMNS) mixing matrix.

Notice that for the right-handed fields $\nu_{lR}(x)$ we have

$$\nu_{lR}(x) = \sum_{i=1}^{3} V_{li} \, \nu_{iR}(x) \quad (l = e, \mu, \tau). \tag{4.9}$$

From (4.2), (4.7) and (4.8) follows that the total Lagrangian is invariant under the global transformations

$$\nu'_{iL}(x) = e^{i\Lambda} \, \nu_i(x), \quad \nu'_{iR}(x) = e^{i\Lambda} \, \nu_{iR}(x), \, l'(x) = e^{i\Lambda} \, l(x), \quad q'(x) = q(x), \tag{4.10}$$

where Λ is an arbitrary constant.

From invariance under the transformation (4.10) follows that the total lepton number L is conserved and that $\nu_i(x)$ is the Dirac field of neutrinos ν_i and antineutrinos $\bar{\nu}_i$, particles with the same mass m_i and different lepton numbers: $L(\nu_i) = 1$, $L(\bar{\nu}_i) = -1$. The mass term (4.5) is the standard *Dirac mass term*.

4.3 Majorana Mass Term

Is it possible to built the neutrino mass in which only left-handed flavor fields ν_{lL} enter? If we require the conservation of the total lepton number L the answer to this question is negative. In fact, we have

$$\gamma_5 \nu_{lL} = -\nu_{lL}, \quad \bar{\nu}_{lL} \gamma_5 = \bar{\nu}_{lL}. \tag{4.11}$$

From these relations follow that product of neutrino fields invariant under global transformation $\nu'_{lL} = e^{i\Lambda} \nu_{lL}$ is equal to zero:

$$\bar{\nu}_{l'L} \nu_{lL} = \bar{\nu}_{l'L} \gamma_5 \gamma_5 \nu_{lL} = -\bar{\nu}_{l'L} \nu_{lL} = 0. \tag{4.12}$$

However, if we assume that the total lepton number is violated we can built the neutrino mass term in which only flavor neutrino fields ν_{lL} enter.[1] In fact, from (4.11) we find

$$\gamma_5^T \bar{\nu}_{lL}^T = \bar{\nu}_{lL}^T. \tag{4.13}$$

---

[1] In the case of two flavor neutrinos this was shown for the first time by Gribov and Pontecorvo [54].

Let us multiply this relation by the unitary matrix of the charge conjugation C, which satisfies the relations

$$C \gamma_\alpha^T C^{-1} = -\gamma_\alpha, \quad C^T = -C. \tag{4.14}$$

Taking into account that $C \gamma_5^T C^{-1} = \gamma_5$ we obtain

$$\gamma_5 (\nu_{lL})^c = (\nu_{lL})^c, \tag{4.15}$$

where $(\nu_{lL})^c = C \bar{\nu}_{lL}^T$ is the conjugated flavor field.

From (4.11) and (4.15) we find

$$\bar{\nu}_{l'L}(\nu_{lL})^c = \bar{\nu}_{l'L}\gamma_5\gamma_5(\nu_{lL})^c = \bar{\nu}_{l'L}(\nu_{lL})^c. \tag{4.16}$$

It is obvious that the product $\bar{\nu}_{l'L} C \bar{\nu}_{lL}^T$ is not invariant under global transformations.

The most general neutrino mass term in which only flavor fields enter has the form

$$\mathscr{L}^M = -\frac{1}{2} \sum_{l',l=e,\mu,\tau} \bar{\nu}_{l'L} M_{l'l}^M (\nu_{lL})^c + \text{h.c.} = -\frac{1}{2} \bar{\nu}_L M^M (\nu_L)^c + \text{h.c.} \tag{4.17}$$

Here

$$\nu_L = \begin{pmatrix} \nu_{eL} \\ \nu_{\mu L} \\ \nu_{\tau L} \end{pmatrix} \tag{4.18}$$

and M^M is a complex 3×3 nondiagonal matrix.

From the Fermi-Dirac statistics of neutrino fields and (4.14) follows that M^M is a symmetrical matrix. In fact, we have

$$\bar{\nu}_L M^M (\nu_L)^c = \bar{\nu}_L M^M C \bar{\nu}_L^T = -\bar{\nu}_L (M^M)^T C^T \bar{\nu}_L^T = \bar{\nu}_L (M^M)^T (\nu_L)^c. \tag{4.19}$$

From (4.19) we have

$$M^M = (M^M)^T. \tag{4.20}$$

The symmetrical matrix M^M can be presented in the form (see Appendix C)

$$M^M = U \, m \, U^T, \tag{4.21}$$

where U is an unitary matrix and m is a diagonal matrix ($m_{ik} = m_i \, \delta_{ik}, \quad m_i > 0$). From (4.17) and (4.21) we find

$$\mathscr{L}^{\mathrm{M}} = -\frac{1}{2} \, \bar{v}_L \, U \, m \, U^T \, C \, \bar{v}_L^T + \text{h.c.} = -\frac{1}{2} \, \overline{U^\dagger v_L} \, m \, (U^\dagger v_L)^c + \text{h.c.} = -\frac{1}{2} \, \bar{v}^{\mathrm{M}} \, m \, v^{\mathrm{M}}.$$

(4.22)

Here

$$v^{\mathrm{M}} = U^\dagger v_L + (U^\dagger v_L)^c = \begin{pmatrix} v_1 \\ v_2 \\ v_3 \end{pmatrix}, \quad m = \begin{pmatrix} m_1 & 0 & 0 \\ 0 & m_2 & 0 \\ 0 & 0 & m_3 \end{pmatrix},$$ (4.23)

From (4.22) and (4.23) we have

$$\mathscr{L}^{\mathrm{M}} = -\frac{1}{2} \sum_{i=1}^{3} m_i \, \bar{v}_i \, v_i.$$ (4.24)

Thus, $v_i(x)$ is the field of the neutrino with mass m_i. From (4.23) we obviously have

$$(v^{\mathrm{M}}(x))^c = v^{\mathrm{M}}(x).$$ (4.25)

Thus, the field of neutrinos with definite mass $v_i(x)$ satisfy the Majorana condition

$$v_i^c(x) = v_i(x).$$ (4.26)

The mass term (4.17) is called the Majorana mass term.

The Majorana field $v_i(x)$ is *the field of neutrinos*. There is no notion of neutrinos and antineutrinos in the case of the Majorana field $v_i(x)$ (or $\bar{v}_i \equiv v_i$). This is connected with the fact that in the case of the mass term (4.22) the Lagrangian is not invariant under the global transformations and there is no conserved lepton number which allows to distinguish neutrino and antineutrino. For the Majorana field $v_i(x)$ we have the following expansion

$$v(x) = \int \frac{1}{(2\pi)^{3/2} \sqrt{2 \, p^0}} \left(a_r(p) \, u^r(p) \, e^{-i \, px} + a_r^\dagger(p) \, C \, (\bar{u}^r(p))^T \, e^{i \, px} \right) d^3 p.$$

(4.27)

Here $a_r(p)$ ($a_r^\dagger(p)$) is the operator of the absorption of neutrino (creation of antineutrino) with momentum p and helicity r, the spinor $u^r(p)$ describes the state with momentum p and helicity r.

From (4.23) we find

$$v_L(x) = U \, v_L^M(x), \quad v_{lL}(x) = \sum_{i=1}^{3} U_{li} \, v_{iL}(x) \tag{4.28}$$

Here U is the unitary 3×3 mixing matrix. Thus, in the case of the Majorana mass term (4.22) the left-handed flavor fields v_{lL}, which enter into CC and NC of the Standard Model, are connected with the left-handed components of the Majorana fields v_{iL} by the mixing relation (4.28).

In conclusion let us stress an important difference between Majorana and Dirac fields. Any fermion field is the sum of the left-handed and right-handed components. We have

$$v^M(x) = v_L^M(x) + v_R^M(x). \tag{4.29}$$

Comparing (4.23) and (4.29), we find

$$v_L^M(x) = U^\dagger v_L(x), \quad v_R^M(x) = (U^\dagger v_L(x))^c. \tag{4.30}$$

Thus, right-handed and left-handed components of the Majorana field are connected by the relation

$$v_R^M(x) = (v_L^M(x))^c, \quad v_{iR}(x) = (v_{iL}(x))^c \tag{4.31}$$

It is obvious that this relation is a direct consequence of the Majorana condition:

$$(v_{iL})^c = (\frac{1 - \gamma_5}{2} v_i)^c = C \frac{1 + \gamma_5^T}{2} \bar{v}_i^T = \frac{1 + \gamma_5}{2} v_i^c = \frac{1 + \gamma_5}{2} v_i = v_{iR} \tag{4.32}$$

Thus, right-handed and left-handed components of the Majorana field are connected by the relation (4.31). The right-handed and left-handed components of the Dirac field are independent.

4.4   Dirac and Majorana Mass Term

The most general neutrino mass term is not invariant under the global transformations (does not conserve the total lepton number L) includes the left-handed flavor fields v_{lL} ($l = e, \mu, \tau$) and right-handed sterile fields v_{sL} which do not enter into SM interaction Lagrangian ($s = s_1, s_2, s_3$). We have

$$\mathscr{L}^{D+M} = -\frac{1}{2} \bar{v}_L \, M_L^M (v_L)^c - \bar{v}_L \, M^D \, v_R - \frac{1}{2} \overline{(v_R)^c} \, M_R^M v_R + \text{h.c.} \tag{4.33}$$

Here M_L^M and M_R^M are complex non-diagonal symmetrical 3×3 matrices, M^D is a complex non-diagonal 3×3 matrix, v_L is given by (4.18) and

$$v_R = \begin{pmatrix} v_{s_1 R} \\ v_{s_2 R} \\ v_{s_3 R} \end{pmatrix}.$$
(4.34)

The mass term (4.33) is the sum of the left-handed Majorana mass term, the Dirac mass term and right-handed Majorana mass term. It is called the Dirac and Majorana mass term.

The mass term $\mathcal{L}^{D+M}$ can be written in the following matrix form

$$\mathcal{L}^{D+M} = -\frac{1}{2} \bar{n}_L M^{D+M} (n_L)^c + \text{h.c.}$$
(4.35)

Here

$$n_L = \begin{pmatrix} v_L \\ (v_R)^c \end{pmatrix}$$
(4.36)

and

$$M^{D+M} = \begin{pmatrix} M_L^M & M^D \\ (M^D)^T & M_R^M \end{pmatrix}$$
(4.37)

is a symmetrical 6×6 matrix. Notice that in (4.35) we took into account the following relation

$$\bar{v}_L M^D v_R = -(v_R)^T (M^D)^T (\bar{v}_L)^T = \overline{(v_R)^c} (M^D)^T (v_L)^c.$$
(4.38)

The matrix M^{D+M} can be presented in the following diagonal form

$$M^{D+M} = U m U^T,$$
(4.39)

where U is an unitary 6×6 matrix and $m_{ik} = m_i \delta_{ik}$, $m_i > 0$ $(i,k = 1, \ldots 6)$.

From (4.35) and (4.39) we have

$$\mathcal{L}^{D+M} = -\frac{1}{2} \overline{U^\dagger n_L} \, m \, (U^\dagger n_L)^c + \text{h.c.} = -\frac{1}{2} \bar{v}^M m \, v^M = -\frac{1}{2} \sum_{i=1}^{6} m_i \bar{v}_i v_i.$$
(4.40)

Here

$$v^M = v_L^M + (v_L^M)^c = \begin{pmatrix} v_1 \\ \vdots \\ v_6 \end{pmatrix}, \tag{4.41}$$

where

$$v_L^M = U^\dagger n_L. \tag{4.42}$$

From (4.41) we have

$$(v^M)^c = v^M \quad \text{and} \quad v_i^c = v_i \ \ (i = 1, 2, \dots .6). \tag{4.43}$$

From (4.40) and (4.43) follow that $v_i(x)$ is the field of Majorana particles with mass m_i.

It is obvious from (4.42) that v_{lL} and $(v_{lR})^c$ are connected with left-handed components of the Majorana fields v_{iL} by a unitary transformation. In fact, we have

$$n_L = U v_L^M. \tag{4.44}$$

From (4.44) we obtain the following relations

$$v_{lL}(x) = \sum_{i=1}^{6} U_{li} \, v_{iL}(x), \quad (v_{sR}(x))^c = \sum_{i=1}^{6} U_{si} \, v_{iL}(x), \tag{4.45}$$

where U is the unitary 6×6 mixing matrix. Thus, in the case of the Dirac and Majorana mass term, flavor fields v_{lL} are "mixture" of the six left-handed fields of Majorana particles with mass m_i. The sterile fields $(v_{sR})^c$, which do not enter into charged and neutral currents, are "mixture" of the same left-handed components of the Majorana fields.

Let us notice that there are no special reasons to assume that the index s takes three values. In the general case the index s takes n_{ster} values ($n_{ster} \geq 3$). In this case for the mass term we have

$$\mathscr{L}^{D+M} = -\frac{1}{2} \sum_{i=1}^{3+n_{ster}} m_i \, \bar{v}_i \, v_i, \quad v_i^c = v_i \tag{4.46}$$

The mixing relations take the form

$$v_{lL}(x) = \sum_{i=1}^{3+n_{ster}} U_{li} \, v_{iL}(x), \quad (v_{sR}(x))^c = \sum_{i=1}^{3+n_{ster}} U_{si} \, v_{iL}(x). \tag{4.47}$$

Here U is a unitary $(3 + n_{ster}) \times (3 + n_{ster})$ matrix.

4.5   Neutrino Mass Term in the Simplest Case of Two Neutrino Fields

It is instructive to consider a neutrino mass term in the simplest case of two neutrino fields. Let us consider the Dirac and Majorana mass term in the case of one generation. We have

$$\mathscr{L}^{D+M} = -\frac{1}{2} m_L \, \bar{\nu}_L \, (\nu_L)^c - m_D \, \bar{\nu}_L \, \nu_R - \frac{1}{2} m_R \, \overline{(\nu_R)^c} \, \nu_R + \text{h.c.}$$

$$= -\frac{1}{2} \bar{n}_L \, M^{D+M}(n_L)^c + \text{h.c.} \tag{4.48}$$

Here m_L, m_D and m_R are real parameters (we assume CP invariance) and

$$M^{D+M} = \begin{pmatrix} m_L & m_D \\ m_D & m_R \end{pmatrix}, \quad n_L = \begin{pmatrix} \nu_L \\ (\nu_R)^c \end{pmatrix}. \tag{4.49}$$

It is convenient to present the matrix M^{D+M} in the form

$$M^{D+M} = \frac{1}{2} \operatorname{Tr} M^{D+M} + M, \tag{4.50}$$

where $\operatorname{Tr} M^{D+M} = m_L + m_R$ and $\operatorname{Tr} M = 0$. We have

$$M = \begin{pmatrix} -\frac{1}{2}(m_R - m_L) & m_D \\ m_D & \frac{1}{2}(m_R - m_L) \end{pmatrix}. \tag{4.51}$$

The matrix M can be easily diagonalized by the orthogonal transformation (see Appendix A). We have

$$M = O \, \bar{m} \, O^T, \tag{4.52}$$

where

$$O = \begin{pmatrix} \cos\theta & \sin\theta \\ -\sin\theta & \cos\theta \end{pmatrix}, \quad \bar{m} = \begin{pmatrix} \bar{m}_1 & 0 \\ 0 & \bar{m}_1 \end{pmatrix} \tag{4.53}$$

Here

$$\bar{m}_{1,2} = \mp \frac{1}{2} \sqrt{(m_R - m_L)^2 + 4 m_D^2} \tag{4.54}$$

From (4.52)–(4.54) we find

$$\tan 2\theta = \frac{2m_D}{m_R - m_L}, \quad \cos 2\theta = \frac{m_R - m_L}{\sqrt{(m_R - m_L)^2 + 4\,m_D^2}}. \tag{4.55}$$

For the matrix M^{D+M} from (4.50), (4.52) and (4.54) we have

$$M^{D+M} = O\, m'\, O^T, \tag{4.56}$$

where

$$m'_{1,2} = \frac{1}{2}\,(m_R + m_L) \mp \frac{1}{2}\sqrt{(m_R - m_L)^2 + 4\,m_D^2} \tag{4.57}$$

are eigenvalues of the matrix M^{D+M} which can be positive or negative. Let us write down

$$m'_i = m_i\, \eta_i \tag{4.58}$$

where $m_i = |m'_i|$ and $\eta_i = \pm 1$. Taking into account (4.57) and (4.58) we have

$$M^{D+M} = O\, m\, \eta\, O^T = U\, m\, U^T, \tag{4.59}$$

where $U = O\,\eta^{1/2}$ is unitary matrix.

From (4.48) and (4.59) we obtain the following expression for the mass term

$$\mathscr{L}^{D+M} = -\frac{1}{2}\,\overline{\nu^M}\, m\, \nu^M = -\frac{1}{2}\sum_{i=1,2} m_i\, \bar{\nu}_i\, \nu_i. \tag{4.60}$$

Here

$$\nu^M = U^\dagger n_L + (U^\dagger n_L)^c = \begin{pmatrix} \nu_1 \\ \nu_2 \end{pmatrix}. \tag{4.61}$$

It is obvious from (4.61) that

$$\nu_i^c = \nu_i. \tag{4.62}$$

Thus, ν_1 and ν_2 are fields of Majorana neutrino with masses m_1 and m_2, respectively.

From (4.49), (4.53) and (4.61) we obtain the following mixing relations in the case of the Dirac and Majorana mass term for one neutrino family

$$\nu_L = \cos\theta\,\sqrt{\eta_1}\,\nu_{1L} + \sin\theta\,\sqrt{\eta_2}\,\nu_{2L}$$
$$(\nu_R)^c = -\sin\theta\,\sqrt{\eta_1}\,\nu_{1L} + \cos\theta\,\sqrt{\eta_2}\,\nu_{2L}. \tag{4.63}$$

The neutrino masses m_1 and m_2 and the mixing angle θ are determined by three real parameters m_L, m_R and m_D (see relations (4.55) and (4.57)). The parameter η_i ($i = 1, 2$) determines the CP parity of the Majorana neutrino ν_i (see the next section).

For the Majorana mass term in the case of two flavor fields (say, ν_μ and ν_τ) we have

$$\mathscr{L}^M = -\frac{1}{2}\,\bar{\nu}_L\,M^M (\nu_L)^c + \text{h.c.} \tag{4.64}$$

Here

$$M^M = \begin{pmatrix} m_{\mu\mu} & m_{\mu\tau} \\ m_{\mu\tau} & m_{\tau\tau.} \end{pmatrix}, \quad \nu_L = \begin{pmatrix} \nu_{\mu L} \\ \nu_{\tau L.} \end{pmatrix} \tag{4.65}$$

It is obvious that if we change $m_{\mu\mu} \to m_L$, $m_{\tau\tau} \to m_R$, $m_{\mu\tau} \to m_D$ we can use the relations obtained for the Dirac and Majorana mass term. For the masses of the Majorana neutrinos ν_1 and ν_2 we have

$$m_{1,2} = \left| \frac{1}{2}\,(m_{\tau\tau} + m_{\mu\mu}) \mp \frac{1}{2}\sqrt{(m_{\tau\tau} - m_{\mu\mu})^2 + 4\,m_{\mu\tau}^2} \right|. \tag{4.66}$$

The flavor fields $\nu_{\mu L}$ and $\nu_{\tau L}$ are given by the relations

$$\nu_{\mu L} = \cos\theta\sqrt{\eta_1}\,\nu_{1L} + \sin\theta\sqrt{\eta_2}\,\nu_{2L}$$
$$\nu_{\tau L} = -\sin\theta\sqrt{\eta_1}\,\nu_{1L} + \cos\theta\sqrt{\eta_2}\,\nu_{2L}, \tag{4.67}$$

where for the mixing angle θ we have

$$\tan 2\theta = \frac{2m_{\mu\tau}}{m_{\tau\tau} - m_{\mu\mu}}, \quad \cos 2\theta = \frac{m_{\tau\tau} - m_{\mu\mu}}{\sqrt{(m_{\tau\tau} - m_{\mu\mu})^2 + 4\,m_{\mu\tau}^2}}. \tag{4.68}$$

Two extreme cases are of interest:

- $m_{\mu\tau} = 0$.

  In this case there is no mixing:

$$\theta = 0, \quad m_1 = m_{\mu\mu}, \quad m_2 = m_{\tau\tau}, \quad \nu_{\mu L} = \nu_{1L}, \quad \nu_{\tau L} = \nu_{2L}$$

- $m_{\mu\mu} = m_{\tau\tau}$

  This is the case of the maximal mixing:

$$\theta = \frac{\pi}{4}, \quad m_{1,2} = m_{\mu\mu} \pm m_{\mu\tau}, \quad \nu_{\mu L} = \frac{1}{\sqrt{2}}(\nu_{1L} + \nu_{2L}), \quad \nu_{\tau L} = \frac{1}{\sqrt{2}}(-\nu_{1L} + \nu_{2L})$$

4.6 Concluding Remarks

There are three phenomenologically possible neutrino mass terms.

1. *The Dirac mass term* in which flavor left-handed and sterile right-handed fields enter. The total lepton number L is conserved in the case of the Dirac mass term.
2. *The Majorana mass term* in which only flavor left-handed field enter. The total lepton number L is violated in the case of the Majorana mass term.
3. *The most general Dirac and Majorana mass term* in which flavor left-handed and sterile right-handed fields enter. The total lepton number L is not conserved in the case of the Dirac and Majorana mass term.

In the case of the Dirac mass term fields of neutrinos with definite masses are Dirac fields of neutrinos ($L(\nu_i) = 1$) and antineutrinos ($L(\bar{\nu}_i) = -1$). In the case of the Majorana mass term fields of neutrinos with definite masses are Majorana fields of neutrinos ($\nu_i \equiv \bar{\nu}_i$). In the case of the Dirac mass term and the Majorana mass term the number of massive neutrinos ν_i is equal to three (the number of neutrino flavors) and mixing have the form

$$\nu_{lL}(x) = \sum_{i=1}^{3} U_{li} \nu_{iL}(x).$$

Here U is the unitary PMNS mixing matrix and $\nu_i(x)$ is the field of the Dirac (Majorana) neutrino with the mass m_i.

In the case of the Dirac and Majorana mass term fields of neutrinos with definite masses are Majorana fields and the number of massive neutrinos depends on the number of sterile fields n_{ster} and is larger than three. For the mixing we have in this case

$$\nu_{lL}(x) = \sum_{i=1}^{3+n_{\text{ster}}} U_{li}\, \nu_{iL}(x), \quad (\nu_{sR}(x))^c = \sum_{i=1}^{3+n_{\text{ster}}} U_{si}\, \nu_{iL}(x). \tag{4.69}$$

The discussion in this chapter was purely phenomenological: neutrino masses and elements of the mixing matrix are considered here as parameters which must be determined from experimental data. In the following chapters we will consider in details neutrino oscillations, major consequence of the neutrino masses and mixing, and neutrino oscillation experiments which allows to determine neutrino mixing angles and neutrino mass-squared differences. We will also discuss experiments which allow to establish the nature of the massive neutrinos (experiments on the search for neutrinoless double β decay) and briefly discuss experiments on the search for transitions of flavor neutrinos into sterile states.

Chapter 5
Seesaw Mechanism of the Neutrino Mass Generation

5.1 Introduction

Information about neutrino masses was obtained from neutrino oscillation experiments, experiments on the precise measurement of the end-point part of the β-spectrum of ^{3}H and from cosmological data. From analysis of the data of the neutrino oscillation experiments the values of two neutrino mass-squared differences were determined. From ^{3}H experiments the upper bound on the effective neutrino mass was obtained. From cosmological measurements an upper bound of the sum of the neutrino masses can be inferred. In spite absolute values of neutrino masses at present are not known from these data we can conclude

- Neutrino masses are different from zero.
- The mass of the heaviest neutrino is in the range

$$(5 \cdot 10^{-2} \leq m_h \leq 3 \cdot 10^{-1}) \, \text{eV}$$

Thus, *neutrino masses are many orders of magnitude smaller than masses of other fundamental fermions* (leptons and quarks). In this chapter we will consider the most viable beyond the Standard Model seesaw mechanism of the neutrino mass generation which connect the smallness of neutrino masses with a new lepton number violating physics at a scale which is much larger than electroweak scale determined by $v \simeq 246 \, \text{GeV}$.

We will discuss first problems connected with the generation of neutrino masses by the standard Brout-Englert-Higgs mechanism. Let us consider the following $SU_L(2) \times U_Y(1)$ invariant Lagrangian of the Yukawa interaction of the leptons and the Higgs boson

$$\mathscr{L}_Y = -\sqrt{2} \sum_{l_1, l_2} \overline{\psi}^{\text{lep}}_{l_1 L} \, Y'_{l_1 l_2} \, \nu'_{l_2 R} \, \tilde{H} + \text{h.c.} \tag{5.1}$$

© Springer International Publishing AG, part of Springer Nature 2018
S. Bilenky, *Introduction to the Physics of Massive and Mixed Neutrinos*,
Lecture Notes in Physics 947, https://doi.org/10.1007/978-3-319-74802-3_5

Here ψ_{lL}^{lep} is the left-handed lepton doublet (see (3.66)), $\tilde{H} = i\tau_2 H^*$ is the conjugated Higgs doublet, v'_{lR} are right-handed singlets of the $SU_L(2) \times U(1)_Y$ group and $Y'_{ll'}$ are dimensionless, complex Yukawa constants.

After the spontaneous symmetry braking from (3.130) and (5.1) we find

$$\mathscr{L}_Y = -\sum_{l_1, l_2} \bar{v}'_{l_1 L} \, Y'_{l_1 l_2} \, v'_{l_2 R} \, (v + h) + \text{h.c.} = -\bar{v}'_L \, Y' \, v'_R \, (v + h) + \text{h.c.} \qquad (5.2)$$

Here

$$v'_L = \begin{pmatrix} v'_{eL} \\ v'_{\mu L} \\ v'_{\tau L} \end{pmatrix} \qquad v'_R = \begin{pmatrix} v'_{eR} \\ v'_{\mu R} \\ v'_{\tau R} \end{pmatrix}. \qquad (5.3)$$

In terms of the flavor neutrino fields (see (3.164)) the proportional to v term of the Lagrangian (5.2) takes the form

$$\mathscr{L}^D = -v \, \bar{v}_L \, Y \, v_R + \text{h.c.}, \qquad (5.4)$$

where $Y = U_L^\dagger \, Y'$, $v_R \equiv v'_R$. Taking into account the results of the previous chapter we conclude that *the Brout-Englert-Higgs mechanism generates the Dirac neutrino mass term.*

For the complex matrix Y we have

$$Y = U \, y \, V^\dagger, \qquad (5.5)$$

where U and V are unitary matrices and $y_{ik} = y_i \, \delta_{ik}, \quad y_i > 0$.

From (5.4) and (5.5) for the neutrino mass term we find the following standard expression

$$\mathscr{L}^D(x) = -\sum_{i=1}^{3} m_i \, \bar{v}_i(x) v_i(x). \qquad (5.6)$$

The Dirac fields of neutrinos with definite masses $v_i(x)$ are determined by the relation

$$U^\dagger v_L(x) + V^\dagger v_R(x) = \begin{pmatrix} v_1(x) \\ v_2(x) \\ v_3(x) \end{pmatrix} \qquad (5.7)$$

and the Dirac neutrino masses are given by the relation

$$m_i = y_i \, v. \qquad (5.8)$$

For the neutrino mixing we have

$$\nu_{lL}(x) = \sum_{i=1}^{3} U_{li} \, \nu_{iL}(x). \tag{5.9}$$

From (5.8) follows that neutrino masses generated by the Brout-Englert-Higgs mechanism, like all other SM masses, are proportional to the vacuum expectation value of the Higgs field v. For the heaviest neutrino we find the following inequality

$$y_3 = \frac{m_3}{v} \lesssim 10^{-12}. \tag{5.10}$$

For other particles of the third family we have $y_\tau \simeq 0.7 \cdot 10^{-2}$, $y_b \simeq 1.7 \cdot 10^{-2}$, $y_t \simeq 0.7$. Thus, in order to generate neutrino masses by the SM mechanism we need to assume that dimensionless neutrino Yukawa coupling constants are many orders of magnitude smaller than Yukawa coupling constants of leptons and quarks. *It is very unlikely that neutrino masses are of the SM origin.* It is very plausible that the SM neutrinos are massless.

5.2   The Generation of the Neutrino Masses by Effective Lagrangian Method

If the Standard Model neutrinos are massless particles, small neutrino masses are generated by a beyond the Standard Model (BSM) mechanism. A general method which allows to reveal a BSM effects by investigation of relatively low-energy phenomena is the method of the effective Lagrangian.

Effective Lagrangian is dimension five or more nonrenormalizable Lagrangian, invariant under $SU_L(2) \times U_Y(1)$ transformations and built from SM fields. In order to build the effective Lagrangian which generate a neutrino mass term let us consider the $SU_L(2) \times U_Y(1)$ scalar

$$(\bar{\psi}_{lL}^{\mathrm{lep}} \tilde{H}),$$

where ψ_{lL}^{lep} is the lepton doublet (see (3.66)), $\tilde{H} = i \, \tau_2 H^*$ is the conjugated Higgs doublet.

After the spontaneous symmetry breaking (SSB) we have

$$(\bar{\psi}_{lL}^{\mathrm{lep}} \tilde{H}) = \frac{v + h}{\sqrt{2}} \bar{\nu}'_{lL}. \tag{5.11}$$

From this expression it is obvious that the only possible $SU_L(2) \times U_Y(1)$ invariant effective Lagrangian which generate the neutrino mass term (after SSB) has a form (Weinberg)

$$\mathscr{L}_I^{\text{eff}} = -\frac{1}{\Lambda} \sum_{l_1,l_2} (\bar{\psi}_{l_1 L}^{\text{lep}} \tilde{H}) \, Y'_{l_1 l_2} \, C(\bar{\psi}_{l_2 L}^{\text{lep}} \tilde{H})^T + \text{h.c.} \tag{5.12}$$

Here C is matrix of the charge conjugation, $Y'_{l_1 l_2} = Y'_{l_2 l_1}$ are dimensionless constants which are not constrained by the $SU_L(2) \times U_Y(1)$ symmetry.

The operator in the expression (5.12) has a dimension M^5. Because the Lagrangian must have the dimension M^4 the constant Λ has a dimension of a mass.

The Lagrangian (5.12) does not conserve the total lepton number L. Thus, the constant Λ characterizes the scale of a L-violating, beyond the SM physics. After spontaneous symmetry breaking from (5.12) we obtain the following neutrino mass term

$$\mathscr{L}^M = -\frac{1}{2}\left(\frac{v^2}{\Lambda}\right) \sum_{l_1,l_2} \bar{\nu}'_{l_1 L} \, Y'_{l_1 l_2} \, C(\bar{\nu}'_{l_2 L})^T + \text{h.c.} = -\frac{1}{2}\left(\frac{v^2}{\Lambda}\right) \bar{\nu}'_L \, Y' \, C(\bar{\nu}'_L)^T + \text{h.c.}$$

$$\tag{5.13}$$

The fields ν'_{lL} are connected with the flavor neutrino fields ν_{lL} by the relation

$$\nu'_L = U_L \, \nu_L, \tag{5.14}$$

where U_L is an unitary matrix (see 3.164). From (5.13) and (5.14) for the neutrino mass term we find the following expression

$$\mathscr{L}^M = -\frac{1}{2}\left(\frac{v^2}{\Lambda}\right) \bar{\nu}_L \, Y \, C(\bar{\nu}_L)^T + \text{h.c.} = -\frac{1}{2}\left(\frac{v^2}{\Lambda}\right) \sum_{l_1,l_2} \bar{\nu}_{l_1 L} \, Y_{l_1 l_2} \, (\nu_{l_2 L})^c + \text{h.c.}$$

$$\tag{5.15}$$

Here $Y = U_L^\dagger Y' (U_L^\dagger)^T$ is a symmetrical 3×3 matrix. The matrix Y can be presented in the form

$$Y = U \, y \, U^T, \quad U^\dagger U = 1, \quad y_{ik} = y_i \delta_{ik}, \quad y_i > 0. \tag{5.16}$$

From (5.15) and (5.16) we find the following standard expression for the Majorana neutrino mass term

$$\mathscr{L}^M = -\frac{1}{2} \sum_{i=1}^{3} m_i \bar{\nu}_i \, \nu_i. \tag{5.17}$$

Here $\nu_i = \nu_i^c$ is the Majorana field with the mass

$$m_i = \frac{v^2}{\Lambda}y_i = \left(\frac{v}{\Lambda}\right)(vy_i). \tag{5.18}$$

The flavor fields ν_{lL} are connected with the left-handed components of the fields of neutrinos with definite mass ν_{iL} by the standard mixing relation

$$\nu_{lL} = \sum_{i=1}^{3} U_{li}\,\nu_{iL}. \tag{5.19}$$

Thus, *from the Weinberg effective Lagrangian (5.12) we obtained the Majorana mass term.*

We showed in the previous chapter that the Majorana mass term is the most economical, L-violating mass term in which only left-handed flavor fields ν_{lL} enter. Neutrino masses considered in Sect. 4.3 were parameters which must be determined from experiments. There were no any theoretical hints in favor of the smallness of neutrino masses.

By the effective Lagrangian method we obtained for the neutrino masses the relation (5.18). The factor (vy_i) is "a typical fermion mass" in the SM. Thus, neutrino masses generated by the BSM effective Lagrangian (5.12) are suppressed with the respect to the "SM fermion masses" by the factor

$$\frac{v}{\Lambda} = \frac{\text{SM scale}}{\text{scale of a BSM physics}}. \tag{5.20}$$

Let us stress, however, that the scale of a new L-violating physics Λ is unknown. From (5.20) we have the relation

$$\Lambda = y_i\,\frac{v^2}{m_i}. \tag{5.21}$$

Absolute values of neutrino masses are unknown at present. However, if we assume neutrino mass hierarchy ($m_1 \ll m_2 \ll m_3$) from neutrino oscillation data we find that $m_3 \simeq 5 \cdot 10^{-2}$ eV. We have

$$\frac{v^2}{m_3} \simeq 1.2 \cdot 10^{15}\ \text{GeV}. \tag{5.22}$$

If we assume that $\Lambda \simeq$ TeV in this case for the dimensionless quantity y_3 we find too small value: $y_3 \simeq 10^{-12}$. If we assume that y_3 is of the order of one (like the Yukawa constant for the t-quark) in this case $\Lambda \simeq 10^{15}$. These are extreme cases. Large Λ ($\Lambda \gg v$) is a plausible possibility.

We will consider now a possible origin of the effective Lagrangian (5.12). Let us notice that the typical effective Lagrangian is the four-fermion Fermi Lagrangian of the β-decay

$$\mathscr{L}_I^F = -G_F \bar{p}\gamma^\alpha n \,\bar{e}\gamma_\alpha \nu + \text{h.c.} \tag{5.23}$$

The Lagrangian (5.23) has dimension M^6 and the Fermi constant G_F has dimension M^{-2}.

We know that the modern effective Lagrangian of the β-decay, of the process $\bar{\nu} + p \rightarrow e^+ + n$ and other connected processes

$$\mathscr{L}_I = -\frac{G_F}{\sqrt{2}} 4 \bar{p}_L \gamma^\alpha n_L \,\bar{e}_L \gamma_\alpha \nu_L + \text{h.c.} \tag{5.24}$$

is generated by the exchange of the virtual vector W boson between $(p - n)$ and $(e - \nu)$ vertices.

The Weinberg effective Lagrangian (5.12) can be generated by the exchange of virtual heavy Majorana leptons between lepton-Higgs vertices. In fact, let us assume that exist heavy Majorana leptons N_k ($k = 1, 2, \ldots$), singlets of the $SU_L(2) \times U_Y(1)$ group, which interact with lepton-Higgs pairs via $SU_L(2) \times U_Y(1)$ invariant Yukawa interaction

$$\mathscr{L}_I^Y = -\sqrt{2} \sum_{l,k} \bar{\psi}_{lL}^{\text{lep}} \tilde{H} \, Y_{lk}' N_{kR} + \text{h.c.} \tag{5.25}$$

where Y_{lk}' is a dimensionless Yukawa coupling constant. We are interested in the processes with virtual leptons N_k in the region $Q^2 \ll M_k^2$ (M_k is the mass of the lepton N_k). Taking into account that $N_k^T(x) = -\bar{N}_k(x)C$ for the propagator of the heavy Majorana lepton we have in the small Q^2 region

$$\langle 0|N_{kR}(x_1) N_{kR}^T(x_2)|0\rangle = -\left(\frac{1+\gamma_5}{2}\right) \langle 0|N_k(x_1)\bar{N}_k(x_2)|0\rangle C \left(\frac{1+\gamma_5}{2}\right)^T$$

$$\simeq \frac{i}{M_k}\delta(x_1 - x_2)C \left(\frac{1+\gamma_5}{2}\right)^T. \tag{5.26}$$

Using this relation, for the local effective Lagrangian, induced by the interaction (5.25), in the second order of the perturbation theory we find the following expression

$$\mathscr{L}_I^{\text{eff}} = -(\bar{\psi}_L^{\text{lep}} \tilde{H}) \, \bar{Y}' \frac{1}{M} \bar{Y}'^T C (\bar{\psi}_L^{\text{lep}} \tilde{H})^T + \text{h.c.}$$

$$= -\sum_{l_1,l_2} (\bar{\psi}_{l_1L}^{\text{lep}} \tilde{H}) \sum_k \bar{Y}_{l_1k}' \frac{1}{M_k} \bar{Y}_{kl_2}'^T C (\bar{\psi}_{l_2L}^{\text{lep}} \tilde{H})^T + \text{h.c.} \tag{5.27}$$

Comparing (5.12) and (5.27) we conclude that

$$\frac{1}{\Lambda} Y'_{l_1 l_2} = \sum_k \bar{Y}'_{l_1 k} \frac{1}{M_k} \bar{Y}'^T_{k l_2}. \tag{5.28}$$

Thus, the scale of L-violating physics Λ is determined by the masses of heavy Majorana leptons.

Let us consider the Lagrangian (5.25) and the mass term of the heavy Majorana leptons. After the spontaneous symmetry breaking in terms of the flavor neutrino fields we have

$$\mathcal{L} = -\bar{\nu}_L M^D N_R - \frac{1}{2} \bar{N}_L M N_R + \text{h.c.} \tag{5.29}$$

Here $M^D = v\, U_L^\dagger \bar{Y}'$ is a 3×3 complex matrix and $M_{ik} = M_i\, \delta_{ik}$.

Comparing (5.28) with (4.33) we conclude that (5.29) is the Dirac and Majorana mass term in which

- there is no left-handed Majorana mass term;
- the right-handed Majorana mass term has diagonal form; its elements are much larger than the elements of the matrix M^D.

Diagonalization of the Dirac and Majorana mass term (5.29) leads to the same results as the effective Lagrangian: small Majorana neutrino masses. We will show that in the next section.

5.3 Diagonalization of the Seesaw Matrix

The seesaw mechanism, proposed at the end of the seventies, is apparently the most natural and viable mechanism of neutrino mass generation. In the previous section we have considered equivalent mechanism based on the effective Lagrangian. Here we will discuss the seesaw mechanism in the framework of the Dirac and Majorana mass term.

In order to expose the main idea of the mechanism let us consider the simplest case of one family. General case of one family was considered in Sect. 4.5. We assume that

- there is no left-handed Majorana mass term, i.e. $m_L = 0$,
- the Dirac mass term m_D is generated by the Standard Higgs mechanism, i.e. it is of the order of a mass of quark or lepton,
- the lepton number is violated at a scale which is much larger than the electroweak scale, i.e. $m_R \gg m_D$.

From (4.57) we find in this case

$$m_1 = \frac{1}{2}|m_R - \sqrt{m_R^2 + 4m_D^2}| \simeq \frac{m_D^2}{m_R} \ll m_D \tag{5.30}$$

and

$$m_2 = \frac{1}{2}|m_R + \sqrt{m_R^2 + 4m_D^2}| \simeq m_R \gg m_D. \tag{5.31}$$

For the mixing angle from (4.55) we have

$$\theta \simeq \frac{m_D}{m_R} \ll 1. \tag{5.32}$$

Let us consider now the case of three families. The Dirac and Majorana seesaw matrix has the form

$$M = \begin{pmatrix} 0 & m_D \\ m_D^T & M_R \end{pmatrix}, \tag{5.33}$$

where m_D and M_R are 3×3 matrices and $M_R = M_R^T$. We will assume that M_R is a diagonal matrix.

Let us introduce the matrix m by the relation

$$V^T M V = m, \tag{5.34}$$

where V is a unitary matrix. We will show that the matrix V can be chosen in such a form that the matrix m has a block-diagonal form.

In the one family case in the linear over the mixing angle θ approximation for the mixing angle we have

$$U \simeq \begin{pmatrix} 1 & \frac{m_D}{m_R} \\ -\frac{m_D}{m_R} & 1 \end{pmatrix}. \tag{5.35}$$

By analogy we will present the unitary matrix V in the form

$$V = \begin{pmatrix} 1 & a^\dagger \\ -a & 1 \end{pmatrix}, \tag{5.36}$$

where $a \ll 1$. It is easy to check that in the linear over a approximation $V^\dagger V = 1$. From (5.34) and (5.36) follows that the matrix m takes the block-diagonal form if we choose

$$a = M_R^{-1} m_D^T. \tag{5.37}$$

We have

$$m \simeq \begin{pmatrix} -m_D M_R^{-1} m_D^T & 0 \\ 0 & M_R \end{pmatrix}. \tag{5.38}$$

From (5.34) and (5.38) for the Dirac and Majorana mass term we find the following expression

$$\mathcal{L}^{\text{seesaw}} = -\frac{1}{2}\bar{\nu}_L m_L (\nu_L)^c - \frac{1}{2}\overline{(\nu_R)^c} M_R \nu_R + \text{h.c.} \tag{5.39}$$

where left-handed Majorana matrix is given by the relation

$$m_L = -m_D \, M_R^{-1} \, m_D^T. \tag{5.40}$$

The first term of the expression (5.39) is the Majorana mass term of light neutrinos and the second term is the mass term of the heavy Majorana leptons. The structure of the relation (5.40) with large M_R in denominator ensure the smallness of neutrino masses with respect to masses of leptons and quarks.

5.4 Concluding Remarks

If the SM neutrinos are left-handed, massless, Weil particles, neutrino masses and mixing are generated by a beyond the Standard Model mechanism. We considered here the seesaw mechanism which is the most viable mechanism of the generation of small neutrino masses. The seesaw mechanism is based on the assumption that exist heavy Majorana leptons which interact with lepton-Higgs pairs. This interaction induces L-violating effective Lagrangian which (after spontaneous symmetry breaking) generates the most economical Majorana mass term in which only left-handed flavor neutrino fields ν_{lL} enter. The seesaw neutrino masses are suppressed with respect to the SM masses of quarks and leptons by a factor which is the ratio $\frac{v}{\Lambda}$ ($v \simeq 246\,\text{GeV}$ is the scale of the SM physics and Λ is a scale of a new L-violating physics). From the measurement of the absolute values of the neutrino masses the scale Λ cannot be determined. The value of the parameter Λ is an open problem. $\Lambda \gg v$ is a plausible possibility. If the seesaw mechanism is realized in nature in this case

- Neutrino with definite masses ν_i are Majorana particles. Observation of L-violating neutrinoless double β-decay of heavy even-even nuclei would be a crucial test of this prediction. We will consider this process in details in Chap. 9.
- The number of neutrinos with definite masses ν_i is equal to the number of flavor neutrinos (three). The experiments on the search for transitions of flavor neutrinos into sterile states would allow to check this prediction. We will discuss these experiments later.

- Exist heavy Majorana leptons. If masses of the Majorana leptons are much larger that v they cannot be produced at modern accelerators. However, they can be created in the early Universe. In Chap. 12. we will consider a popular mechanism of the generation of the barion asymmetry of the Universe induced by CP-violating decays of the heavy Majorana leptons (leptogenesis).

In conclusion let us notice that the effective Lagrangian (5.12) can be induced by three different beyond the SM interactions:

1. The interaction of lepton-Higgs pairs with a heavy Majorana $SU_l(2) \times U_Y(1)$ singlet leptons N_{kR} which considered in this chapter. The Lagrangian $\mathscr{L}_I^{\text{eff}}$ is induced in this case by the exchange of a virtual Majorana leptons between lepton-Higgs pairs. This scenarios is called type I seesaw.
2. An interaction of lepton pairs and Higgs pair with triplet heavy scalar boson. The effective Lagrangian $\mathscr{L}_I^{\text{eff}}$ is generated in this case by the exchange of a virtual scalar boson between lepton and Higgs pairs (type II seesaw).
3. An interaction of lepton-Higgs pairs with heavy Majorana triplet fermions. The effective Lagrangian $\mathscr{L}_I^{\text{eff}}$ is generated in this case by the exchange of a virtual Majorana triplet fermions between the lepton-Higgs pairs (type III seesaw).

Chapter 6
Neutrino Mixing Matrix

6.1 Introduction

As we have seen in the previous chapters, if in the total Lagrangian there is a
neutrino mass term, the flavor neutrino fields ν_{lL} ($l = e, \mu, \tau$) which enter into
CC lepton and NC neutrino currents

$$j_\alpha^{CC} = 2 \sum_{l=e,\mu,\tau} \bar\nu_{lL}\gamma_\alpha l_L, \quad j_\alpha^{NC} = \sum_{l=e,\mu,\tau} \bar\nu_{lL}\gamma_\alpha \nu_{lL} \qquad (6.1)$$

are mixtures of left-handed components of the fields of neutrinos with definite
masses $\nu_{iL}(x)$:

$$\nu_{lL}(x) = \sum_{i=1}^{3} U_{li}\, \nu_{iL}(x). \qquad (6.2)$$

Here U is a unitary 3×3 PMNS mixing matrix and $\nu_i(x)$ is the field of neutrino
(Dirac or Majorana) with the mass m_i.

The matrix U is the object of central interest of theory and experiment. In this
chapter we will consider the general properties of a unitary $n \times n$ mixing matrix in
the Dirac and Majorana cases. We will consider the standard parametrization of the
3×3 mixing matrix and obtain conditions which the mixing matrix must satisfy
if there are CP invariance in the lepton sector. In the last section of this chapter
we will briefly discuss models of neutrino mixing based on assumptions of flavor
symmetry.

© Springer International Publishing AG, part of Springer Nature 2018   89
S. Bilenky, *Introduction to the Physics of Massive and Mixed Neutrinos*,
Lecture Notes in Physics 947, https://doi.org/10.1007/978-3-319-74802-3_6

6.2   The Number of Angles and Phases in the Matrix U

A unitary mixing matrix U in the general $n \times n$ case can be presented in the form $U = e^{iH}$, where H is a hermitian $n \times n$ matrix. From the condition $H_{ik} = H_{ki}^*$ follows that the matrix H is characterized by n (diagonal elements) + $2\left(\frac{n^2-n}{2}\right)$ (non diagonal elements) $= n^2$ real parameters.

The number of angles which characterizes a unitary $n \times n$ matrix coincides with the number of parameters which characterizes a real orthogonal $n \times n$ matrix O which satisfies the condition

$$O^T O = 1. \tag{6.3}$$

An orthogonal matrix O can be presented in the form $O = e^A$, where $A^T = -A$. The diagonal elements of the matrix A are equal to zero. The number of real non diagonal elements is equal to $\frac{n(n-1)}{2}$. Thus, the number of angles which characterize a unitary matrix U is equal to

$$n_{\text{angles}} = \frac{n(n-1)}{2}. \tag{6.4}$$

Other parameters of the unitary matrix U are phases. The number of phases is equal to

$$n_{\text{phases}} = n^2 - \frac{n(n-1)}{2} = \frac{n(n+1)}{2}. \tag{6.5}$$

The number of *physical phases*, which characterize the unitary mixing matrix, is smaller than n_{phases}. This is connected with the fact that the mixing matrix enters into the charged current together with Dirac fields of charged leptons and Dirac (or Majorana) fields of neutrinos.

In fact, a unitary $n \times n$ matrix can be presented in the form

$$S(\beta)\, U^D\, S^\dagger(\alpha). \tag{6.6}$$

Here

$$S_{l'l}(\beta) = e^{i\beta_l}\, \delta_{l'l}, \quad S_{i'i}(\alpha) = e^{i\alpha_i}\, \delta_{i'i}, \tag{6.7}$$

where β_l and α_i are real phases. There are $(2n-1)$ independent parameters in phase matrices in (6.6). In fact, we can present the matrix $S(\alpha)$ in the form

$$S(\alpha) = e^{i\alpha_1} S(\bar{\alpha}), \tag{6.8}$$

where $\bar{\alpha}_1 = 0$, $\bar{\alpha}_i = \alpha_i - \alpha_1$ $(i \geq 2)$. From (6.6) and (6.8) we obviously find

$$S(\beta) \, U^D \, S^\dagger(\alpha) = S(\bar{\beta}) \, U^D \, S^\dagger(\bar{\alpha}), \tag{6.9}$$

where $\bar{\beta}_i = \beta_i - \alpha_1$. Thus, the matrix U^D is characterized by $\frac{n(n+1)}{2} - (2n - 1) = \frac{(n-1)(n-2)}{2}$ phases.

In the case of the neutrino mixing the leptonic charged current has a form

$$j_\alpha^{CC\dagger}(x) = 2 \sum_l \bar{l}_L(x) \, \gamma_\alpha \sum_i U_{li} \, \nu_{iL}(x) \tag{6.10}$$

Let us consider first the Dirac neutrinos ν_i. Because phases of the Dirac fields are not physical, arbitrary quantities it is obvious from (6.10) that phase factors $e^{i\bar{\beta}_l}$ and $e^{i\bar{\alpha}_i}$ can be included, correspondingly, into lepton and neutrino fields. Thus, in the case of the Dirac neutrinos neutrino mixing matrix is U^D. It is characterized by

$$n_{\text{phases}}^D = \frac{(n-1)(n-2)}{2} \tag{6.11}$$

physical phases.

In the case of the Majorana neutrinos only phase factors $e^{i\bar{\beta}_l}$ can be included in the Dirac fields $l_L(x)$. The Majorana fields satisfy the Majorana condition $\nu_i(x) = C\bar{\nu}_i^T(x)$ which fix phases of the fields. Thus, the Majorana mixing matrix has the form

$$U^M = U^D \, S(\bar{\alpha}). \tag{6.12}$$

It is characterized by

$$n_{\text{phases}}^M = \frac{(n-1)(n-2)}{2} + (n-1) = \frac{n(n-1)}{2} \tag{6.13}$$

physical phases.

In the simplest case of two families 2×2 mixing matrix for Dirac neutrinos is real. It is characterized by one angle. In the case of the Majorana neutrinos the mixing matrix is characterized by one angle and one phase.

In the most important case of three families 3×3 mixing matrix is characterized by three angles and one phase for the Dirac neutrinos. In the case of the Majorana neutrinos the mixing matrix is characterized by three angles and three phases.

6.3 CP Conservation in the Lepton Sector

In this section we will obtain conditions which the unitary mixing matrix U (in the case of Dirac or Majorana neutrinos) must satisfy if the CP invariance in the lepton sector holds.

If CP is conserved we have

$$V_{CP} \, \mathscr{L}_I^{CC}(x) \, V_{CP}^{-1} = \mathscr{L}_I^{CC}(x'). \tag{6.14}$$

Here V_{CP} is the operator of the CP conjugation, $x' = (x^0, -\mathbf{x})$ and $\mathscr{L}_I^{CC}(x)$ is the Lagrangian of the CC interaction of leptons and W-bosons:

$$\mathscr{L}_I^{CC}(x) = -\frac{g}{\sqrt{2}} \sum_{l,i} \bar{l}_L(x) \, \gamma_\alpha \, U_{li} \, \nu_{iL}(x) \, W^{\alpha\dagger}(x)$$

$$-\frac{g}{\sqrt{2}} \sum_{l,i} \bar{\nu}_{iL}(x) \, \gamma_\alpha \, U_{li}^* \, l_L(x) \, W^\alpha(x). \tag{6.15}$$

We will consider first the case of the Dirac neutrinos. For the Dirac lepton and neutrino fields we have[1]

$$V_{CP} \, l(x) \, V_{CP}^{-1} = \gamma^0 \, C \, \bar{l}^T(x'), \quad V_{CP} \, \nu_i(x) \, V_{CP}^{-1} = \gamma^0 \, C \, \bar{\nu}_i^T(x'), \tag{6.16}$$

where C is the matrix of the charge conjugation. From (6.16) we find

$$V_{CP} \, \bar{l}_L(x) \, V_{CP}^{-1} = -l_L^T(x') \, C^{-1} \gamma^0, \quad V_{CP} \, \bar{\nu}_{iL}(x) \, V_{CP}^{-1} = -\nu_{iL}^T(x') \, C^{-1} \gamma^0. \tag{6.17}$$

From (6.16) and (6.17) we have

$$V_{CP} \, \bar{l}_L(x) \, \gamma_\alpha \, \nu_{iL}(x) \, V_{CP}^{-1} = -l_L^T(x') \, C^{-1} \gamma^0 \, \gamma_\alpha \, \gamma^0 \, C \, \bar{\nu}_{iL}^T(x')$$

$$= \delta_\alpha \, l_L^T(x') \, \gamma_\alpha^T \, \bar{\nu}_{iL}^T(x')$$

$$= -\delta_\alpha \, \bar{\nu}_{iL}(x') \, \gamma_\alpha \, l_L(x'). \tag{6.18}$$

Here $\delta_\alpha = (1, -1, -1, -1)$ is the sign factor.[2]

For the field of the $W^\pm$ bosons we have

$$V_{CP} \, W^{\alpha\dagger}(x) \, V_{CP}^{-1} = -\delta_\alpha \, W^\alpha(x'). \tag{6.19}$$

---

[1] By a redefinition of arbitrary phases of the lepton fields we can always put the CP phase factors of the Dirac fields equal to one.

[2] Notice that in (6.18) no sum over α is assumed and minus sign in the last term is due to the transposition of the fermion fields.

From (6.18) and (6.19) we find

$$V_{CP}\, \mathscr{L}_I^{CC}(x)\, V_{CP}^{-1} = -\frac{g}{\sqrt{2}} \sum_{l,i} \bar{\nu}_{iL}(x')\, \gamma_\alpha\, U_{li}\, l_L(x')\, W^\alpha(x')$$

$$-\frac{g}{\sqrt{2}} \sum_{l,i} \bar{l}_L(x')\, \gamma_\alpha\, U_{li}^*\, \nu_{iL}(x')\, W^{\alpha\dagger}(x')$$

$$= \mathscr{L}_I^{CC}(x'). \tag{6.20}$$

Comparing (6.20) and (6.15) we conclude that in the case of the Dirac neutrinos from the CP invariance follows that the neutrino mixing matrix is real:

$$U_{li} = U_{li}^*. \tag{6.21}$$

Let us now consider the case of the Majorana neutrinos. Because of the Majorana condition $\nu_i(x) = C\bar{\nu}_i^T(x)$ the CP phase factors cannot be included in the Majorana fields. We have

$$V_{CP}\, \nu_i(x)\, V_{CP}^{-1} = \eta_i\, \gamma^0\, C\, \bar{\nu}_i^T(x') = \eta_i\, \gamma^0\, \nu_i(x'), \tag{6.22}$$

where η_i is the CP phase factor of the Majorana field $\nu_i(x)$ $(|\eta_i|^2 = 1)$.

We will show that $\eta_i = \pm i$. In fact, we have

$$V_{CP}^2\, \nu_i(x)\, (V_{CP}^{-1})^2 = \nu_i(x) = \eta_i\, V_{CP}\, \gamma^0\, \nu_i(x')\, V_{CP}^{-1} = \eta_i^2\, \gamma^0\, C\, \overline{(\gamma^0 \nu_i(x))}^T$$

$$= \eta_i^2\, \gamma^0\, C\, \gamma^{0T}\, \bar{\nu}_i^T(x) = -\eta_i^2\, \nu_i(x). \tag{6.23}$$

From this relation follows that

$$\eta_i^2 = -1, \quad \eta_i = \pm i. \tag{6.24}$$

Taking into account the Majorana condition, from (6.17) and (6.22) we find

$$V_{CP}\, \bar{l}_L(x)\gamma_\alpha \nu_{iL}(x)\, V_{CP}^{-1} = -\eta_i l_L^T(x')C^{-1}\gamma^0\gamma_\alpha\gamma^0 C\bar{\nu}_i^T(x') = -\eta_i \delta_\alpha \bar{\nu}_i(x')\gamma_\alpha l_L(x'). \tag{6.25}$$

Further, from (6.14), (6.19) and (6.25) we have

$$V_{CP}\, \mathscr{L}_I^{CC}(x)\, V_{CP}^{-1} = -\frac{g}{\sqrt{2}} \sum_{l,i} \bar{\nu}_{iL}(x')\, \gamma_\alpha\, U_{li}^M\, \eta_i\, l_L(x')\, W^\alpha(x')$$

$$-\frac{g}{\sqrt{2}} \sum_{l,i} \bar{l}_L(x')\, \gamma_\alpha\, U_{li}^{M*}\, \eta_i^*\, \nu_{iL}(x')\, W^{\alpha\dagger}(x')$$

$$= \mathscr{L}_I^{CC}(x'). \tag{6.26}$$

From this relation we conclude that if there is the CP invariance in the lepton sector Majorana neutrino mixing matrix satisfies the relation

$$U_{li}^{M}\, \eta_i = U_{li}^{M*}. \tag{6.27}$$

6.4   Standard Parametrization of 3 × 3 Mixing Matrix

We will consider here the unitary 3×3 mixing matrix for Dirac neutrinos and introduce the standard parameters which characterizes it: three mixing angles and one phase. We will show in the next chapter that from the neutrino mixing

$$\nu_{lL}(x) = \sum_{i=1}^{3} U_{li}\, \nu_{iL}(x) \tag{6.28}$$

follows that state of the flavor neutrinos $|\nu_l\rangle$ ($l = e, \mu, \tau$) is connected with states $|\nu_i\rangle$ of neutrinos with masses m_i ($i = 1, 2, 3$) and momentum $\mathbf{p}$ by the relation

$$|\nu_l\rangle = \sum_{i=1}^{3} U_{li}^{*}\, |\nu_i\rangle, \tag{6.29}$$

where

$$\langle \nu_i | \nu_k \rangle = \delta_{ik}. \tag{6.30}$$

In the matrix form the relation (6.29) can be written as follows

$$|\nu_f\rangle = U^{*}\, |\nu\rangle, \tag{6.31}$$

where

$$|\nu_f\rangle = \begin{pmatrix} |\nu_e\rangle \\ |\nu_\mu\rangle \\ |\nu_\tau\rangle \end{pmatrix}, \quad |\nu\rangle = \begin{pmatrix} |\nu_1\rangle \\ |\nu_2\rangle \\ |\nu_3\rangle \end{pmatrix}. \tag{6.32}$$

In order to parameterize the matrix U we perform three Euler rotations. The first rotation will be performed at the angle θ_{12} around the vector $|\nu_3\rangle$. The new vectors are

$$|\nu_1\rangle' = c_{12}\, |\nu_1\rangle + s_{12}\, |\nu_2\rangle$$
$$|\nu_2\rangle' = -s_{12}\, |\nu_1\rangle + c_{12}\, |\nu_2\rangle$$
$$|\nu_3\rangle' = |\nu_3\rangle, \tag{6.33}$$

where $c_{12} = \cos\theta_{12}$ and $s_{12} = \sin\theta_{12}$. In the matrix form (6.33) can be written as follows

$$|\nu\rangle' = R_{12}(\theta_{12}) |\nu\rangle. \tag{6.34}$$

Here

$$R_{12}(\theta_{12}) = \begin{pmatrix} c_{12} & s_{12} & 0 \\ -s_{12} & c_{12} & 0 \\ 0 & 0 & 1 \end{pmatrix}. \tag{6.35}$$

The second rotation will be performed at the angle θ_{13} around the vector $|\nu_2\rangle'$. At this step we will introduce the CP phase δ, connected with the rotation of the vector of the third family $|\nu_3\rangle$. We will obtain the following three orthogonal and normalized vectors:

$$
\begin{aligned}
|\nu_1\rangle'' &= c_{13} |\nu_1\rangle' + s_{13}e^{i\delta} |\nu_3\rangle' \\
|\nu_2\rangle'' &= |\nu_2\rangle' \\
|\nu_3\rangle'' &= -s_{13}e^{-i\delta} |\nu_1\rangle' + c_{13} |\nu_3\rangle'.
\end{aligned} \tag{6.36}
$$

In the matrix form we have

$$|\nu\rangle'' = R_{13}^*(\theta_{13}) |\nu\rangle' = R_{13}^*(\theta_{13}) R_{12}(\theta_{12}) |\nu\rangle. \tag{6.37}$$

Here

$$R_{13}^*(\theta_{13}) = \begin{pmatrix} c_{13} & 0 & s_{13}e^{i\delta} \\ 0 & 1 & 0 \\ -s_{13}e^{-i\delta} & 0 & c_{13} \end{pmatrix}. \tag{6.38}$$

Finally, after the third rotation at the angle θ_{23} around the vector $|\nu_1\rangle''$ we find

$$
\begin{aligned}
|\nu_e\rangle &= |\nu_1\rangle'' \\
|\nu_\mu\rangle &= c_{23}|\nu_2\rangle'' + s_{23} |\nu_3\rangle'' \\
|\nu_\tau\rangle &= -s_{23} |\nu_2\rangle'' + c_{23} |\nu_3\rangle''.
\end{aligned} \tag{6.39}
$$

In the matrix form we have

$$|\nu_f\rangle = R_{23}(\theta_{23}) |\nu\rangle'' = R_{23}(\theta_{23}) R_{13}^*(\theta_{13}) R_{12}(\theta_{12}) |\nu\rangle = U^* |\nu\rangle. \tag{6.40}$$

Here

$$R_{23}(\theta_{23}) = \begin{pmatrix} 1 & 0 & 0 \\ 0 & c_{23} & s_{23} \\ 0 & -s_{23} & c_{23} \end{pmatrix}. \tag{6.41}$$

From (6.40) we find

$$U^* = R_{23}(\theta_{23})\, R_{13}^*(\theta_{13})\, R_{12}(\theta_{12}). \tag{6.42}$$

Thus, the unitary neutrino mixing matrix U is given by the product of three Euler rotation matrices

$$U = \begin{pmatrix} 1 & 0 & 0 \\ 0 & c_{23} & s_{23} \\ 0 & -s_{23} & c_{23} \end{pmatrix} \begin{pmatrix} c_{13} & 0 & s_{13}e^{-i\delta} \\ 0 & 1 & 0 \\ -s_{13}e^{i\delta} & 0 & c_{13} \end{pmatrix} \begin{pmatrix} c_{12} & s_{12} & 0 \\ -s_{12} & c_{12} & 0 \\ 0 & 0 & 1 \end{pmatrix}. \tag{6.43}$$

From (6.43) we find the following standard expression for the Dirac neutrino mixing matrix

$$U^D = \begin{pmatrix} c_{13}c_{12} & c_{13}s_{12} & s_{13}e^{-i\delta} \\ -c_{23}s_{12} - s_{23}c_{12}s_{13}e^{i\delta} & c_{23}c_{12} - s_{23}s_{12}s_{13}e^{i\delta} & c_{13}s_{23} \\ s_{23}s_{12} - c_{23}c_{12}s_{13}e^{i\delta} & -s_{23}c_{12} - c_{23}s_{12}s_{13}e^{i\delta} & c_{13}c_{23} \end{pmatrix}. \tag{6.44}$$

The mixing angles θ_{12}, θ_{23} and θ_{13} and the phase δ are parameters. They can be determined from neutrino oscillation experiments (see later).

The 3×3 Majorana mixing matrix has the form (see previous section)

$$U^M = U^D\, S^M(\bar{\alpha}) \tag{6.45}$$

where the phase matrix $S^M(\bar{\alpha})$ is characterized by two Majorana phases and has the form

$$S^M(\bar{\alpha}) = \begin{pmatrix} e^{i\bar{\alpha}_1} & 0 & 0 \\ 0 & e^{i\bar{\alpha}_2} & 0 \\ 0 & 0 & 1 \end{pmatrix}. \tag{6.46}$$

and the matrix U^D is given by (6.44).

Under the change $\theta \to \pi + \theta$ the matrix

$$\begin{pmatrix} \cos\theta & \sin\theta \\ -\sin\theta & \cos\theta \end{pmatrix} \tag{6.47}$$

changes sign. In the observable transition probabilities enters the square of the mixing matrix. Thus, angles θ and $\pi + \theta$ cannot be distinguished. Moreover, from the expressions for observable probabilities (see the next chapter) it follows that for the mixing angles θ_{12}, θ_{23}, θ_{13} we can choose the range $0 \leq \theta_{ik} \leq \frac{\pi}{2}$ and for the CP phase δ the range $0 \leq \delta \leq 2\pi$. As it was shown in the previous section, if CP is conserved the Dirac mixing matrix is real: $U^D = U^{D*}$. Thus, in this case we have $\delta = 0, \pi$.

6.5 On Models of Neutrino Mixing

During many years existed the following upper bound on the value of the parameter $\sin^2 \theta_{13}$:

$$\sin^2 \theta_{13} \leq 5 \cdot 10^{-2}. \tag{6.48}$$

This bound was obtained from analysis of the data of the CHOOZ reactor neutrino experiment (1997–1998). Thus the CHOOZ data were compatible with a popular at that time idea that $\theta_{13} = 0$.

It is obvious from the previous subsection that if $\theta_{13} = 0$ the general unitary mixing matrix can be obtained by (1,2) and (2,3) Euler rotations. It has the following form

$$U = \begin{pmatrix} 1 & 0 & 0 \\ 0 & c_{23} & s_{23} \\ 0 & -s_{23} & c_{23} \end{pmatrix} \begin{pmatrix} c_{12} & s_{12} & 0 \\ -s_{12} & c_{12} & 0 \\ 0 & 0 & 1 \end{pmatrix} = \begin{pmatrix} c_{12} & s_{12} & 0 \\ -c_{23}s_{12} & c_{23}c_{12} & s_{23} \\ s_{23}s_{12} & -s_{23}c_{12} & c_{23} \end{pmatrix}. \tag{6.49}$$

From global analysis of the data of atmospheric, accelerator and solar experiments and the reactor experiment KamLAND, performed in 2003 the following 2σ ranges were obtained

$$0.25 \leq \sin^2 \theta_{12} \leq 0.36, \quad 0.36 \leq \sin^2 \theta_{23} \leq 0.67. \tag{6.50}$$

For the best fit values of these parameters it was found

$$\sin^2 \theta_{12} = 0.30, \quad \sin^2 \theta_{23} = 0.52. \tag{6.51}$$

These data were compatible with the simplest assumption

$$\sin^2 \theta_{12} = \frac{1}{3}, \quad \sin^2 \theta_{23} = \frac{1}{2}. \tag{6.52}$$

From (6.52) follows that the neutrino mixing matrix can be chosen in the following form

$$
U_{TB} = \begin{pmatrix} \sqrt{\frac{2}{3}} & \frac{1}{\sqrt{3}} & 0 \\ -\frac{1}{\sqrt{6}} & \frac{1}{\sqrt{3}} & -\frac{1}{\sqrt{2}} \\ -\frac{1}{\sqrt{6}} & \frac{1}{\sqrt{3}} & \frac{1}{\sqrt{2}} \end{pmatrix}.
\tag{6.53}
$$

The matrix U_{TB} is called tri-bimaximal mixing matrix. Notice that the elements of the second column of the matrix (6.53) are equal. We have

$$
\nu_{2L} = \frac{1}{\sqrt{3}}(\nu_{eL} + \nu_{\mu L} + \nu_{\tau L}).
\tag{6.54}
$$

Idea of the tri-bimaximal mixing inspired many models of neutrino mixing based on broken finite flavor symmetries. For illustration we will consider the tri-bimaximal mixing from the point of view of the broken A4 symmetry.

We will assume that neutrino with definite masses are Majorana particles. In this case for the mass matrix M we have

$$
M = U^M \, m \, (U^M)^T,
\tag{6.55}
$$

where U^M is the mixing matrix and $m_{ik} = m_i \, \delta_{ik}$. As we have seen in the previous section $U^M = (e^{i\alpha_1}, e^{i\alpha_2}, 1)$ where the matrix U^D is given by (6.44) and $\bar{\alpha}_{1,2}$ are Majorana phases.

Let us assume that $U^D = U_{TB}$ and $\bar{\alpha}_{1,2} = 0$. In this case the neutrino mass matrix is given by the expression

$$
M_{TB} = U_{TB} \, m \, U_{TB}^T.
\tag{6.56}
$$

From (6.53) and (6.56) for the neutrino mixing matrix we easily find the following expression

$$
M_{TB} = \begin{pmatrix} (\frac{2}{3}m_1 + \frac{1}{3}m_2) & (-\frac{1}{3}m_1 + \frac{1}{3}m_2) & (-\frac{1}{3}m_1 + \frac{1}{3}m_2) \\ (-\frac{1}{3}m_1 + \frac{1}{3}m_2) & (\frac{1}{6}m_1 + \frac{1}{3}m_2 + \frac{1}{2}m_3) & (\frac{1}{6}m_1 + \frac{1}{3}m_2 - \frac{1}{2}m_3) \\ (-\frac{1}{3}m_1 + \frac{1}{3}m_2) & (\frac{1}{6}m_1 + \frac{1}{3}m_2 - \frac{1}{2}m_3) & (\frac{1}{6}m_1 + \frac{1}{3}m_2 + \frac{1}{2}m_3) \end{pmatrix}
\tag{6.57}
$$

The tri-bimaximal mass matrix depends on three parameters. Let us introduce

$$
x = \frac{2}{3}m_1 + \frac{1}{3}m_2, \quad y = -\frac{1}{3}m_1 + \frac{1}{3}m_2, \quad \upsilon = -\frac{1}{2}m_1 + \frac{1}{2}m_3.
\tag{6.58}
$$

From (6.57) we have

$$M_{\mathrm{TB}} = \begin{pmatrix} x & y & y \\ y & x+v & y-v \\ y & y-v & x+v \end{pmatrix}. \tag{6.59}$$

The group A4 is the group of even permutations of four objects. All elements of the group are products of two generators S and T which satisfy the relations

$$S^2 = T^3 = (ST)^3 = 1. \tag{6.60}$$

The number of elements in the A4 group is equal to $\frac{4!}{2} = 12$. Taking into account the relations (6.60) we see that elements of the A4 group (all possible products of S and T) are given by

$$1, T, S, ST, TS, T^2, T^2S, TST, ST^2, STS, T^2ST, TST^2. \tag{6.61}$$

The group A4 has four irreducible representations: one triplet and three singlets

$$3, 1, 1', 1''. \tag{6.62}$$

For one-dimensional unitary representations we can choose $S = 1, T = e^{i\alpha}$. From the relations (6.60) follows that $3\alpha = 2\pi n$, where n is an integer number. Thus, we have

$$1 : S = 1 \ T = 1; \quad 1' : S = 1 \ T = \omega^2 : \quad 1'' : S = 1 \ T = \omega, \tag{6.63}$$

where $\omega = e^{\frac{2\pi}{3}i}$.

In the basis in which the matrix T is diagonal, the 3×3 unitary matrix T has the following general form

$$T = \begin{pmatrix} e^{i\beta_1} & 0 & 0 \\ 0 & e^{i\beta_2} & 0 \\ 0 & 0 & e^{i\beta_3} \end{pmatrix}, \tag{6.64}$$

where β_i is a real phase. Taking into account that $T^3 = 1$, we have $\beta_i = \frac{2\pi}{3}n_i$ $(n_i = 0, 1, 2, \ldots)$ We can choose the three-dimensional representation of the generator T in the form

$$T = \begin{pmatrix} 1 & 0 & 0 \\ 0 & \omega^2 & 0 \\ 0 & 0 & \omega \end{pmatrix}. \tag{6.65}$$

The real unitary matrix S satisfies the condition $S^T S = 1$. Taking into account that $S^2 = 1$ we have $S^T = S$. It is easy to check that the three-dimensional orthogonal symmetrical matrix

$$S = \frac{1}{3} \begin{pmatrix} -1 & 2 & 2 \\ 2 & -1 & 2 \\ 2 & 2 & -1 \end{pmatrix}. \qquad (6.66)$$

satisfies the relations (6.60).

Let us assume that the lepton doublets L_{lL} ($l = e, \mu, \tau$) are A4 triplets. In order to generate the neutrino mass term we will further assume that exist a scalar triplet Higgs-like flavon fields ϕ_T, ϕ_S and a singlet field ξ which enter into the A4 invariant Yukawa interactions together with lepton doublets.

If we assume the vacuum alignments

$$< \phi_T >= (v_T, 0, 0), \quad < \phi_S >= (v_S, v_S, v_S), \quad < \xi >= u \qquad (6.67)$$

the A4 symmetry will be broken down to the G_S symmetry generated by the operator S. The mass matrix M satisfies in this case the relation

$$S M S = M, \qquad (6.68)$$

where S is given by (6.66). From (6.68) we find that the mass matrix M is given by

$$M = \begin{pmatrix} x & y & z \\ y & x + v + y - z & w \\ z & w & x + v \end{pmatrix}. \qquad (6.69)$$

This matrix is characterized by four real parameters and does not have the tri-bimaximal form (6.59).

We will come to the tri-bimaximal mixing if we further assume $\mu - \tau$ symmetry of the neutrino mixing matrix

$$S_{\mu\tau} M S_{\mu\tau} = M, \qquad (6.70)$$

where

$$S_{\mu\tau} = \begin{pmatrix} 1 & 0 & 0 \\ 0 & 0 & 1 \\ 0 & 1 & 0 \end{pmatrix}. \qquad (6.71)$$

From (6.70) and (6.71) we find that

$$z = y. \qquad (6.72)$$

With the relation (6.72) we come to the tri-bimaximal neutrino mixing matrix (6.59). Let us stress that in order to implement $\mu - \tau$ symmetry we need to assume that there are no $1'$ and $1''$ scalar flavon fields in the Yukawa interactions.

In 2012 it was discovered in the Daya Bay, RENO and Double Chooz reactor neutrino experiments that the angle θ_{13} is different from zero. From the global analysis of neutrino oscillation data it was found that

$$\sin^2 \theta_{12} = 0.304^{+0.013}_{-0.012}, \quad \sin^2 \theta_{23} = 0.452^{+0.052}_{-0.028}, \quad \sin^2 \theta_{13} = 0.0218^{+0.0010}_{-0.0010}.$$
$$(6.73)$$

Thus, the tri-bimaximal mixing can be considered only as a first approximation valid with accuracy of a few %.

Numerous models, which are based on different broken discrete flavor symmetries and take into account the fact that the angle θ_{13} is different from zero, were proposed. It is not our aim to discuss here these models. References to original papers can be found in reviews which we present at the end of the book.

For illustration of some ideas and results we will briefly discuss one class of the models. As we saw in Chap. 3, the leptonic charged current is given by the expression

$$j_\alpha^{CC}(x) = 2 \bar{\nu}'_L(x) \gamma_\alpha L'_L(x) = 2 \sum_{l=e,\mu,\tau} \bar{\nu}'_{lL}(x) \gamma_\alpha l'_L(x). \tag{6.74}$$

After the diagonalization of the charged lepton and neutrino mass terms we have

$$L'_L(x) = U_L^{\text{lep}} L_L(x), \quad \nu'_L(x) = U_L^\nu \nu_L(x). \tag{6.75}$$

where U_L^{lep} and U_L^ν are unitary 3×3 matrices and

$$L_L(x) = \begin{pmatrix} e_L(x) \\ \mu_L(x) \\ \tau_L(x) \end{pmatrix}, \quad \nu_L(x) = \begin{pmatrix} \nu_{1L}(x) \\ \nu_{2L}(x) \\ \nu_{1L}(x) \end{pmatrix}. \tag{6.76}$$

Here $l(x)$ ($l = e, \mu, \tau$) is the field of a lepton with mass m_l and $\nu_i(x)$ ($i = 1, 2, 3$) is the field of neutrino with mass m_i.

From (6.74) and (6.75) we have

$$j_\alpha^{CC}(x) = 2 \bar{\nu}_L(x) U_L^{\nu\dagger} U_L^{\text{lep}} \gamma_\alpha L_L(x) = 2 \bar{\nu}_L(x) U^\dagger \gamma_\alpha L_L(x) = 2 \sum_{l,i} \bar{\nu}_i(x) U_{li}^* \gamma_\alpha l_L(x). \tag{6.77}$$

Thus the PMNS mixing matrix is given by the relation

$$U = U_L^{\text{lep}\dagger} U_L^\nu. \tag{6.78}$$

It was assumed that the matrix U_L^ν is determined by a broken finite flavor symmetry (A4, S4 and others) and $e'3$ element of this matrix is equal to zero. The tri-bimaximal matrix (6.53) is one of the possible examples. In this approach nonzero value of the angle θ_{13} is due to the matrix U_L^{lep}.

So let us assume that the PMNS mixing matrix is given by the expression

$$U = U_{\text{TB}} R_{23}(\alpha), \tag{6.79}$$

where U_{TB} is the tri-bimaximal matrix (6.53) and the matrix $R_{23}(\alpha)$ has the form

$$R_{23}(\alpha) = \begin{pmatrix} 1 & 0 & 0 \\ 0 & \cos\alpha & \sin\alpha \\ 0 & -\sin\alpha & \cos\alpha \end{pmatrix}, \tag{6.80}$$

where the parameter α is determined by θ_{13}. The relation (6.79) leads to the following sum rules

$$\sin^2\theta_{23} = \frac{1}{2} - \sqrt{2}\sin\theta_{13}\cos\delta + O(\sin^2\theta_{13}), \quad \sin^2\theta_{12} = \frac{1}{3} - \frac{2}{3}\sin^2\theta_{13} + O(\sin^4\theta_{13}). \tag{6.81}$$

Another possibility is to assume

$$U = U_{\text{TB}} R_{13}(\alpha), \tag{6.82}$$

where

$$R_{13}(\alpha) = \begin{pmatrix} \cos\alpha & 0 & \sin\alpha \\ 0 & 1 & 0 \\ -\sin\alpha & 0 & \cos\alpha \end{pmatrix}, \tag{6.83}$$

The relation (6.82) leads to another sum rules

$$\sin^2\theta_{23} = \frac{1}{2} + \frac{1}{\sqrt{2}}\sin\theta_{13}\cos\delta + O(\sin^2\theta_{13}), \quad \sin^2\theta_{12} = \frac{1}{3} + \frac{1}{3}\sin^2\theta_{13} + O(\sin^4\theta_{13}). \tag{6.84}$$

The sum rules (6.81) and (6.84) will be tested in future neutrino oscillation experiments. Many other sum rules, based on different symmetry groups, were obtained in different papers.

Chapter 7
Neutrino Oscillations in Vacuum

7.1 Introduction

Discovery of neutrino oscillations in the atmospheric Super Kamiokande, solar SNO, reactor KamLAND, solar Homestake, GALLEX-GNO and SAGE experiments was one of the most important recent discovery in the elementary particle physics. This discovery was confirmed by the accelerator K2K, MINOS, T2K, NOvA, reactor Daya Bay, RENO, Double Chooz and solar BOREXINO experiments in which neutrino oscillations were studied in details. In 2015 T. Kajita and A. McDonald were awarded by the Nobel prize "for the discovery of neutrino oscillations, which shows that neutrinos have mass". Neutrino oscillations are driven by small neutrino mass-squared differences and neutrino mixing. From all existing data it follows that neutrino masses are many orders of magnitude smaller than masses of other fundamental fermions (quarks and leptons). Small neutrino masses cannot be naturally explained by the SM Higgs mechanism of the mass generation. A new beyond the SM mechanism of the mass generation is required.

Neutrino oscillation parameters are known at present with accuracy from about 3% to about 10%. Investigation of neutrino oscillations enter now into a new era, era of precision measurements. Main questions which will be addressed in the future neutrino oscillation experiments are the following:

- What is the character of neutrino mass spectrum (normal or inverted).
- What is the value of the CP phase δ.
- Are there transitions of flavor neutrinos into sterile states.

In this chapter we will consider in some details neutrino oscillations in vacuum.

Neutrinos and antineutrinos are emitted in weak decays of pions and kaons, which are produced at accelerators and in the processes of interaction of cosmic rays with the atmosphere, in decays of muons, products of decays of pions and

© Springer International Publishing AG, part of Springer Nature 2018
S. Bilenky, *Introduction to the Physics of Massive and Mixed Neutrinos*,
Lecture Notes in Physics 947, https://doi.org/10.1007/978-3-319-74802-3_7

kaons, in β-decays of nuclei, products of the fission of uranium and plutonium in reactors, in nuclear reactions in the sun etc. The first question which we will address here will be the following: in the case of the neutrino mixing *what are the states of neutrinos and antineutrinos* which are produced in weak interaction processes and take part in CC and NC neutrino processes.

7.2   Flavor Neutrino States. Oscillations of Flavor Neutrinos

Let us consider a decay

$$a \rightarrow b + l^+ + \nu_l. \tag{7.1}$$

Here a and b are some hadrons (in the case of two-body decays b is vacuum).

The leptonic part of the SM Lagrangian of the CC interaction is given by the expression

$$\mathscr{L}_I^{CC} = -\frac{g}{\sqrt{2}} \sum_{l=e,\mu,\tau} \bar{\nu}_{lL} \, \gamma_\alpha \, l_L \, W^\alpha + \text{h.c.} \tag{7.2}$$

In the case of the three-neutrino mixing,

$$\nu_{lL} = \sum_{i=1}^{3} U_{li} \nu_{iL}, \tag{7.3}$$

where U is a unitary PMNS mixing matrix and ν_i is the field of neutrino (Dirac or Majorana) with the mass m_i.

The matrix element of the transition $a \rightarrow b + l^+ + \nu_i$ is given by the expression

$$\langle \nu_i \, l^+ \, b \, | S | \, a \rangle = U_{li}^* \, (-i) \frac{G_F}{\sqrt{2}} \, N \, 2 \, \bar{u}_L(p_i) \, \gamma_\alpha \, u_L(-p_{\,l}) \, \langle b | \, J^\alpha(0) \, | a \rangle \, (2\pi)^4 \, \delta(P' - P). \tag{7.4}$$

Here p_i and p_l are momenta of ν_i and l^+, P and P' are the total initial and final momenta, J^α is the hadronic charged current and N is the product of the standard normalization factors.

In neutrino oscillation experiments neutrino energies E are much larger than neutrino masses. In reactor experiments $E \simeq$ a few MeV, in the atmospheric and accelerator experiments $E \simeq$ a few GeV etc. For neutrino masses from exiting data we have $m_i \lesssim 1\,\text{eV}$.

For the ultrarelativistic neutrinos we have

$$|\mathbf{p_i} - \mathbf{p_k}| \simeq \frac{\Delta m_{ik}^2}{2E}, \quad (i \neq k) \tag{7.5}$$

where $\Delta m_{ik}^2 = m_k^2 - m_i^2$. From the three-neutrino analysis of the neutrino oscillation data for two mass-squared differences the following values were obtained

$$\Delta m_{12}^2 \simeq 8 \cdot 10^{-5} \text{ eV}^2, \quad \Delta m_{23}^2 \simeq 2.4 \cdot 10^{-3} \text{ eV}^2 \tag{7.6}$$

Thus for the reactor (KamLAND) and the accelerator (atmospheric) neutrino experiments we have

$$\frac{\Delta m_{12}^2 \, c^4}{2E \, \hbar c} \simeq \frac{2}{10 \text{ km}}, \quad \frac{\Delta m_{23}^2 \, c^4}{2E \, \hbar c} \simeq \frac{0.6}{10^2 \text{ km}} \tag{7.7}$$

On the other side from the Heisenberg uncertainty relation for the uncertainty of the neutrino momentum he have

$$(\Delta p)_{QM} \simeq \frac{1}{d}, \tag{7.8}$$

where d is a microscopic size of a wave packet of a neutrino source.

From (7.5), (7.7) and (7.8) we have

$$|\mathbf{p_i} - \mathbf{p_k}| \ll (\Delta p)_{QM}. \tag{7.9}$$

Thus from the Heisenberg uncertainty relation follows that it is impossible to resolve production of ultrarelativistic neutrinos with small mass-squared differences in weak decays. This means that in weak decays the flavor lepton numbers L_e, L_μ, L_τ are conserved. Neutrinos ν_e, ν_μ, ν_τ (antineutrinos $\bar{\nu}_e$, $\bar{\nu}_\mu$, $\bar{\nu}_\tau$) which can be produced in weak decays together with e^+, μ^+, τ^+ (e^-, μ^-, τ^-), respectively, are called *flavor neutrinos (antineutrinos)*.[1]

What is the state of the flavor neutrino? From (7.4) we find

$$\langle \nu_i \, l^+ \, b \, | S | \, a \rangle \simeq U_{li}^* \, \langle \nu_l \, l^+ \, b \, | S | \, a \rangle_{SM}, \tag{7.10}$$

Here $\langle \nu_l \, l^+ \, b \, | S | \, a \rangle_{SM}$ is the SM matrix element of the decay (7.1).

---

[1]Similarly, flavor neutrinos ν_e, ν_μ, ν_τ (antineutrinos $\bar{\nu}_e$, $\bar{\nu}_\mu$, $\bar{\nu}_\tau$) produce e^-, μ^-, τ^- (e^+, μ^+, τ^+) in CC neutrino reactions.

From (7.10) follows that the coherent state of neutrino produced in the decay (7.1) is given by

$$\sum_{i=1}^{3} |\nu_i\rangle U_{li}^* \langle \nu_l \, l^+ \, b \, |S| \, a \rangle_{SM},\tag{7.11}$$

where $|\nu_i\rangle$ is normalized state of the left-handed neutrino with the mass m_i, momentum $\mathbf{p}$ and the energy $E_i = \sqrt{|\mathbf{p}|^2 + m_i^2}$. From (7.11) we find that *the normalized coherent state of the flavor neutrino ν_l is given by the expression*

$$|\nu_l\rangle = \sum_{i=1}^{3} U_{li}^* \, |\nu_i\rangle.\tag{7.12}$$

Analogously for the state of the flavor antineutrino $\bar{\nu}_l$ we have[2]

$$|\bar{\nu}_l\rangle = \sum_{i=1}^{3} U_{li} \, |\bar{\nu}_i\rangle.\tag{7.13}$$

As we have discussed in the previous section *due to the Heisenberg uncertainty relation in weak decays and neutrino reactions flavor neutrinos (antineutrinos), i.e. coherent superpositions of states of neutrinos (antineutrinos) with definite masses, are produced and detected.* In other words in weak decays and reactions we cannot resolve productions and absorption of neutrinos (antineutrinos) with small mass-squared differences. Small neutrino mass-squared differences can be revealed in special experiments with large distances between neutrino sources and detectors. Physical basis for such type of experiments is the time-energy uncertainty relation

$$\Delta E \, \Delta t \gtrsim 1.\tag{7.14}$$

Here Δt is the time interval during which the state with the energy uncertainty ΔE is significantly changed.

In the case of the neutrino beams $\Delta E = |E_i - E_k| \simeq \frac{|\Delta m_{ki}^2|}{2E}$ and $\Delta t \simeq L$, where L is the distance between a neutrino source and neutrino detector. From the time-energy uncertainty relation (7.14) follows that in order to reveal mass-squared difference $|\Delta m_{ki}^2|$ the quantity $\frac{L}{E}$ must satisfy the condition

$$|\Delta m_{ki}^2| \frac{L}{2E} \gtrsim 1.\tag{7.15}$$

---

[2]In the case of the Majorana neutrino ν_i we need to change $|\bar{\nu}_i\rangle$ by the state of the right-handed Majorana neutrino with the mass m_i.

We will come to the same condition later when we will consider probabilities on the flavor neutrino transitions in vacuum.

We will consider now the theory of the neutrino oscillations in vacuum. Let us assume that at $t = 0$ in a weak decay ($\pi^+ \to \mu^+ + \nu_\mu$, $\mu^+ \to e^+ + \nu_e + \bar{\nu}_\mu$ etc.) the flavor neutrino ν_l was produced. At the time $t > 0$ for the neutrino state we have

$$|\nu_l\rangle_t = e^{-iHt} |\nu_l\rangle. \tag{7.16}$$

Here H is the free Hamiltonian. In a neutrino detector at the time t flavor neutrinos are detected. Let us expand the state $|\nu_l\rangle_t$ over the complete system of the states of flavor neutrinos $|\nu_{l'}\rangle$. We have

$$|\nu_l\rangle_t = \sum_{l'=e,\mu,\tau} A(\nu_l \to \nu_{l'}) |\nu_{l'}\rangle \tag{7.17}$$

Here

$$A(\nu_l \to \nu_{l'}) = \langle \nu_{l'}| \, e^{-iHt} \, |\nu_l\rangle = \sum_{i=1}^{3} U_{l'i} \, e^{-i E_i t} \, U_{li}^*. \tag{7.18}$$

In (7.18) we take into account that

$$H |\nu_i\rangle = E_i |\nu_i\rangle, \quad E_i = \sqrt{|\mathbf{p}|^2 + m_i^2} \quad \langle \nu_{l'}|i\rangle = U_{l'i} \quad \langle i|\nu_l\rangle = U_{li}^*. \tag{7.19}$$

From (7.18) follows that $A(\nu_l \to \nu_{l'})$ is the amplitude of the probability to find at the time t the flavor neutrino $\nu_{l'}$ in the beam of neutrinos which originally at $t = 0$ was a beam of ν_l. The expression (7.18) has the simple meaning: the factor U_{li}^* is the amplitude of the transition from the initial flavor state $|\nu_l\rangle$ into the state of a neutrino with definite mass $|\nu_i\rangle$; the factor $e^{-i E_i t}$ describes the propagation in the state with definite mass and energy; the factor $U_{l'i}$ is the amplitude of the transition from the state $|\nu_i\rangle$ to the flavor state $|\nu_{l'}\rangle$. The sum over all states with definite masses must be performed.

The probabilities of the transitions $\nu_l \to \nu_{l'}$ and $\bar{\nu}_l \to \bar{\nu}_{l'}$ are equal, respectively,

$$P(\nu_l \to \nu_{l'}) = |\sum_{i=1}^{3} U_{l'i} \, e^{-i E_i t} \, U_{li}^*|^2 \tag{7.20}$$

and

$$P(\bar{\nu}_l \to \bar{\nu}_{l'}) = |\sum_{i=1}^{3} U_{l'i}^* \, e^{-i E_i t} \, U_{li}|^2. \tag{7.21}$$

From the unitarity of the mixing matrix it follows that (7.20) is *normalized transition probability*. In fact, we have

$$
\sum_{l'} P(\nu_l \rightarrow \nu_{l'}) = \sum_{i,k} \sum_{l'} (U_{l'i} U^*_{l'k}) \, e^{-i\,(E_i - E_k)\,t} \, U^*_{li} U_{lk}
$$

$$
= \sum_{i,k} \delta_{ik} \, e^{-i\,(E_i - E_k)\,t} \, U^*_{li} U_{lk}
$$

$$
= \sum_i U^*_{li} U_{li} = 1. \tag{7.22}
$$

Analogously we find

$$
\sum_{l'} P(\bar{\nu}_l \rightarrow \bar{\nu}_{l'}) = 1 \tag{7.23}
$$

and also

$$
\sum_l P(\nu_l \rightarrow \nu_{l'}) = 1, \qquad \sum_l P(\bar{\nu}_l \rightarrow \bar{\nu}_{l'}) = 1. \tag{7.24}
$$

Further, the expression (7.20) for the $\nu_l \rightarrow \nu_{l'}$ transition probability can be written in the form

$$
P(\nu_l \rightarrow \nu_{l'}) = |\sum_{i=1}^{3} U^*_{li} \, e^{-i\, E_i\, t} \, U_{l'i}|^2 \tag{7.25}
$$

Comparing this expression with (7.21) we conclude that the following relation holds

$$
P(\nu_l \rightarrow \nu_{l'}) = P(\bar{\nu}_{l'} \rightarrow \bar{\nu}_l). \tag{7.26}
$$

This relation is a consequence of the CPT invariance. From (7.26) follows that ν_l and $\bar{\nu}_l$ survival probabilities are equal

$$
P(\nu_l \rightarrow \nu_l) = P(\bar{\nu}_l \rightarrow \bar{\nu}_l). \tag{7.27}
$$

Let us assume now that the CP invariance in the lepton sector holds. In this case, as it was shown in the previous chapter,

$$
U_{li} = U^*_{li} \tag{7.28}
$$

in the case of the Dirac neutrinos ν_i and

$$U_{li}\,\eta_i = U_{li}^* \tag{7.29}$$

in the case of the Majorana neutrinos ($\eta_i = \pm i$ is the CP parity of the neutrino ν_i). For the Majorana neutrinos we have

$$U_{l'i}\,U_{li}^* = U_{l'i}^*\,\eta_i^*\eta_i\,U_{li} = U_{l'i}^*\,U_{li}. \tag{7.30}$$

From (7.20), (7.21), (7.28) and (7.30) we find that in the case of the CP invariance

$$P(\nu_l \to \nu_{l'}) = P(\bar{\nu}_l \to \bar{\nu}_{l'}). \tag{7.31}$$

The probabilities of the transitions $\nu_l \to \nu_l$ and $\bar{\nu}_l \to \bar{\nu}_l$ are equal due to CPT invariance. Thus, in order to check whether CP is conserved in the lepton sector we need to test the relations (7.31) at $l' \neq l$.

Let us notice that from relations (7.26) and relations (7.31) we find

$$P(\nu_l \to \nu_{l'}) = P(\nu_{l'} \to \nu_l), \quad P(\bar{\nu}_l \to \bar{\nu}_{l'}) = P(\bar{\nu}_{l'} \to \bar{\nu}_l). \tag{7.32}$$

These relations are a consequence of the invariance under time reversal T which obviously holds if both CP invariance and CPT invariance take place.

Finally, we will show that investigation of the neutrino oscillations does not allow to reveal the nature of neutrinos with definite masses (Dirac or Majorana?). In fact, the neutrino mixing matrix in the case of the Majorana neutrinos has the form (see (6.45))

$$U_{li}^M = U_{li}^D\,S_{ii}(\bar{\alpha}), \tag{7.33}$$

where $S_{ii}(\bar{\alpha}) = e^{i\bar{\alpha}_i}$ ($\bar{\alpha}_3 = 0, \bar{\alpha}_{1,2}$ are Majorana phases) and U^D is the mixing matrix in the case of the Dirac neutrinos. We have

$$U_{l'i}^M\,U_{li}^{M*} = U_{l'i}^D\,U_{li}^{D*}. \tag{7.34}$$

From (7.34), (7.20) and (7.21) follows that the Majorana phases do not enter into transition probabilities. In other words neutrino transition probabilities have the same form in the Majorana and Dirac cases.

7.3 Standard Expression for $\nu_l \rightarrow \nu_{l'}$ ($\bar{\nu}_l \rightarrow \bar{\nu}_{l'}$) Transition Probability

Let us consider now the expression (7.20) for the $\nu_l \rightarrow \nu_{l'}$ transition probability. We have

$$
\begin{aligned}
P(\nu_l \rightarrow \nu_{l'}) &= \sum_{i,k} U_{l'i} U_{l'k}^* U_{li}^* U_{lk} e^{-i\,(E_i - E_k)t} \\
&= \sum_i |U_{l'i}|^2 |U_{li}|^2 \\
&\quad + 2\,\text{Re} \sum_{i>k} U_{l'i} U_{l'k}^* U_{li}^* U_{lk} e^{-i\,(E_i - E_k)t}.
\end{aligned} \tag{7.35}
$$

This expression can be presented in an another form if we take into account the unitarity condition

$$
\sum_i U_{l'i} U_{li}^* = \delta_{l'l}. \tag{7.36}
$$

From this relation we find

$$
\sum_i |U_{l'i}|^2 |U_{li}|^2 = \delta_{l'l} - 2\,\text{Re} \sum_{i>k} U_{l'i} U_{l'k}^* U_{li}^* U_{lk}. \tag{7.37}
$$

With the help of this relation we obtain the following expression for the $\nu_l \rightarrow \nu_{l'}$ transition probability

$$
P(\nu_l \rightarrow \nu_{l'}) = \delta_{l'l} - 2\,\text{Re} \sum_{i>k} U_{l'i} U_{l'k}^* U_{li}^* U_{lk} (1 - e^{-i\,(E_i - E_k)t}). \tag{7.38}
$$

From this expression it is obvious that transitions between different flavor neutrinos (described by the second term of the expression (7.38)) can take place only if $E_i \neq E_k$, i.e. $m_i \neq m_k$. The necessary condition for such transitions ($E_i - E_k$) $t \gtrsim 1$) corresponds to the time-energy uncertainty relation (7.15).

For the ultrarelativistic neutrinos we have

$$
t \simeq L, \tag{7.39}
$$

where L is the neutrino detector-source distance. Taking also into account that

$$
E_i - E_k \simeq \frac{\Delta m_{ik}^2}{2E} \tag{7.40}
$$

we find the following general expression for the neutrino transition probability in the case of the three-neutrino mixing into vacuum

$$P(\nu_l \to \nu_{l'}) = \delta_{l'l} - 4 \sum_{i>k} \mathrm{Re}(U_{l'i} U_{l'k}^* U_{li}^* U_{lk}) \sin^2 \Delta_{ki}$$

$$+ 2 \sum_{i>k} \mathrm{Im}(U_{l'i} U_{l'k}^* U_{li}^* U_{lk}) \sin 2\Delta_{ki}, \qquad (7.41)$$

where

$$\Delta_{ki} = \frac{\Delta m_{ik}^2}{4E}. \qquad (7.42)$$

In order to obtain $\bar{\nu}_l \to \bar{\nu}_{l'}$ transition probability we need to make in (7.41) the following change $U_{li} \to U_{li}^*$. We find

$$P(\bar{\nu}_l \to \bar{\nu}_{l'}) = \delta_{l'l} - 4 \sum_{i>k} \mathrm{Re}(U_{l'i} U_{l'k}^* U_{li}^* U_{lk}) \sin^2 \Delta_{ki}$$

$$- 2 \sum_{i>k} \mathrm{Im}(U_{l'i} U_{l'k}^* U_{li}^* U_{lk}) \sin 2\Delta_{ki}, \qquad (7.43)$$

We see from these expressions that transition probabilities depends on the quantity $\frac{L}{E}$ which is determined by experimental conditions. Transition probabilities are determined by the six neutrino oscillation parameters: two neutrino mass-squared differences, three mixing angles and one CP violating phase.

From the standard general expressions for transition probabilities (7.41) and (7.43) it is easy to obtain transition probabilities which are investigated in neutrino oscillation experiments ($\nu_\mu \to \nu_\mu$, $\nu_\mu \to \nu_e$, $\bar{\nu}_e \to \bar{\nu}_e$ etc.). Let us notice that

1. Neutrino mass-squared differences in the expressions (7.41) and (7.43) are not independent. They are connected by the relation $\Delta m_{13}^2 = \Delta m_{23}^2 + \Delta m_{12}^2$.
2. The quantities $\mathrm{Im}\, U_{l'i} U_{li}^* U_{l'k}^* U_{lk}$ for different a i and k are not independent (they are connected by the relations which are based on the unitarity of the mixing matrix).

We will obtain now a simple alternative expressions for the neutrino transition probabilities in which only independent mass-squared differences enter and the unitarity of the mixing matrix is fully employed. The method which we will use can be easily applied to transitions of the flavor neutrinos into sterile ones.

7.4 Alternative Expression for Neutrino (Antineutrino) Transition Probability in Vacuum

The expression (7.20) for the $\nu_l \to \nu_{l'}$ transition probability $(l, l' = e, \mu, \tau)$ can be presented in the form

$$P(\nu_l \to \nu_{l'}) = |\sum_{i=1}^{3} U_{l'i}\, e^{-i(E_i-E_p)\,t}\, U_{li}^*|^2 = |\delta_{l'l} + \sum_{i\neq p} U_{l'i}\, (e^{-i2\Delta_{pi}} - 1)\, U_{li}^*|^2$$

$$= |\delta_{l'l} - 2i \sum_{i\neq p} U_{l'i}\, U_{li}^*\, e^{-i\Delta_{pi}}\, \sin \Delta_{pi}|^2. \tag{7.44}$$

Here p is a fixed, arbitrary index and Δ_{pi} is given by (7.42). Let us notice that in the expression (7.44)

- we took into account that transition probability does not depend on a common phase $(e^{-iE_p t})$;
- we used the unitarity of the mixing matrix in the transition amplitude.

From (7.44) we find the following expression for the $\nu_l \to \nu_{l'}$ transition probability

$$P(\nu_l \to \nu_{l'}) = \delta_{l'l} - 4 \sum_{i} |U_{li}|^2(\delta_{l'l} - |U_{l'i}|^2) \sin^2 \Delta_{pi} \tag{7.45}$$

$$+8\, \mathrm{Re} \sum_{i>k} e^{-i(\Delta_{pi}-\Delta_{pk})}\, U_{l'i} U_{li}^* U_{l'k}^* U_{lk}\, \sin \Delta_{pi} \sin \Delta_{pk}.$$

From this expression we obviously have

$$P(\nu_l \to \nu_{l'}) = \delta_{l'l} - 4 \sum_{i} |U_{li}|^2(\delta_{l'l} - |U_{l'i}|^2)\, \sin^2 \Delta_{pi}$$

$$+8 \sum_{i>k} [\mathrm{Re}\, (U_{l'i} U_{li}^* U_{l'k}^* U_{lk})\, \cos(\Delta_{pi} - \Delta_{pk})$$

$$+\, \mathrm{Im}\, (U_{l'i} U_{li}^* U_{l'k}^* U_{lk})\, \sin(\Delta_{pi} - \Delta_{pk})]\, \sin \Delta_{pi} \sin \Delta_{pk}. \tag{7.46}$$

Let us stress that in the second term of this expression $i \neq p$ and in the third term $i, k \neq p$.

In order to obtain $\bar{\nu}_l \to \bar{\nu}_{l'}$ transition probability we will change $U_{li} \to U_{li}^*$ in (7.46). We have

$$P(\bar{\nu}_l \to \bar{\nu}_{l'}) = \delta_{l'l} - 4 \sum_i |U_{li}|^2(\delta_{l'l} - |U_{l'i}|^2)\, \sin^2 \Delta_{pi}$$

$$+ 8 \sum_{i>k} [\mathrm{Re}\, (U_{l'i} U_{li}^* U_{l'k}^* U_{lk})\, \cos(\Delta_{pi} - \Delta_{pk})$$

$$- \mathrm{Im}\, (U_{l'i} U_{li}^* U_{l'k}^* U_{lk})\, \sin(\Delta_{pi} - \Delta_{pk})]\sin \Delta_{pi} \sin \Delta_{pk}.$$

$$(7.47)$$

7.4.1   Two-Neutrino Oscillations

Let us consider the simplest case of the two-neutrino mixing

$$\nu_l = \sum_{i=1,2} U_{li}\, \nu_i. \tag{7.48}$$

Here U is a orthogonal 2×2 matrix

$$U = \begin{pmatrix} \cos \theta & \sin \theta \\ -\sin \theta & \cos \theta, \end{pmatrix}, \tag{7.49}$$

where θ is the mixing angle.

We will choose in (7.46) and (7.47) $p = 1$. In this case $i = 2$ and there is no interference terms in the transition probability. We find

$$P(\nu_l \to \nu_{l'}) = P(\bar{\nu}_l \to \bar{\nu}_{l'}) = \delta_{l'l} - 4\, |U_{l2}|^2(\delta_{l'l} - |U_{l'2}|^2)\, \sin^2 \Delta. \tag{7.50}$$

where $\Delta = \frac{\Delta m^2 L}{4E}$, $\Delta m^2 = m_2^2 - m_1^2$.

From this expression for the probability of ν_l to survive we obtain the following expression

$$P(\nu_l \to \nu_l) = 1 - 4\, |U_{l2}|^2(1 - |U_{l2}|^2)\, \sin^2 \Delta. \tag{7.51}$$

The $\nu_l \to \nu_{l'}$ transition probability is given by the expression

$$P(\nu_l \to \nu_{l'}) = 4\, |U_{l2}|^2 |U_{l'2}|^2\, \sin^2 \Delta \quad (l' \neq l). \tag{7.52}$$

From this expression it is obvious that

$$P(\nu_l \to \nu_{l'}) = P(\nu_{l'} \to \nu_l). \tag{7.53}$$

Further, from the conservation of the probability we have

$$P(\nu_l \to \nu_l) = 1 - P(\nu_l \to \nu_{l'}), \quad P(\nu_{l'} \to \nu_{l'}) = 1 - P(\nu_{l'} \to \nu_l) \qquad (7.54)$$

From (7.53) and (7.54) we conclude that in the two-neutrino case the following equation holds

$$P(\nu_l \to \nu_l) = P(\nu_{l'} \to \nu_{l'}) \quad (l' \neq l). \qquad (7.55)$$

From the unitarity of the mixing matrix we have

$$|U_{l'2}|^2 = 1 - |U_{l2}|^2. \qquad (7.56)$$

Thus from (7.52) and (7.56) we find for the appearance probability the expression

$$P(\nu_l \to \nu_{l'}) = 4 \, |U_{l2}|^2 (1 - |U_{l2}|^2) \, \sin^2 \Delta \quad (l' \neq l). \qquad (7.57)$$

which obviously follows from (7.51) and the conservation of the probability.

Let us also notice that in the case of the two-neutrino mixing the CP relation

$$P(\nu_l \to \nu_{l'}) = P(\bar{\nu}_l \to \bar{\nu}_{l'}) \quad (l' \neq l), \qquad (7.58)$$

is always valid.[3] Thus in order to check whether in the lepton sector CP is violated we need to study effects of the three-neutrino mixing.

From (7.51) and (7.49) for the survival probability we find the following standard two-neutrino expression

$$P(\nu_l \to \nu_l) = 1 - \sin^2 2\theta \, \sin^2 \frac{\Delta m^2 L}{4E} = 1 - \frac{1}{2} \sin^2 2\theta \, (1 - \cos \frac{\Delta m^2 L}{2E}). \qquad (7.59)$$

The $\nu_l \to \nu_{l'}$ appearance probability is given by

$$P(\nu_l \to \nu_{l'}) = \sin^2 2\theta \, \sin^2 \frac{\Delta m^2 L}{4E} = \frac{1}{2} \sin^2 2\theta \, (1 - \cos \frac{\Delta m^2 L}{2E}) \quad (l' \neq l). \qquad (7.60)$$

---

[3]This is also obvious from the results of the previous chapter: for $n = 2$ the number of CP phases in the mixing matrix is equal to zero.

All previous formulas were written in the $\hbar = c = 1$ system. For the survival probability in the CGS system we have[4]

$$P(\nu_l \to \nu_l) = 1 - \sin^2 2\theta \, \sin^2 1.27 \, \frac{\Delta m^2 L}{E} = 1 - \frac{1}{2} \sin^2 2\theta \, (1 - \cos 2.54 \, \frac{\Delta m^2 L}{E}). \tag{7.61}$$

Appearance probability is given by the expression

$$P(\nu_l \to \nu_{l'}) = \sin^2 2\theta \, \sin^2 1.27 \, \frac{\Delta m^2 L}{E} = \frac{1}{2} \sin^2 2\theta \, (1 - \cos 2.54 \, \frac{\Delta m^2 L}{E}). \tag{7.62}$$

In expressions (7.61) and (7.62) Δm^2 is the neutrino mass-squared difference in eV^2, L is the source-detector distance in m (km) and E is the neutrino energy in MeV (GeV).

We will now introduce *the oscillation length*

$$L^{osc} = 4\pi \, \frac{E}{\Delta m^2}. \tag{7.63}$$

In the CGS units we have

$$L^{osc}_{12} = 4\pi \, \frac{E \, \hbar c}{\Delta m^2 c^4} \simeq 2.47 \, \frac{E}{\Delta m^2} \, m \, (km), \tag{7.64}$$

where E is the neutrino energy in MeV (GeV) and Δm^2_{12} is neutrino mass-squared difference in eV^2.

From (7.59), (7.60) and (7.63) for the survival and appearance probabilities we have

$$P(\nu_l \to \nu_l) = 1 - \frac{1}{2} \sin^2 2\theta \, (1 - \cos 2\pi \frac{L}{L^{osc}}), \tag{7.65}$$

and

$$P(\nu_l \to \nu_{l'}) = \frac{1}{2} \sin^2 2\theta \, (1 - \cos 2\pi \frac{L}{L^{osc}}) \tag{7.66}$$

Thus, the oscillation length is the period of the oscillations. For neutrino oscillations to be observed it is necessary that

$$L \gtrsim L^{osc}. \tag{7.67}$$

---

[4]We take into account that in the CGS system the argument of the cosine in (7.59) is equal to $\frac{\Delta m^2_{12} c^4 L}{2E \, \hbar c}$.

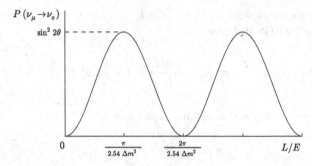

Fig. 7.1 Transition probability $P(\nu_\mu \rightarrow \nu_e)$ as a function of $\frac{L}{E}$

Notice that the same estimate we obtained from the time-energy uncertainty relation (see (7.15)).

In Fig. 7.1 the transition probability $P(\nu_\mu \rightarrow \nu_e)$ as a function of $\frac{L}{E}$ (in units $\frac{m}{MeV}$) is plotted. At the points

$$\frac{L}{E} = \frac{2\pi\, n}{2.54\, \Delta m^2} \quad \text{and} \quad \frac{L}{E} = \frac{2\pi\, (n + \frac{1}{2})}{2.54\, \Delta m^2} \quad (n = 0, 1, 2, ...)$$

only ν_μ or, correspondingly, ν_e can be observed. At other values of $\frac{L}{E}$ both ν_μ and ν_e can be found (ν_e with the probability $P(\nu_\mu \rightarrow \nu_e)$ and ν_μ with the probability $P(\nu_\mu \rightarrow \nu_\mu) = 1 - P(\nu_\mu \rightarrow \nu_e)$). From Fig. 7.1 it is clear why phenomenon, we are considering, is called neutrino oscillations.

7.4.2 Three-Neutrino Oscillations

Existing neutrino oscillation data are perfectly described under the assumption of the three-neutrino mixing

$$\nu_{lL} = \sum_{i=1}^{3} U_{li}\nu_{iL}, \quad l = e, \mu, \tau. \tag{7.68}$$

We will consider here in details neutrino oscillations in this most important case.

In the case of the three-neutrino mixing neutrino (antineutrino) transition probabilities depend on two mass-squared differences. From analysis of the existing data it follows that one mass-squared difference is much smaller than the other one.

We will assume that the mass of neutrino ν_2 is larger than the mass of ν_1 ($m_2 > m_1$) and that the small ("solar") mass-squared difference is given by

$$\Delta m_S^2 \equiv \Delta m_{12}^2 = m_2^2 - m_1^2 > 0. \tag{7.69}$$

For the mass of the third neutrino ν_3 there are two possibilities.

NO. The mass of ν_3 is larger than the mass of ν_2

$$m_1 < m_2 < m_3. \tag{7.70}$$

Such ordering of the neutrino masses is called the normal ordering (NO).

IO. The mass of ν_3 is smaller than the mass of ν_1

$$m_3 < m_1 < m_2. \tag{7.71}$$

Such ordering of the neutrino masses is called the inverted ordering (IO).

The character of the neutrino mass spectrum at present is not known. Its determination is a major aim of modern and future accelerator, reactor and atmospheric neutrino oscillation experiments.

Neutrino mixing matrix U is determined by the relation (7.68) and does not depend on the neutrino mass spectrum. In the standard parametrization the matrix U is characterized by three mixing angles $\theta_{12}, \theta_{23}, \theta_{13}$ and one CP phase δ (see (6.44)). For both possible neutrino mass-spectra the small (solar) mass-squared difference is determined by the relation (7.69). It is natural to determine the sixth neutrino oscillation parameter, large ("atmospheric") mass-squared difference, in such a way that it does not depend on the neutrino mass spectrum.

We will determine the atmospheric mass-squared difference Δm_A^2 as follows

$$\Delta m_A^2 = \Delta m_{23}^2 \,(NO) \quad \Delta m_A^2 = |\Delta m_{13}^2| \,(IO). \tag{7.72}$$

Thus Δm_A^2 is the mass-squared difference between the largest and the intermediate masses (NO) or between the intermediate and the smallest masses (IO). It is obvious that Δm_A^2, determined by the relations (7.72), does not depend on the neutrino mass spectrum.[5]

For the NO neutrino mass spectrum we will choose in (7.46) $p = 2$. In this case the index i takes values $i = 1, 3$ and in the interference term $i = 3, k = 1$. Taking into account that $\Delta m_{23}^2 = \Delta m_A^2$ and $\Delta m_{21}^2 = -\Delta m_S^2$ for the $\nu_l \to \nu_{l'}$ ($\bar{\nu}_l \to \bar{\nu}_{l'}$) transition probability we find the following general expression

$$
\begin{aligned}
\mathrm{P}^{\mathrm{NO}}&(\nu_l \to \nu_{l'})(P^{\mathrm{IO}}(\bar{\nu}_l \to \bar{\nu}_{l'})) \\
&= \delta_{l'l} - 4\,|U_{l3}|^2(\delta_{l'l} - |U_{l'3}|^2)\,\sin^2 \Delta_A \\
&\quad -4\,|U_{l1}|^2(\delta_{l'l} - |U_{l'1}|^2)\,\sin^2 \Delta_S - 8\,[\mathrm{Re}\,(U_{l'3}U_{l3}^*U_{l'1}^*U_{l1})\,\cos(\Delta_A + \Delta_S) \\
&\quad \pm \mathrm{Im}\,(U_{l'3}U_{l3}^*U_{l'1}^*U_{l1})\,\sin(\Delta_A + \Delta_S)]\sin \Delta_A \sin \Delta_S. \tag{7.73}
\end{aligned}
$$

---

[5]Other possibility: $\Delta m_A^2 = \Delta m_{13}^2$ (NO) and $\Delta m_A^2 = |\Delta m_{23}^2|$ (IO). In this case Δm_A^2 is the mass-squared difference between the largest and the smallest masses.

Here

$$\Delta_{A,S} = \frac{\Delta m_{A,S}^2 L}{4E}.$$ (7.74)

For the IO neutrino mass spectrum we choose $p = 1$. The index i in this case takes values $2, 3$ and in the interference term $i = 3, k = 2$. Taking into account that $\Delta m_{13}^2 = -\Delta m_A^2$ and $\Delta m_{12}^2 = \Delta m_S^2$ for the $\nu_l \to \nu_{l'}$ ($\bar{\nu}_l \to \bar{\nu}_{l'}$) transition probability in the IO case we have

$$
\begin{aligned}
P^{IO}(\nu_l &\to \nu_{l'})(P^{IO}(\bar{\nu}_l \to \bar{\nu}_{l'})) \\
&= \delta_{l'l} - 4\,|U_{l3}|^2(\delta_{l'l} - |U_{l'3}|^2)\,\sin^2 \Delta_A \\
&\quad -4\,|U_{l2}|^2(\delta_{l'l} - |U_{l'2}|^2)\,\sin^2 \Delta_S - 8\,[\mathrm{Re}\,(U_{l'3}U_{l3}^* U_{l'2}^* U_{l2})\,\cos(\Delta_A + \Delta_S) \\
&\quad \mp \mathrm{Im}\,(U_{l'3}U_{l3}^* U_{l'2}^* U_{l2})\,\sin(\Delta_A + \Delta_S)]\sin \Delta_A \sin \Delta_S.
\end{aligned}
$$ (7.75)

Let us stress that in (7.73) and (7.75) there is only one interference term. Transition probabilities for the normal and inverted ordering differ by the change $U_{l1} \leftrightarrows U_{l2}$, $U_{l'1} \leftrightarrows U_{l'2}$ and $(\pm) \to (\mp)$ in the last term.

The last terms of the expressions (7.73) and (7.75) are different from zero if CP is violated in the lepton sector. Let us determine the CP asymmetry

$$A_{ll'}^{CP} = P(\nu_l \to \nu_{l'}) - P(\bar{\nu}_l \to \bar{\nu}_{l'}).$$ (7.76)

The CP asymmetries in different channels satisfy the following relations

$$A_{e\tau}^{CP} = -A_{\mu\tau}^{CP} = A_{\mu e}^{CP}.$$ (7.77)

In fact, from the relation (7.26), which is based on the CPT invariance, follows that

$$A_{ll'}^{CP} = -A_{l'l}^{CP}.$$ (7.78)

Further, from the conservation of the probability we have

$$\sum_{l'} A_{ll'}^{CP} = \sum_{l'} P(\nu_l \to \nu_{l'}) - \sum_{l'} P(\bar{\nu}_l \to \bar{\nu}_{l'}) = 0.$$ (7.79)

Taking into account that $A_{ll}^{CP} = 0$ from (7.79) we obtain the following relations

$$A_{e\mu}^{CP} + A_{e\tau}^{CP} = 0,\quad A_{\mu e}^{CP} + A_{\mu\tau}^{CP} = 0,\quad A_{\tau e}^{CP} + A_{\tau\mu}^{CP} = 0.$$ (7.80)

From (7.78) and (7.80) we can easily find the relations (12.201).

For the normal and inverted ordering we obtain the following expressions for the CP asymmetry

$$(A_{ll'}^{CP})_{NO} = -16 \, \mathrm{Im} \, U_{l'3} U_{l3}^* U_{l'1}^* U_{l1} \, \sin(\Delta_A + \Delta_S) \sin \Delta_A \sin \Delta_S \qquad (7.81)$$

and

$$(A_{ll'}^{CP})_{IO} = 16 \, \mathrm{Im} \, U_{l'3} U_{l3}^* U_{l'2}^* U_{l2} \, \sin(\Delta_A + \Delta_S) \sin \Delta_A \sin \Delta_S. \qquad (7.82)$$

7.4.3 Neutrino Oscillations in Leading Approximation

From analysis of the neutrino oscillation data it follows that the solar mass-squared difference is much smaller than the atmospheric one

$$\Delta m_S^2 \simeq 3 \cdot 10^{-2} \, \Delta m_A^2 \qquad (7.83)$$

and that the parameter $\sin^2 \theta_{13}$ is small:

$$\sin^2 \theta_{13} \simeq 2.3 \cdot 10^{-2}. \qquad (7.84)$$

If we neglect the contribution of small parameters to the neutrino transition probabilities we will obtain simple two-neutrino expressions which describe basic features of neutrino oscillations valid with accuracies of a few %.

Let us consider $\nu_\mu \to \nu_\mu$ transition in the atmospheric and long baseline accelerator experiments. In these experiments $\Delta_A \simeq 1$ and $\Delta_S \ll 1$. Neglecting the contribution of the parameter Δ_S, from (7.73) and (7.75) for the probability of ν_μ to survive (for NO and IO spectra) we find the following expression

$$P(\nu_\mu \to \nu_\mu) \simeq 1 - 4 \, |U_{\mu 3}|^2 (1 - |U_{\mu 3}|^2) \, \sin^2 \Delta_A. \qquad (7.85)$$

If we further neglect the contribution of $\sin^2 \theta_{13}$ we have

$$P(\nu_\mu \to \nu_\mu) \simeq 1 - \sin^2 2\theta_{23} \sin^2 \frac{\Delta m_A^2 L}{4E}. \qquad (7.86)$$

Thus in the leading approximation ν_μ ($\bar{\nu}_\mu$) survival probability in the atmospheric range of the parameter $\frac{L}{E}$ is characterized by two parameters: $\sin^2 2\theta_{23}$ and Δm_A^2.

It follows from (7.73) and (7.75) that in the leading approximation we have $P(\nu_\mu \to \nu_e) \simeq 0$. This means that in this approximation ν_μ disappearance is due to $\nu_\mu \to \nu_\tau$ transitions. From the conservation of the probability we have

$$P(\nu_\mu \to \nu_\tau) \simeq 1 - P(\nu_\mu \to \nu_\mu) \simeq \sin^2 2\theta_{23} \sin^2 \frac{\Delta m_A^2 L}{4E}. \qquad (7.87)$$

We come to the conclusion that in the atmospheric range of L/E *in the leading approximation neutrino oscillations are two-neutrino* $\nu_\mu \leftrightarrows \nu_\tau$ *oscillations*.

Let us consider now $\bar{\nu}_e$ disappearance in the long baseline reactor experiment KamLAND. In this experiment

$$\Delta_S \simeq 1, \quad \Delta_A \gg 1. \tag{7.88}$$

Due to the averaging over neutrino energy resolution the term proportional to $\sin \Delta_A$ do not give contribution to the survival probability. Neglecting the contribution of $\sin^2 \theta_{13}$, for the $\bar{\nu}_e$ survival probability (for NO and IO mass spectra) from (7.73) and (7.75) we find the following expression

$$P(\bar{\nu}_e \to \bar{\nu}_e) \simeq 1 - \sin^2 2\theta_{12} \sin^2 \frac{\Delta m_S^2 L}{4E}. \tag{7.89}$$

Thus the study of neutrino oscillations in the reactor KamLAND experiment allows to determine the parameters $\sin^2 2\theta_{12}$ and Δm_S^2.

The disappearance of $\bar{\nu}_e$'s in the KamLAND experiment is due to $\bar{\nu}_e \to \bar{\nu}_\mu$ and $\bar{\nu}_e \to \bar{\nu}_\tau$ transitions in the region determined by the condition (7.88). In fact, from (7.73) and (7.75) in the leading approximation we find

$$P(\bar{\nu}_e \to \bar{\nu}_\mu) \simeq \cos^2 \theta_{23} \sin^2 2\theta_{12} \sin^2 \frac{\Delta m_S^2 L}{4E} \tag{7.90}$$

and

$$P(\bar{\nu}_e \to \bar{\nu}_\tau) \simeq \sin^2 \theta_{23} \sin^2 2\theta_{12} \sin^2 \frac{\Delta m_S^2 L}{4E}. \tag{7.91}$$

From analysis of the neutrino oscillation data follows that $\sin^2 \theta_{23} \simeq \cos^2 \theta_{23} \simeq \frac{1}{2}$. From (7.90) and (7.91) we have

$$P(\bar{\nu}_e \to \bar{\nu}_\mu) \simeq P(\bar{\nu}_e \to \bar{\nu}_\tau) \simeq \frac{1}{2} (1 - P(\bar{\nu}_e \to \bar{\nu}_e)). \tag{7.92}$$

In analysis of the first data of the atmospheric and accelerator neutrino oscillation experiments (Super-Kamiokande, K2K, MINOS) and the first data of the long-baseline reactor experiment KamLAND the expressions (7.86) and (7.89) for ν_μ and $\bar{\nu}_e$ survival probabilities in the leading approximation were used. These data allowed to obtain the first information about Δm_A^2 and $\sin^2 2\theta_{23}$, Δm_S^2 and $\sin^2 2\theta_{12}$. In the recent years the value of the parameter $\sin^2 \theta_{13}$ was measured in the T2K, Daya Bay, RENO, Double Chooz and NOvA experiments. In analysis of the modern data expressions for the three-neutrino oscillations are used. The leading approximation allows us, however, to understand the basic picture of neutrino oscillations in the regions sensitive to Δm_A^2 and to Δm_S^2. In order to study beyond the leading approximation effects such as CP violation in the lepton sector or the neutrino mass ordering we need to perform high precision neutrino oscillation experiments. In the next sections we will present exact expressions for three-neutrino transition probabilities in different channels.

7.4.4 $\bar{\nu}_e \to \bar{\nu}_e$ Survival Probability

In reactor neutrino oscillation experiments $\bar{\nu}_e$ survival probability is measured. From (7.73) and (7.75) for the normal and inverted mass ordering we have, respectively,

$$P^{NO}(\bar{\nu}_e \to \bar{\nu}_e) = 1 - 4\,|U_{e3}|^2(1 - |U_{e3}|^2)\,\sin^2 \Delta_A$$
$$- 4\,|U_{e1}|^2(1 - |U_{e1}|^2)\,\sin^2 \Delta_S$$
$$- 8\,|U_{e3}|^2|U_{e1}|^2\,\cos(\Delta_A + \Delta_S)\sin \Delta_A \sin \Delta_S. \tag{7.93}$$

and

$$P^{IO}(\bar{\nu}_e \to \bar{\nu}_e) = 1 - 4\,|U_{e3}|^2(1 - |U_{e3}|^2)\,\sin^2 \Delta_A$$
$$- 4\,|U_{e2}|^2(1 - |U_{e2}|^2)\,\sin^2 \Delta_S$$
$$- 8\,|U_{e3}|^2|U_{e2}|^2\,\cos(\Delta_A + \Delta_S)\sin \Delta_A \sin \Delta_S. \tag{7.94}$$

Using the standard parametrization of the PMNS matrix (see (6.44)) from (7.93) and (7.94) we

$$P^{NO}(\bar{\nu}_e \to \bar{\nu}_e) = 1 - \sin^2 2\theta_{13} \sin^2 \Delta_A$$
$$- (\cos^4 \theta_{13} \sin^2 2\theta_{12} + \cos^2 \theta_{12} \sin^2 2\theta_{13})\,\sin^2 \Delta_S$$
$$- 2\sin^2 2\theta_{13} \cos^2 \theta_{12}\,\cos(\Delta_A + \Delta_S)\sin \Delta_A \sin \Delta_S. \tag{7.95}$$

and

$$P^{IO}(\bar{\nu}_e \to \bar{\nu}_e) = 1 - \sin^2 2\theta_{13} \sin^2 \Delta_A$$
$$- (\cos^4 \theta_{13} \sin^2 2\theta_{12} + \sin^2 \theta_{12} \sin^2 2\theta_{13})\,\sin^2 \Delta_S$$
$$- 2\sin^2 2\theta_{13} \sin^2 \theta_{12}\,\cos(\Delta_A + \Delta_S)\sin \Delta_A \sin \Delta_S. \tag{7.96}$$

Notice that $P^{NO}(\bar{\nu}_e \to \bar{\nu}_e)$ and $P^{IO}(\bar{\nu}_e \to \bar{\nu}_e)$ differ by the change $\cos^2 \theta_{12} \leftrightarrows \sin^2 \theta_{12}$.

7.4.5　$\nu_\mu \rightarrow \nu_e$ and $\bar{\nu}_\mu \rightarrow \bar{\nu}_e$ Appearance Probabilities

In this section we will present expressions for $\nu_\mu \rightarrow \nu_e$ ($\bar{\nu}_\mu \rightarrow \bar{\nu}_e$) vacuum transition probabilities in the case of the normal and inverted neutrino mass ordering. From general expressions (7.73) and (7.75) we find

$$P^{NO}(\nu_\mu \rightarrow \nu_e)(P^{NO}(\bar{\nu}_\mu \rightarrow \bar{\nu}_e))$$

$$= 4 \, |U_{e3}|^2 |U_{\mu 3}|^2 \, \sin^2 \Delta_A$$

$$+ 4 \, |U_{e1}|^2 |U_{\mu 1}|^2 \, \sin^2 \Delta_S$$

$$- 8 \, \text{Re} \, (U_{e3} U_{\mu 3}^* U_{e1}^* U_{\mu 1}) \, \cos(\Delta_A + \Delta_S) \sin \Delta_A \sin \Delta_S$$

$$\mp 8 \, \text{Im} \, (U_{e3} U_{\mu 3}^* U_{e1}^* U_{\mu 1}) \, \sin(\Delta_A + \Delta_S) \sin \Delta_A \sin \Delta_S. \qquad (7.97)$$

and

$$P^{IO}(\nu_\mu \rightarrow \nu_e)(P^{IO}(\bar{\nu}_\mu \rightarrow \bar{\nu}_e))$$

$$= 4 \, |U_{e3}|^2 |U_{\mu 3}|^2 \, \sin^2 \Delta_A$$

$$+ 4 \, |U_{e2}|^2 |U_{\mu 2}|^2 \, \sin^2 \Delta_S$$

$$- 8 \, \text{Re} \, (U_{e3} U_{\mu 3}^* U_{e2}^* U_{\mu 2}) \, \cos(\Delta_A + \Delta_S) \sin \Delta_A \sin \Delta_S$$

$$\pm 8 \, \text{Im} \, (U_{e3} U_{\mu 3}^* U_{e2}^* U_{\mu 2}) \, \sin(\Delta_A + \Delta_S) \sin \Delta_A \sin \Delta_S. \qquad (7.98)$$

Using the standard parameterizations of the PMNS mixing matrix in the case of the normal ordering we have

$$P^{NO}(\nu_\mu \rightarrow \nu_e)(P^{NO}(\bar{\nu}_\mu \rightarrow \bar{\nu}_e)) = \sin^2 2\theta_{13} \sin^2 \theta_{23} \sin^2 \Delta_A$$

$$+ (\sin^2 2\theta_{12} \cos^2 \theta_{13} \cos^2 \theta_{23} + \sin^2 2\theta_{13} \cos^4 \theta_{12} \sin^2 \theta_{23}$$

$$+ K \cos^2 \theta_{12} \cos \delta) \sin^2 \Delta_S$$

$$+ (2 \sin^2 2\theta_{13} \sin^2 \theta_{23} \cos^2 \theta_{12} + K \cos \delta) \, \cos(\Delta_A + \Delta_S) \sin \Delta_A \sin \Delta_S$$

$$\mp 8 J_{CP} \, \sin(\Delta_A + \Delta_S) \sin \Delta_A \sin \Delta_S. \qquad (7.99)$$

Here

$$K = \sin 2\theta_{12} \sin 2\theta_{13} \sin 2\theta_{23} \cos \theta_{13} \qquad (7.100)$$

and

$$J_{CP} = \frac{1}{8} \sin 2\theta_{12} \sin 2\theta_{13} \sin 2\theta_{23} \cos \theta_{13} \sin \delta \qquad (7.101)$$

is the Jarlskog invariant.

In the case of the inverted ordering we find

$$P^{IO}(\nu_\mu \to \nu_e)(P^{IO}(\bar{\nu}_\mu \to \bar{\nu}_e)) = \sin^2 2\theta_{13} \sin^2 \theta_{23} \sin^2 \Delta_A$$
$$+ (\sin^2 2\theta_{12} \cos^2 \theta_{13} \cos^2 \theta_{23} + \sin^2 2\theta_{13} \sin^4 \theta_{12} \sin^2 \theta_{23}$$
$$- K \sin^2 \theta_{12} \cos \delta) \sin^2 \Delta_S$$
$$+ (2 \sin^2 2\theta_{13} \sin^2 \theta_{23} \sin^2 \theta_{12} - K \cos \delta) \cos(\Delta_A + \Delta_S) \sin \Delta_A \sin \Delta_S$$
$$\mp 8 J_{CP} \sin(\Delta_A + \Delta_S) \sin \Delta_A \sin \Delta_S. \tag{7.102}$$

From (7.99) and (7.102) follows that the CP asymmetry

$$A_{e\mu}^{CP} = P(\nu_\mu \to \nu_e) - P(\bar{\nu}_\mu \to \bar{\nu}_e) \tag{7.103}$$

does not depend on the neutrino mass ordering. We have

$$A_{e\mu}^{CP} = -16 J_{CP} \sin(\Delta_A + \Delta_S) \sin \Delta_A \sin \Delta_S. \tag{7.104}$$

The CP asymmetry is different from zero if all three mixing angles $\theta_{23}, \theta_{12}, \theta_{13}$ are not equal to zero and both mass-squared differences solar Δm_S^2 and atmospheric Δm_A^2 are relevant.

7.4.6 $\nu_\mu \to \nu_\mu$ Survival Probability

From general expressions (7.73) and (7.75) for the $\nu_\mu \to \nu_\mu$ survival probability in the case of the normal and inverted mass ordering we have, correspondingly,

$$P^{NO}(\nu_\mu \to \nu_\mu) = 1 - 4 |U_{\mu3}|^2 (1 - |U_{\mu3}|^2) \sin^2 \Delta_A$$
$$- 4 |U_{\mu1}|^2 (1 - |U_{\mu1}|^2) \sin^2 \Delta_S$$
$$- 8 |U_{\mu3}|^2 |U_{\mu1}|^2 \cos(\Delta_A + \Delta_S) \sin \Delta_A \sin \Delta_S. \tag{7.105}$$

and

$$P^{IO}(\nu_\mu \to \nu_\mu) = 1 - 4 |U_{\mu3}|^2 (1 - |U_{\mu3}|^2) \sin^2 \Delta_A$$
$$- 4 |U_{\mu2}|^2 (1 - |U_{\mu2}|^2) \sin^2 \Delta_S$$
$$- 8 |U_{\mu3}|^2 |U_{\mu2}|^2 \cos(\Delta_A + \Delta_S) \sin \Delta_A \sin \Delta_S. \tag{7.106}$$

Using standard parametrization of the PMNS matrix we find in the case of the normal ordering

$$
\begin{aligned}
P^{NO}(\nu_\mu \to \nu_\mu) = {} & 1 - (\sin^2 2\theta_{23} \cos^4 \theta_{13} + \sin^2 2\theta_{13} \sin^2 \theta_{23}) \ \sin^2 \Delta_A \\
& - 4\,|U_{\mu 1}|^2 (1 - |U_{\mu 1}|^2) \ \sin^2 \Delta_S \\
& - 2(\sin^2 2\theta_{23} \cos^2 \theta_{13} \sin^2 \theta_{12} + \sin^2 2\theta_{13} \cos^2 \theta_{12} \sin^4 \theta_{23} \\
& + K \sin^2 \theta_{23} \cos \delta) \ \cos(\Delta_A + \Delta_A) \sin \Delta_A \sin \Delta_S. \tag{7.107}
\end{aligned}
$$

Here

$$
|U_{\mu 1}|^2 = \cos^2 \theta_{23} \sin^2 \theta_{12} + \sin^2 \theta_{23} \cos^2 \theta_{12} \sin^2 \theta_{13} + \frac{K \cos \delta}{4 \cos^2 \theta_{13}} \tag{7.108}
$$

and K is given by the relation (7.100).

For the inverted ordering we have

$$
\begin{aligned}
P^{IO}(\nu_\mu \to \nu_\mu) = {} & 1 - (\sin^2 2\theta_{23} \cos^4 \theta_{13} + \sin^2 2\theta_{13} \sin^2 \theta_{23}) \ \sin^2 \Delta_A \\
& - 4\,|U_{\mu 2}|^2 (1 - |U_{\mu 2}|^2) \ \sin^2 \Delta_S \\
& - 2(\sin^2 2\theta_{23} \cos^2 \theta_{13} \cos^2 \theta_{12} + \sin^2 2\theta_{13} \sin^2 \theta_{12} \sin^4 \theta_{23} \\
& - K \sin^2 \theta_{23} \cos \delta) \ \cos(\Delta_A + \Delta_A) \sin \Delta_A \sin \Delta_S, \tag{7.109}
\end{aligned}
$$

where

$$
|U_{\mu 2}|^2 = \cos^2 \theta_{23} \cos^2 \theta_{12} + \sin^2 \theta_{23} \sin^2 \theta_{12} \sin^2 \theta_{13} - \frac{K \cos \delta}{4 \cos^2 \theta_{13}}. \tag{7.110}
$$

7.4.7 Transitions of Flavor Neutrinos into Sterile States

From existing experimental data indications in favor of transitions of flavor ν_μ and ν_e into sterile neutrino states were obtained. These indications were obtained in LSND, MiniBooNE, short baseline reactor and neutrino source experiments which are sensitive to neutrino mass-squared difference(s) much larger than Δm_A^2.

Transitions of the flavor neutrinos into sterile states are possible if the number of neutrinos with definite masses ν_i is larger than the number of the flavor neutrinos (three). For the neutrino mixing we have in this case

$$
\nu_{lL} = \sum_{i=1}^{3+n_{ster}} U_{li} \nu_{iL}, \qquad \nu_{sL} = \sum_{i=1}^{3+n_{ster}} U_{si} \nu_{iL}, \tag{7.111}
$$

where the index s takes n_{ster} values $(s_1, s_2, \ldots s_{n_{ster}})$ and U is a unitary $(3 + n_{ster}) \times (3 + n_{ster})$ matrix.

We will assume that all mass-squared differences are small enough so that due to the Heisenberg uncertainty relation production (and absorption) of neutrinos with different masses cannot be resolved in weak processes. For neutrino states (flavor and sterile) we have in this case

$$|\nu_\alpha\rangle = \sum U_{\alpha i}^* |\nu_i\rangle, \quad \alpha = e, \mu, \tau, s_1, s_2, ... \tag{7.112}$$

Here $|\nu_i\rangle$ is the state of neutrino with mass m_i, momentum $\mathbf{p}$ and the energy $E_i \simeq E + \frac{m_i^2}{2E}$, $E = |\mathbf{p}|$. From the unitarity of the mixing matrix we have

$$\langle\nu_{\alpha'}|\nu_\alpha\rangle = \delta_{\alpha'\alpha}, \quad |\nu_i\rangle = \sum_{\alpha'} U_{\alpha' i}|\nu_{\alpha'}\rangle. \tag{7.113}$$

If at $t = 0$ a flavor neutrino ν_α with momentum $\mathbf{p}$ was produced, at the time t the neutrino state is given by

$$|\nu_\alpha\rangle_t = e^{-iHt}|\nu_\alpha\rangle = \sum_{\alpha'} |\nu_{\alpha'}\rangle\langle\nu_{\alpha'}|e^{-iHt}|\nu_\alpha\rangle = \sum_{\alpha'} |\nu_{\alpha'}\rangle(\sum_i U_{\alpha' i}e^{-iE_i t}U_{\alpha i}^*), \tag{7.114}$$

where H is the free Hamiltonian.

The transition probability in the general case of the transitions into flavor and sterile states has the form

$$P(\nu_\alpha \to \nu_{\alpha'}) = | \sum_{i=1}^{3+n_{ster}} U_{\alpha' i}^* \, e^{-iE_i t} \, U_{\alpha i}|^2 \tag{7.115}$$

It is obvious that the results obtained before for the case of the transitions between flavor neutrinos can be applied to the transitions of flavor neutrinos into flavor and sterile states. We have

$$P(\nu_\alpha \to \nu_{\alpha'})(P(\bar{\nu}_\alpha \to \bar{\nu}_{\alpha'})) = \delta_{\alpha\alpha'} - 4 \sum_i |U_{\alpha i}|^2(\delta_{\alpha'\alpha} - |U_{\alpha' i}|^2) \, \sin^2 \Delta_{pi}$$

$$+8 \sum_{i>k}[\text{Re} \, (U_{\alpha' i}U_{\alpha i}^* U_{\alpha' k}^* U_{\alpha k}) \, \cos(\Delta_{pi} - \Delta_{pk})$$

$$\pm \text{Im} \, (U_{\alpha' i}U_{\alpha i}^* U_{\alpha' k}^* U_{\alpha k}) \, \sin(\Delta_{pi} - \Delta_{pk})] \sin \Delta_{pi} \sin \Delta_{pk}. \tag{7.116}$$

It is easy to check that $P(\nu_\alpha \to \nu_{\alpha'})$ $(P(\bar{\nu}_\alpha \to \bar{\nu}_{\alpha'}))$ transition probabilities are correctly normalized. In fact, taking into account the unitarity of the $(3 + n_{n_{ster}}) \times (3 + n_{n_{ster}})$ mixing matrix U from (7.114) we find

$$\sum_{\alpha'} P(\nu_\alpha \to \nu_{\alpha'}) = \sum_{i,k}(\sum_{\alpha'} U_{\alpha' i} U_{\alpha' k}^*) \, e^{-i\,(E_i - E_k)\,t} \, U_{\alpha i}^* \, U_{\alpha k} = \sum_i |U_{\alpha i}|^2 = 1 \tag{7.117}$$

Similarly we have

$$\sum_{\alpha'} P(\bar{\nu}_\alpha \to \bar{\nu}_{\alpha'}) = 1 \qquad (7.118)$$

Sterile neutrinos cannot be detected in the standard weak processes. Transitions of flavor neutrinos into the sterile states can be revealed if we can proof that there exist flavor neutrino oscillations $\nu_l \leftrightarrows \nu_{l'}$ in the short baseline region of $\frac{L}{E}$ which corresponds mass-squared difference(s) larger than $\Delta^2 m_A$. Other possibility to obtain an information about transitions of flavor neutrinos into sterile states is based on the relation (7.115). From this relation we have

$$\sum_{l'=e,\mu,\tau} P(\nu_l \to \nu_{l'}) = 1 - \sum_{s=s_1,s_2,\dots,s_{n_{ster}}} P(\nu_l \to \nu_s) \qquad (7.119)$$

The left-handed part of this relation is the total probability of the transition of a flavor neutrino ν_l into all flavor neutrinos (ν_e, ν_μ and ν_τ). This probability can be determined if neutrinos are detected at some distance from a source by an observation of a NC process. If the probability $\sum_{l'=e,\mu,\tau} P(\nu_l \to \nu_{l'})$ is less than one (and depends on $\frac{L}{E}$) it would be a proof of the transitions of an active neutrino into sterile states.

Let us consider the simplest 3+1 scheme with four massive mixed neutrinos ν_i and with $\Delta m_{14}^2 \gg \Delta m_A^2$, m_4 and m_1 being masses of the heaviest and the lightest neutrinos.[6]

For the probability of the $\nu_l \to \nu_{l'}$ transition in an experiment sensitive to Δm_{14}^2 ($\Delta_{14} = \frac{\Delta m_{14}^2 L}{4E} \simeq 1$) from (7.116) we find the following approximate expression

$$P(\nu_l \to \nu_{l'}) \simeq P(\bar{\nu}_l \to \bar{\nu}_{l'}) \simeq \delta_{l'l} - 4 |U_{l4}|^2 (\delta_{l'l} - |U_{l'4}|^2) \sin^2 \Delta_{14}. \qquad (7.120)$$

Notice that we have neglected contributions of small mass-squared differences Δm_{12}^2 and Δm_{13}^2 and chose $p = 1$.

From (7.120) for $\bar{\nu}_\mu \to \bar{\nu}_e$, $\bar{\nu}_e \to \bar{\nu}_e$ and $\nu_\mu \to \nu_\mu$ transition probabilities we find

$$P(\bar{\nu}_\mu \to \bar{\nu}_e) = \sin^2 2\theta_{e\mu} \sin^2 \Delta_{14},$$

$$P(\bar{\nu}_e \to \bar{\nu}_e) = 1 - \sin^2 2\theta_{ee} \sin^2 \Delta_{14},$$

$$P(\nu_\mu \to \nu_\mu) = 1 - \sin^2 2\theta_{\mu\mu} \sin^2 \Delta_{14}. \qquad (7.121)$$

---

[6]From analysis of the existing data it follows that $\Delta m_{14}^2 \simeq 1\,\text{eV}^2$.

Here

$$\sin^2 2\theta_{e\mu} = 4|U_{e4}|^2 |U_{\mu4}|^2, \quad \sin^2 2\theta_{ee} = 4|U_{e4}|^2(1 - |U_{e4}|^2),$$
$$\sin^2 2\theta_{\mu\mu} = 4|U_{\mu4}|^2(1 - |U_{\mu4}|^2). \tag{7.122}$$

From analysis of the existing data in favor of transitions into sterile neutrinos follows that the amplitudes $\sin^2 2\theta_{\mu\mu}$ and $\sin^2 2\theta_{ee}$ are small. Thus we have

$$\sin^2 2\theta_{ee} \simeq 4|U_{e4}|^2, \quad \sin^2 2\theta_{\mu\mu} \simeq 4|U_{\mu4}|^2. \tag{7.123}$$

From (7.122) and (7.123) we find the following relation between transition amplitudes

$$\sin^2 2\theta_{e\mu} \simeq \frac{1}{4} \sin^2 2\theta_{ee} \sin^2 2\theta_{\mu\mu}. \tag{7.124}$$

Thus if we know $\sin^2 2\theta_{e\mu}$ and $\sin^2 2\theta_{ee}$ we can predict $\sin^2 2\theta_{\mu\mu}$. Notice that the existing data are not in agreement with this prediction.

Chapter 8
Neutrino in Matter

8.1 Introduction

Up to now we have considered the propagation of mixed neutrinos in vacuum. We have seen that due to neutrino mass-squared differences and neutrino mixing the flavor content of the neutrino beam in vacuum depends on distance (time). As it was shown by Wolfenstein in the case of matter not only neutrino masses and mixing but also the coherent scattering of neutrinos in matter must be taken into account. The contribution of the coherent scattering into the Hamiltonian of neutrino in matter is proportional to the electron number-density. If the electron density depends on the distance (as in the case of the sun) the $\nu_e \to \nu_e$ transition probability can have a resonance character (MSW effect). We will consider in this chapter the propagation of the mixed neutrino in matter.

8.2 Evolution Equation of Neutrino in Matter

The state of neutrino with momentum $\mathbf{p}$ in matter $|\Psi(t)\rangle$ satisfies the Schrödinger equation

$$i \frac{\partial |\Psi(t)\rangle}{\partial t} = H |\Psi(t)\rangle. \tag{8.1}$$

Here $H = H_0 + H_I$ is the total Hamiltonian, where H_0 is the free Hamiltonian and H_I is the effective interaction Hamiltonian.

© Springer International Publishing AG, part of Springer Nature 2018
S. Bilenky, *Introduction to the Physics of Massive and Mixed Neutrinos*,
Lecture Notes in Physics 947, https://doi.org/10.1007/978-3-319-74802-3_8

Let us expand the state $|\Psi(t)\rangle$ over the total system of states of the flavor neutrinos ν_l with momentum $\mathbf{p}$

$$|\Psi(t)\rangle = \sum_{l=e,\mu,\tau} a_{\nu_l}(t) |\nu_l\rangle. \tag{8.2}$$

Here

$$|\nu_l\rangle = \sum_{i=1}^{3} U_{li}^* |\nu_i\rangle, \tag{8.3}$$

where

$$H_0 |\nu_i\rangle = E_i |\nu_i\rangle, \quad E_i \simeq p + \frac{m_i^2}{2p}. \tag{8.4}$$

In the flavor representation the evolution equation takes the form

$$i \frac{\partial a_{\nu_l}(t)}{\partial t} = \sum_{l'} [\langle \nu_l | H_0 | \nu_{l'} \rangle + \langle \nu_l | H_I | \nu_{l'} \rangle] a_{l'}(t). \tag{8.5}$$

For the free Hamiltonian we have

$$\langle \nu_{l'} | H_0 | \nu_l \rangle = \sum_i \langle \nu_{l'} | i \rangle E_i \langle i | \nu_l \rangle \simeq p \delta_{\nu_{l'} \nu_l} + \sum_i U_{li} \frac{m_i^2}{2 p} U_{l'i}^*. \tag{8.6}$$

It is easy to see that the first, proportional to the unite matrix term, can be removed from the Hamiltonian by the redefinition of the unphysical common phase.[1]

The second term of the evolution equation (8.5) is due to the coherent scattering of neutrinos in matter. Because of the coherent scattering the refraction index of neutrino in matter is different from one (vacuum value) and for the flavor neutrino ν_l at the point x is given by the following classical expression

$$n_{\nu_l;\nu_l}(x) = 1 + \frac{2\pi}{p^2} \sum_a f_{\nu_l a \to \nu_l a}(0) n_a(x). \tag{8.7}$$

Here $f_{\nu_l a \to \nu_l a}(0)$ is the amplitude of the elastic $\nu_l - a$ scattering in the forward direction and $n_a(x)$ is the number density of particles a. The potential generated by

---

[1]In fact, let us consider the equation $i \frac{\partial a(t)}{\partial t} = (H(t) + \alpha(t) \cdot I) a(t)$ where $\alpha(t)$ is an arbitrary function. For the function $a'(t) = e^{i\alpha(t)} a(t)$, which cannot be distinguished from $a(t)$, we have $i \frac{\partial a'(t)}{\partial t} = H(t) a'(t)$.

the coherent scattering is given by the difference of neutrino energy in matter and neutrino energy in vacuum. We have

$$\langle \nu_l | H_I | \nu_l \rangle = p\, n_{\nu_l;\nu_l}(x) - p = \frac{2\pi}{p} \sum_a f_{\nu_l a \to \nu_l a}(0)\, n_a(x). \tag{8.8}$$

We will consider the propagation of neutrino in a neutral medium of electrons, protons and neutrons. The amplitude of the elastic $\nu_e - e$ scattering is due to the $W^\pm$ exchange (CC) and Z^0 exchange (NC). The amplitudes of the elastic $\nu_{\mu,\tau} - e$ scattering and $\nu_{e,\mu,\tau} - N$ scattering are due to the Z^0 exchange (NC).

The corresponding CC and NC effective Hamiltonians are given by the expressions

$$\mathcal{H}^{CC}(x) = \frac{G_F}{\sqrt{2}} 4\, \bar{\nu}_{eL}(x)\gamma^\alpha \nu_{eL}(x)\, \bar{e}_L(x)\gamma_\alpha e_L(x) \tag{8.9}$$

and

$$\mathcal{H}^{NC}(x) = \frac{G_F}{\sqrt{2}} 2 \sum_{l=e,\mu,\tau} \bar{\nu}_{lL}(x)\gamma^\alpha \nu_{lL}(x)\, j_\alpha^{NC}(x), \tag{8.10}$$

where $j_\alpha^{NC}(x)$ is the NC of electrons and quarks.

Because of the $\nu_e - \nu_\mu - \nu_\tau$ universality, potentials which are due to the NC elastic $\nu_{e,\mu,\tau} - e$ and $\nu_{e,\mu,\tau} - N$ scattering are proportional to unit matrices. As we saw before such terms can be removed from the Hamiltonian by the redefinition of a common phase. The only term which can change the flavor content of the neutrino beam is due to the CC elastic $\nu_e - e$ scattering. From (8.9) we have

$$f_{\nu_e e \to \nu_e e}^{CC}(0) = \frac{G_F\, p}{\sqrt{2}\pi} \tag{8.11}$$

and corresponding effective potential is equal

$$\langle \nu_e | H_I | \nu_e \rangle = \sqrt{2}\, G_F\, n_e(x) \tag{8.12}$$

Finally, the Wolfenstein evolution equation for the mixed neutrino in matter has the form

$$i\frac{\partial a(t)}{\partial t} = \left(U \frac{m^2}{2E} U^\dagger + \sqrt{2} G_F\, n_e(t)\beta \right) a(t). \tag{8.13}$$

Here $\beta_{ee} = 1$ and all other elements of the matrix β are equal to zero. We took into account that for the ultrarelativistic neutrino $x \simeq t$ where t is the neutrino propagation time in matter.[2] We will finish this section with two remarks.

1. The electron neutrino current can be written in the form

$$\bar{v}_{eL}\gamma^{\alpha} v_{eL} = v_{eL}^T \gamma^{\alpha} \bar{v}_{eL}^T = -v_{eL}^T C^{-1} C\gamma^{\alpha}C^{-1} C \bar{v}_{eL}^T = -\bar{v}_{eR}^c \gamma^{\alpha} v_{eR}^c,$$
(8.14)

where C is the charge conjugation matrix. It follows from this expression that the amplitude of the elastic scattering of the right-handed antineutrinos on electrons differs by the sign from the amplitude of the CC $v_e - e$ scattering. For the effective interaction Hamiltonian of the antineutrino in matter we have

$$H_I(t) = -\sqrt{2}\, G_F\, n_e(t)\beta.$$
(8.15)

2. The evolution equations of neutrino in matter in the case of the Majorana and Dirac neutrinos v_i are the same.

   In fact, the mixing matrices for Majorana and Dirac neutrinos are connected by the relation (see (6.45))

$$U^M = U^D S^M(\bar{\alpha})$$
(8.16)

where $S^M(\bar{\alpha})$ is the diagonal phase matrix. We obviously have

$$U^M \frac{m^2}{2E} U^{M\dagger} = U^D \frac{m^2}{2E} U^{D\dagger}.$$
(8.17)

We have shown in the previous section that via the investigation of the neutrino oscillations in vacuum it is impossible to reveal the nature of the massive neutrinos. From (8.17) it follows that the same is true for the neutrino transitions in matter.

---

[2] Let us notice that in the case of flavor and sterile neutrinos both CC and NC interactions contribute to the effective potential of neutrino in matter. We have in this case

$$H_I(x) = \sqrt{2}G_F\, n_e(x)\beta^e + \sqrt{2}G_F \frac{1}{2}\, n_n(x)\beta^s$$

where $n_n(x)$ is the number density of neutrons, $\beta_{ee}^e = 1$ and $\beta_{ss}^s = 1$. Other elements of the matrices β^e and β^s are equal to zero.

8.3 Propagation of Neutrino in Matter with Constant Density

We will consider first the case of a matter with a constant density. The hermitian effective Hamiltonian of neutrino in a matter in the flavor representation can be diagonalized by the unitary transformation

$$H = U^{\mathrm{m}} E^{\mathrm{m}} U^{\mathrm{m}\dagger}. \tag{8.18}$$

Here $E^{\mathrm{m}}_{ik} = E^{\mathrm{m}}_i \, \delta_{ik}$, E^{m}_i being the eigenvalue of the matrix H and $U^{\mathrm{m}} U^{\mathrm{m}\dagger} = 1$.
We have

$$\langle \nu_{l'} | H | \nu_l \rangle = \sum_i \langle \nu_{l'} | i \rangle^{\mathrm{m}} \, E^{\mathrm{m}}_i \, {}^{\mathrm{m}}\langle i | \nu_l \rangle, \tag{8.19}$$

where

$$H | i \rangle^{\mathrm{m}} = E^{\mathrm{m}}_i | i \rangle^{\mathrm{m}}. \tag{8.20}$$

Comparing (8.18) and (8.19) we conclude that

$$\langle \nu_l | i \rangle^{\mathrm{m}} = U^{\mathrm{m}}_{li}, \quad {}^{\mathrm{m}}\langle i | \nu_l \rangle = U^{\mathrm{m}*}_{li}. \tag{8.21}$$

The states of the flavor neutrinos ν_l are connected with states of neutrinos with definite energies in matter by the mixing relations

$$| \nu_l \rangle = \sum_{i=1}^{3} U^{\mathrm{m}*}_{li} | i \rangle^{\mathrm{m}}, \quad l = e, \mu, \tau. \tag{8.22}$$

From (8.18) follows that the evolution equation in the flavor representation has the form

$$i \frac{\partial \, a(t)}{\partial t} = U^{\mathrm{m}} E^{\mathrm{m}} U^{\mathrm{m}\dagger} a(t). \tag{8.23}$$

Let us introduce the function

$$a'(t) = U^{\mathrm{m}\dagger} a(t). \tag{8.24}$$

From (8.23) and (8.24) we obtain the equation

$$i \frac{\partial a'(t)}{\partial t} = E^{\mathrm{m}} a'(t). \tag{8.25}$$

It is obvious that the solution of this equation has the form

$$a'(t) = e^{-i E^m t} a'(0) , \tag{8.26}$$

where $a'(0)$ is the wave function at the initial time $t = 0$. From (8.24) and (8.26) for the wave function in the flavor representation we have

$$a(t) = U^m e^{-i E^m t} U^{m\dagger} a(0) . \tag{8.27}$$

Let us assume that at initial time the flavor neutrino ν_l was produced. From (8.27) we find that the probability of the $\nu_l \to \nu_{l'}$ transition in matter with the constant density is given by the expression

$$P^m(\nu_l \to \nu_{l'}) = |\sum_i U^m_{l'i} e^{-i E^m_i t} U^{m*}_{li}|^2 . \tag{8.28}$$

Let us consider now in some details the simplest case of two flavor neutrinos ν_e and ν_x ($x = \mu$ or τ). For the vacuum mixing we have

$$\nu_{eL} = \cos\theta \nu_{1L} + \sin\theta \nu_{2L}$$
$$\nu_{xL} = -\sin\theta \nu_{1L} + \cos\theta \nu_{2L} , \tag{8.29}$$

where ν_1 and ν_2 are neutrino fields with masses m_1 and m_2. The effective Hamiltonian of neutrino in a matter has the form

$$H = U \frac{m^2}{2E} U^\dagger + \sqrt{2} G_F n_e \beta \tag{8.30}$$

where U is a 2×2 real orthogonal matrix

$$U = \begin{pmatrix} \cos\theta & \sin\theta \\ -\sin\theta & \cos\theta \end{pmatrix} . \tag{8.31}$$

It is convenient to present the total Hamiltonian in the form

$$H = \frac{1}{2} \operatorname{Tr} H + \tilde{H}. \tag{8.32}$$

Here

$$\frac{1}{2} \operatorname{Tr} H = \frac{m_1^2 + m_2^2}{4E} + \frac{1}{2} \sqrt{2} G_F n_e \tag{8.33}$$

and $\tilde{H}$ is the traceless part of the Hamiltonian.

We have

$$\tilde{H} = \frac{1}{4E} \begin{pmatrix} -\Delta m^2 \cos 2\theta + A & \Delta m^2 \sin 2\theta \\ \Delta m^2 \sin 2\theta & \Delta m^2 \cos 2\theta - A \end{pmatrix}, \tag{8.34}$$

where

$$A = 2\sqrt{2}\, G_F\, n_e\, E \tag{8.35}$$

and $\Delta m^2 = m_2^2 - m_1^2$. We will label neutrino masses in such a way that $m_2 > m_1$ and $\Delta m^2 > 0$.

The first term of (8.32), which is proportional to the unit matrix, can be excluded from the Hamiltonian. The real symmetrical 2×2 matrix $\tilde{H}$ can be diagonalized by orthogonal transformation (see Appendix A). We have

$$\tilde{H} = U^m\, E^m\, U^{m\dagger}. \tag{8.36}$$

Here

$$U^m = \begin{pmatrix} \cos\theta^m & \sin\theta^m \\ -\sin\theta^m & \cos\theta^m \end{pmatrix} \tag{8.37}$$

and

$$E^m = \begin{pmatrix} E_1^m & 0 \\ 0 & E_2^m \end{pmatrix} \tag{8.38}$$

where

$$E_{1,2}^m = \mp \frac{1}{4E} \sqrt{(\Delta m^2 \cos 2\theta - A)^2 + (\Delta m^2 \sin 2\theta)^2} \tag{8.39}$$

are eigenvalues of the matrix $\tilde{H}$.

The mixing angle of neutrino in matter θ^{mat} is given by the relation

$$\tan 2\theta^m = \frac{\Delta m^2 \sin 2\theta}{\Delta m^2 \cos 2\theta - A}. \tag{8.40}$$

From (8.28) and (8.37) for the probability of the $\nu_l \to \nu_{l'}$ ($\nu_{l'} \to \nu_l$) transition in matter we find the following expression

$$\mathrm{P}^m(\nu_l \to \nu_{l'}) = \mathrm{P}^m(\nu_{l'} \to \nu_l) = \frac{1}{2} \sin^2 2\theta^m (1 - \cos \Delta E^m L). \tag{8.41}$$

Here $l' \neq l$, (l or l' is equal to e), $L \simeq t$ is the distance, which neutrino travels in matter and

$$\Delta E^{\mathrm{m}} = E_2^{\mathrm{m}} - E_1^{\mathrm{m}} = \frac{1}{2E} \sqrt{(\Delta m^2 \cos 2\theta - A)^2 + (\Delta m^2 \sin 2\theta)^2} \,. \qquad (8.42)$$

The probability of ν_l ($\nu_{l'}$) to survive can be obtained from the condition of the conservation of the probability. We have

$$\mathrm{P}^{\mathrm{m}}(\nu_l \to \nu_l) = \mathrm{P}^{\mathrm{m}}(\nu_{l'} \to \nu_{l'}) = 1 - \frac{1}{2} \sin^2 2\theta^{\mathrm{m}} \left(1 - \cos \Delta E^{\mathrm{m}} L\right). \qquad (8.43)$$

The expression (8.41) can be written in the form

$$\mathrm{P}^{\mathrm{m}}(\nu_l \to \nu_{l'}) = \mathrm{P}^{\mathrm{m}}(\nu_{l'} \to \nu_l) = \frac{1}{2} \sin^2 2\theta^{\mathrm{m}} \left(1 - \cos 2\pi \frac{L}{L_{\mathrm{osc}}^{\mathrm{m}}}\right), \qquad (8.44)$$

where

$$L_{\mathrm{osc}}^{\mathrm{m}} = \frac{4\pi E}{\sqrt{(\Delta m^2 \cos 2\theta - A)^2 + (\Delta m^2 \sin 2\theta)^2}} \qquad (8.45)$$

is the oscillation length of neutrino in a matter with a constant density.

It is obvious that at $n_e = 0$ we have $\theta^{\mathrm{m}} = \theta$, $\Delta E^{\mathrm{m}} = \Delta m^2$ and expressions (8.41), (8.43) and (8.45) coincide with the standard vacuum two-neutrino transition probabilities and oscillation length, correspondingly.

If $n_e \neq 0$ the neutrino mixing angle in matter can be significantly different from the vacuum value. Let us assume that at some energy E the following equality

$$\Delta m^2 \cos 2\theta = A = 2\sqrt{2}\, G_F\, n_e\, E \qquad (8.46)$$

is satisfied. It follows from (8.40) that in this case $\theta^{\mathrm{m}} = \pi/4$ (maximal mixing) independently on the value of the vacuum mixing angle θ. The condition (8.46) is called MSW resonance condition. We will return to the discussion of this condition later.

If the condition (8.46) is satisfied, the oscillation length in matter can also be significantly different from the oscillation length in vacuum. In fact, we have in this case

$$L_{\mathrm{osc}}^{\mathrm{m}} = \frac{L_{\mathrm{osc}}}{\sin 2\theta} \qquad (8.47)$$

where $L_{\mathrm{osc}} = 4\pi \frac{E}{\Delta m^2}$ is the vacuum oscillation length.

Diagonalization of the Hamiltonian of neutrino in matter in the case of the three-neutrino mixing is a more difficult problem. An expansion over small parameters $\frac{\Delta m_{12}^2}{\Delta m_{13}^2}$ and $\sin^2 \theta_{13}$ simplifies its solution. For analysis of the data of long baseline accelerator experiments the following approximate three-neutrino expression for the $\nu_\mu \to \nu_e$ transition probability in the Earth matter is usually used

$$P^m(\nu_\mu \to \nu_e) = P_0 + P_{\sin \delta} + P_{\cos \delta} + P_3, \qquad (8.48)$$

where

$$P_0 = \sin^2 \theta_{23} \frac{\sin^2 2\theta_{13}}{(a-1)^2} \sin^2[(a-1)\Delta_{13}], \quad P_3 = \alpha^2 \cos^2 \theta_{23} \frac{\sin^2 2\theta_{12}}{a^2} \sin^2[a\Delta_{13}],$$
$$(8.49)$$

$$P_{\sin \delta} = -\alpha \frac{\sin \delta \cos \theta_{13} \sin 2\theta_{12} \sin 2\theta_{23} \sin 2\theta_{13}}{a(1-a)} \sin \Delta_{13} \sin[a\Delta_{13}] \sin[(1-a)\Delta_{13}]$$
$$(8.50)$$

and

$$P_{\cos \delta} = \alpha \frac{\cos \delta \cos \theta_{13} \sin 2\theta_{12} \sin 2\theta_{23} \sin 2\theta_{13}}{a(1-a)} \cos \Delta_{13} \sin[a\Delta_{13}] \sin[(1-a)\Delta_{13}]$$
$$(8.51)$$

Here

$$\alpha = \frac{\Delta m_{12}^2}{\Delta m_{13}^2}, \quad \Delta_{13} = \frac{\Delta m_{13}^2 L}{4E}, \quad a = \frac{2\sqrt{2} G_F n_e E}{\Delta m_{13}^2} \qquad (8.52)$$

The expression (8.48) can be used if the source-detector length L satisfies the condition $L \lesssim 8000\,\text{km}\,(\frac{E}{\text{GeV}})\,(\frac{10^{-4}\,\text{eV}^2}{\Delta m_{12}^2})$.

8.4   Adiabatic Transitions of Neutrino in Matter

We will consider here solar neutrinos. The evolution equation of a mixed neutrino in matter in the flavor representation has the form

$$i \frac{\partial a(t)}{\partial t} = H^m(t)\, a(t), \qquad (8.53)$$

where $H^m(t) = U \frac{m^2}{2E} U^\dagger + H_I(t)$ is the total effective Hamiltonian.

The hermitian matrix $H^{\text{mat}}(t)$ can be diagonalized by a unitary transformation

$$H^{\text{m}}(t) = U^{\text{m}}(t) \, E^{\text{m}}(t) \, U^{\text{m}\dagger}(t) \,, \qquad (8.54)$$

where $U^{\text{m}}(t)U^{\text{m}\dagger}(t) = 1$ and $E^{\text{m}}_{ik}(t) = E^{\text{m}}_i(t)\delta_{ik}$, $E^{\text{m}}_i(t)$ being the eigenvalue of the Hamiltonian.

We have

$$\langle \nu_{l'} \,|H(t)|\, \nu_l \rangle = \sum_i \langle \nu_{l'} \,|\, i(t) \rangle^{\text{m}} \, E^{\text{m}}_i(t) \,\, {}^{\text{m}}\langle i(t) \,|\, \nu_l \rangle, \qquad (8.55)$$

where

$$H(t) \,|\, i(t) \rangle^{\text{m}} = E^{\text{m}}_i(t) \,|\, i(t) \rangle^{\text{m}}. \qquad (8.56)$$

From (8.55) and (8.56) follows that

$$\langle \nu_l \,|\, i(t) \rangle^{\text{m}} = U^{\text{m}}_{li}(t), \quad {}^{\text{m}}\langle i(t) \,|\, \nu_l \rangle = U^{\text{m}*}_{li}(t). \qquad (8.57)$$

The flavor neutrino states ν_l are connected with states of neutrinos with definite energies in matter by the mixing relations

$$|\,\nu_l \rangle = \sum_{i=1}^{3} |\, i(t) \rangle^{\text{m}} \, {}^{\text{m}}\langle i(t) \,|\, \nu_l \rangle = \sum_{i=1}^{3} U^{\text{m}*}_{li} \,|\, i \rangle^{\text{m}}, \quad l = e, \mu, \tau. \qquad (8.58)$$

From (8.53) and (8.54) we have

$$U^{\text{m}\dagger}(t)i\,\frac{\partial a(t)}{\partial t} = E^{\text{m}}(t) \, U^{\text{m}\dagger}(t)a(t) \,. \qquad (8.59)$$

Let us introduce the function

$$a'(t) = U^{\text{m}\dagger}(t) \, a(t) \,. \qquad (8.60)$$

From (8.59) and (8.60) we obtain the following equation for the function $a'(t)$

$$i\,\frac{\partial a'(t)}{\partial t} = (E^{\text{m}}(t) + i\,\frac{\partial U^{\text{m}\dagger}(t)}{\partial t}\, U^{\text{m}}(t) \,)a'(t) \,. \qquad (8.61)$$

If n_e does not depend on t in this case $\frac{\partial U^{\text{m}\dagger}(t)}{\partial t} = 0$. We will assume that the function $n_e(t)$ depends on t so weakly that the second term in Eq. (8.61) can be neglected. In this approximation, which is called the adiabatic approximation, the solution of the evolution equation

$$i \frac{\partial a'(t)}{\partial t} \simeq E^{\mathrm{m}}(t)\, a'(t) \,. \tag{8.62}$$

can be easily found. We have

$$a'(t) = e^{-i \int_0^t E^{\mathrm{m}}(t)dt}\, a'(0). \tag{8.63}$$

From this equation follows that in the adiabatic approximation the neutrino remains on the same energy level during evolution.

From (8.60) and (8.63) we find the following adiabatic solution of the evolution equation in the flavor representation

$$a(t) = U^{\mathrm{m}}(t)\, e^{-i \int_0^t E^{\mathrm{m}}(t)dt}\, U^{\mathrm{mat}\dagger}(0)\, a(0). \tag{8.64}$$

From this expression follows that in the adiabatic approximation the probability of the transition $\nu_l \to \nu_{l'}$ during the time t is given by the expression

$$\mathrm{P}^{\mathrm{m}}(\nu_l \to \nu_{l'}) = |\sum_i U_{l'i}^{\mathrm{m}}(t)\, e^{-i \int_0^t E_i^{\mathrm{m}}(t)dt}\, U_{li}^{\mathrm{m}*}(0)|^2 \,. \tag{8.65}$$

Because in the adiabatic approximation the neutrino remains on the same energy level, the $\nu_l \to \nu_{l'}$ transition amplitude has a very simple structure, similar to the structure of the transition amplitudes in the case of the vacuum and a matter with a constant density : $U_{li}^{\mathrm{m}*}(0)$ is the amplitude of the transition from the state of the initial ν_l to the state with energy $E_i(0)$; the factor $e^{-i \int_0^t E_i^{\mathrm{m}}(t)dt}$ describes the propagation in the state with energy E_i^{m}; $U_{l'i}^{\mathrm{m}}(t)$ is the amplitude of the transition from the state with energy $E_i^{\mathrm{m}}(t)$ to the flavor state $\nu_{l'}$. The coherent sum over i is performed.

From (8.65) we find

$$\mathrm{P}^{\mathrm{m}}(\nu_l \to \nu_{l'}) = \sum_i |U_{l'i}^{\mathrm{m}}(t)|^2\, |U_{li}^{\mathrm{m}}(0)|^2$$

$$+ 2\mathrm{Re} \sum_{i>k} U_{l'i}^{\mathrm{m}}(t)\, U_{l'k}^{\mathrm{m}*}(t) e^{-i \int_0^t (E_i^{\mathrm{m}}(t) - E_k^{\mathrm{m}}(t))dt}\, U_{li}^{\mathrm{m}*}(0)\, U_{lk}^{\mathrm{m}}(0). \tag{8.66}$$

In the case of solar neutrinos, the transition probability must be averaged over the central region of the sun, in which solar ν_e are produced ($\sim 10^5$ km), over the energy resolution etc. After integration over many periods of oscillations, the oscillatory terms in the transition probability disappear. From (8.66) we find the following expression for the averaged probability of the solar ν_e to survive

$$P^m(\nu_e \rightarrow \nu_e) = \sum_{i=1}^{3} |U_{ei}^m(t)|^2 \, |U_{ei}^m(0)|^2 \,. \tag{8.67}$$

In the simplest two-neutrino case the mixing matrix in matter $U^m(t)$ is a real, orthogonal 2×2 matrix:

$$U^m(t) = \begin{pmatrix} \cos\theta_{12}^m(t) & \sin\theta_{12}^m(t) \\ -\sin\theta_{12}^m(t) & \cos\theta_{12}^m(t) \end{pmatrix} \,. \tag{8.68}$$

For the second term of the Hamiltonian (8.61) in the two-neutrino case we easily find

$$i \frac{\partial U^{m\dagger}(t)}{\partial t} U^m(t) = \begin{pmatrix} 0 & -i\dot{\theta}_{12}^m(t) \\ i\dot{\theta}_{12}^m(t) & 0 \end{pmatrix} \,. \tag{8.69}$$

The first term of the Hamiltonian (8.61) can be presented in the form

$$\begin{pmatrix} E_1^m(t) & 0 \\ 0 & E_2^m(t) \end{pmatrix}$$

$$= \frac{1}{2}(E_1^m(t) + E_2^m(t)) + \begin{pmatrix} -\frac{1}{2}(E_2^m(t) - E_1^m(t)) & 0 \\ 0 & \frac{1}{2}(E_2^m(t) - E_1^m(t)) \end{pmatrix} \tag{8.70}$$

The first term of (8.70), proportional to the unit matrix, can be omitted. The Hamiltonian in the evolution equation (8.61) has the form

$$H' = \begin{pmatrix} -\frac{1}{2}(E_2^m(t) - E_1^m(t)) & -i\dot{\theta}_{12}^m(t) \\ i\dot{\theta}_{12}^m(t) & \frac{1}{2}(E_2^m(t) - E_1^m(t)) \end{pmatrix} \,. \tag{8.71}$$

Let us introduce the parameter of adiabaticity

$$\gamma(t) = \frac{(E_2^m(t) - E_1^m(t))}{2\left|\frac{d\theta_{12}^m(t)}{dt}\right|}, \tag{8.72}$$

The solution of the evolution equation (8.61) is adiabatic if nondiagonal elements of the matrix (8.76) are much smaller than the diagonal elements, i.e. if the parameter of adiabaticity is much larger than one:

$$\gamma(t) \gg 1 \,. \tag{8.73}$$

8.5 Two-Neutrino Case

The average survival probability of solar ν_e's in matter is given in the three-neutrino case by the following expression[3]

$$P^m(\nu_e \to \nu_e) = |U_{e3}|^4 + (1 - |U_{e3}|^2)^2 \, P^m_{\nu_e \to \nu_e}(\Delta m^2_{12}, \theta_{12}) , \qquad (8.74)$$

where $P^m_{\nu_e \to \nu_e}(\Delta m^2_{12}, \theta_{12})$ is the two-neutrino transition probability. From the data of the Daya Bay, RENO and Double Chooz reactor experiments follows that $|U_{e3}|^2 = \sin^2\theta_{13} \simeq 2.2 \cdot 10^{-2}$. Thus with an accuracy of a few % the ν_e survival probability is given by the two-neutrino expression $P^m_{\nu_e \to \nu_e}(\Delta m^2_{12}, \theta_{12})$. The two-neutrino $\nu_e \to \nu_e$ survival probability is usually used in the analysis of the solar neutrino data.

We will consider here in some details the propagation of the solar neutrinos in matter in the case of the two-neutrino mixing.

For neutrino fields we have

$$\nu_{eL} = \cos\theta_{12}\nu_{1L} + \sin\theta_{12}\nu_{2L}, \quad \nu_{aL} = -\sin\theta_{12}\nu_{1L} + \cos\theta_{12}\nu_{2L} , \qquad (8.75)$$

where $a = \mu, \tau$. The Hamiltonian of neutrino in matter in the flavor representation has the form (proportional to the unit matrix term is omitted)

$$H^m(x) = \begin{pmatrix} -\Delta m^2_{12}\cos 2\theta_{12} + A(x) & \Delta m^2_{12}\sin 2\theta_{12} \\ \Delta m^2_{12}\sin 2\theta_{12} & \Delta m^2_{12}\cos 2\theta_{12} - A(x) \end{pmatrix} . \qquad (8.76)$$

Here $A(x) = 2\sqrt{2}G_F n_e(x)E$, $\Delta m^2_{12} = m^2_2 - m^2_1 > 0$ and $x \simeq t$.

The real, symmetrical, traceless 2×2 matrix (8.75) can be easily diagonalized (see Appendix A). We have

$$H^m(x) = U^m(x)E^m(x)U^{m\dagger}(x) . \qquad (8.77)$$

Here $E^m_{ik}(x) = E^m_i(x)\delta_{ik}$ and $U^m(x)\,U^{m\dagger}(x) = 1$. The eigenvalues of the matrix $H^m(x)$ are given by

$$E^m_{1,2}(x) = \mp\frac{1}{4E}\sqrt{(\Delta m^2_{12}\cos 2\theta_{12} - A(x))^2 + (\Delta m^2_{12}\sin 2\theta_{12})^2} \qquad (8.78)$$

---

[3]It is instructive to obtain this expression in the vacuum case. We have $P(\nu_e \to \nu_e) = |\sum_{i=1}^3 |U_{ei}|^2 e^{-i\frac{\Delta m^2_{1i}}{2E}}|^2$. For the average probability the effect of interference due to the large Δm^2_{13} disappears. We find $P(\nu_e \to \nu_e) = |U_{e3}|^4 + |\sum_{i=1,2} |U_{ei}|^2 e^{-i\frac{\Delta m^2_{1i}}{2E}}|^2$. Finally, taking into account that $|U_{e1}|^2 = (1 - |U_{e3}|^2)\cos^2\theta_{12}$ and $|U_{e2}|^2 = (1 - |U_{e3}|^2)\sin^2\theta_{12}$ we come to the following expression $P(\nu_e \to \nu_e) = |U_{e3}|^4 + (1 - |U_{e3}|^2)^2(1 - \sin^2 2\theta_{12}\sin^2\frac{\Delta m^2_{12}}{4E})$.

The matrix $U^m(x)$ has the form

$$U^m(x) = \begin{pmatrix} \cos\theta_{12}^m(x) & \sin\theta_{12}^m(x) \\ -\sin\theta_{12}^m(x) & \cos\theta_{12}^m(x) \end{pmatrix}, \tag{8.79}$$

where the mixing angle $\theta^m(x)$ is determined by the equation

$$\tan 2\theta_{12}^m(x) = \frac{\Delta m_{12}^2 \sin 2\theta}{\Delta m_{12}^2 \cos 2\theta_{12} - A(x)}. \tag{8.80}$$

States of the flavor neutrinos are connected with the states of neutrinos with definite energy $|\nu_{1,2}(x)\rangle^m$ by the following mixing relation

$$|\nu_e\rangle = \cos\theta_{12}^m(x)|\nu_1(x)\rangle^m + \sin\theta_{12}^m(x)|\nu_2(x)\rangle^m$$
$$|\nu_a\rangle = -\sin\theta_{12}^m(x)|\nu_1(x)\rangle^m + \cos\theta_{12}^m(x)|\nu_2(x)\rangle^m. \tag{8.81}$$

From (8.78) and (8.80) it follows that at the point x_R at which the condition

$$\Delta m_{12}^2 \cos 2\theta_{12} = A(x_R) = 2\sqrt{2}G_F E\, n_e(x_R) \tag{8.82}$$

is satisfied, the mixing in matter is maximal and the difference of the neutrino energies is minimal:

$$\theta_{12}^m(x_R) = \frac{\pi}{4}, \quad (E_2^m(x_R) - E_1^m(x_R)) = \frac{\Delta m_{12}^2 \sin 2\theta_{12}}{2E}. \tag{8.83}$$

This condition is called *the MSW resonance condition*.

The electron density $n_e(x)$ is maximal in the center of the sun and decreases exponentially to its periphery. Neutrinos, produced in the central region of the sun, are traveling towards the its surface can pass through the resonance region $x = x_R$. The resonance region is the most important one for the neutrino transition in the sun.

Let us calculate the adiabaticity parameter $\gamma(x)$ which is determined by the relation (8.72). From (8.80) we have

$$\frac{d\theta^m(x)}{dx} = \frac{\Delta m_{12}^2 \sin 2\theta_{12}\frac{dA(x)}{dx}}{2[(\Delta m_{12}^2 \cos 2\theta_{12} - A(x))^2 + (\Delta m_{12}^2 \sin 2\theta_{12})^2]} \tag{8.84}$$

Further, from (8.78) and (8.84) we find

$$\gamma(x) = \frac{[(\Delta m_{12}^2 \cos 2\theta_{12} - A(x))^2 + (\Delta m_{12}^2 \sin 2\theta_{12})^2]^{3/2}}{2E\,\Delta m_{12}^2 \sin 2\theta_{12}\,|\frac{dA(x)}{dx}|}. \tag{8.85}$$

From (8.85) we obtain for the value of the parameter of the adiabaticity at the resonance point x_R the following expression

$$\gamma(x_R) = \frac{(\Delta m_{12}^2 \sin 2\theta_{12})^2}{2E \, 2\sqrt{2} G_F E |\frac{dn_e}{dx}|_{x_R}}.$$ (8.86)

Taking into account that $2\sqrt{2} G_F E = \frac{\Delta m_{12}^2 \cos 2\theta_{12}}{n_e(x_R)}$, from this relation we find

$$\gamma(x_R) = \frac{\Delta m_{12}^2 \sin^2 2\theta_{12}}{2E \cos 2\theta_{12} |\frac{d \ln n_e}{dx}|_{x_R}}.$$ (8.87)

The electron density $n_e(x)$ can be presented in the form

$$n_e(x) = Y_e \rho(x) \frac{1}{M}.$$ (8.88)

Here $Y_e = \frac{n_e}{n_N}$ is the ratio of the electron and nucleon number density, $\rho(x)$ is the matter density and M is the nucleon mass.

The sun density $\rho(x)$ is well described by the exponential function

$$\rho(x) \simeq \rho(0) \, e^{-x/r_0}.$$ (8.89)

Here

$$r_0 = \frac{R}{10.54} \simeq 6.6 \cdot 10^4 \text{ km},$$ (8.90)

where R is the solar radius, and $\rho(0) \simeq 150 \text{ g/cm}^3$.

From (8.87) and (8.89) for the adiabaticity parameter at the point x_R we have

$$\gamma(x_R) = \frac{\Delta m_{12}^2 \sin^2 2\theta_{12} \, r_0}{2E \cos 2\theta}.$$ (8.91)

Let us estimate the parameter $\gamma(x_R)$. Taking into account that

$$\Delta m_{12}^2 \simeq 7.4 \cdot 10^{-5} \text{ eV}^2, \quad \sin^2 \theta_{12} \simeq 0.31$$ (8.92)

from (8.90) and (8.91) we obtain the following expression

$$\gamma(x_R) \simeq 2.7 \cdot 10^4 \left(\frac{\text{MeV}}{E}\right).$$ (8.93)

In the solar neutrino experiments neutrinos with energies in the range $(0.2–15)$ MeV are detected. From (8.93) follows that in the whole interval of the detected solar neutrino energies

$$\gamma(x_R) \gg 1 \,. \tag{8.94}$$

This inequality ensures that *transitions of solar neutrinos are adiabatic ones*. From (8.67) for the adiabatic two-neutrino probability of the solar ν_e's to survive we find the following expression

$$P^m(\nu_e \to \nu_e) = \sum_{i=1,2} |U_{ei}^m(x)|^2 \, |U_{ei}^m(0)|^2 = \cos^2 \theta_{12}^m(x) \, \cos^2 \theta_{12}^m(0)$$

$$+ \sin^2 \theta_{12}^m(x) \, \sin^2 \theta_{12}^m(0) = \frac{1}{2}(1 + \cos 2\theta_{12}^m(x) \cos 2\theta_{12}^m(0)) \,. \tag{8.95}$$

The solar ν_e's are produced in the central part of the sun and propagate to its surface. Let us estimate neutrino energies at which the matter term at the production point $A(0)$ is much larger than $\Delta m_{12}^2 \cos \theta_{12}$ term:

$$2\sqrt{2} G_F E n_e(0) \gg \Delta m_{12}^2 \cos \theta_{12} \tag{8.96}$$

From this inequality we have

$$E \gg E_0 = \frac{\Delta m_{12}^2 \cos 2\theta_{12}}{2\sqrt{2}\, G_F\, n_e(0)} = \frac{\Delta m_{12}^2 \cos 2\theta_{12}\, M}{2\sqrt{2}\, G_F\, Y_e \rho(0)} \,. \tag{8.97}$$

Using the values (8.92) of the neutrino oscillation parameters, $\rho_0 \simeq 150\,\mathrm{g/cm^3}$ and $Y_e \simeq 2/3$ we find $E_0 \simeq 1.8\,\mathrm{MeV}$.

From Eq. (8.80) follows that in the high energy region

$$E \gg E_0 \simeq 1.8\,\mathrm{MeV} \tag{8.98}$$

the matter neutrino mixing angle in the production region $\theta^m(0)$ is given by

$$\tan 2\theta_{12}^m(0) \simeq 0, \quad \theta_{12}^m(0) \simeq \frac{\pi}{2} \,. \tag{8.99}$$

On the surface of the sun $A = 0$ and $\theta_{12}^m = \theta_{12}$, where θ_{12} is the vacuum mixing angle. Thus, from (8.95) follows that for the solar ν_e's with energies, which satisfy the condition (8.98), the survival probability is given by the expression

$$P^m(\nu_e \to \nu_e) \simeq \frac{1}{2}(1 - \cos 2\theta_{12}) = \sin^2 \theta_{12} \,. \tag{8.100}$$

From (8.92) and (8.100) we obtain the following value of the ν_e survival probability in the high-energy region:

$$P^m(\nu_e \to \nu_e) \simeq 0.31. \tag{8.101}$$

In the region of neutrino energies significantly smaller than E_0

$$E \ll E_0 \simeq 1.8 \, \text{MeV} \tag{8.102}$$

the matter term in the production region can be neglected and from (8.80) we find $\theta_{12}^m(0) \simeq \theta_{12}$. Thus, in the low-energy region the ν_e survival probability in matter is the same as in the vacuum

$$P^m(\nu_e \to \nu_e) \simeq 1 - \frac{1}{2} \sin^2 2\theta_{12}. \tag{8.103}$$

From (8.92) in the low-energy region we find

$$P^m(\nu_e \to \nu_e) \simeq 0.57. \tag{8.104}$$

In every of the two energy regions, we considered, the ν_e transition probabilities practically do not depend on energy and differ approximately by the factor two. In the transition region a strong energy dependence must be exhibited. Detailed calculations show that the transition region between low-energy and high-energy regimes lies in the interval $(2\text{–}5)\,\text{MeV}$.

In the BOREXINO experiment (see later) the ^7Be neutrinos with energy $E = 0.87\,\text{MeV}$ was detected. For the ν_e survival probability the value

$$P^m(\nu_e \to \nu_e) = 0.51 \pm 0.07 \tag{8.105}$$

was obtained.

We will finish this section with the following remarks

1. It follows from (8.81) and (8.99) that high energy solar neutrinos $(E \gg E_0)$ are produced in the state with definite energy in matter

$$|\nu_e\rangle \simeq |\nu_2(0)\rangle^m. \tag{8.106}$$

In the adiabatic transitions during evolution neutrinos stay on the same energy level. Thus on the surface of the sun the high-energy neutrinos are in the vacuum state $|\nu_2\rangle$.

2. We have considered adiabatic transitions of neutrinos in matter. In the general case for the averaged two-neutrino ν_e survival probability in matter we have

$$P^m(\nu_e \to \nu_e) = \sum_{i,k} |U_{ek}^m(x)|^2 \, P_{ki} \, |U_{ei}^m(0)|^2 \tag{8.107}$$

where P_{ki} is the probability of the transition from the state with energy E_i^m at the point $x = 0$ to the state with energy E_k^m at the point x. From the conservation of the total probability we have

$$P_{11} = 1 - P_{21}; \quad P_{22} = 1 - P_{12} . \tag{8.108}$$

Further, from the T invariance we have $P_{21} = P_{12}$. From (8.107) and (8.108) we obtain the following general expression for the two-neutrino ν_e-survival probability

$$P^m(\nu_e \to \nu_e) = \frac{1}{2} + (\frac{1}{2} - P_{12}) \cos 2\theta \cos 2\theta^m(0) \tag{8.109}$$

Chapter 9
Neutrinoless Double Beta-Decay

9.1 Introduction

Discovery of neutrino oscillations, driven by neutrino mass-squared differences and neutrino mixing, is one of the most important discovery in the particle physics. It is unlikely that small neutrino masses are of the standard Brout-Englert-Higgs mechanism origin. A new, beyond the Standard Model mechanism of the generation of neutrino masses is required.

The most viable, economical and simple effective Lagrangian (seesaw) mechanism of the neutrino mass generation is based on the assumption of the violation of the total lepton number at a large scale. This mechanism predicts that neutrinos with definite masses ν_i are Majorana particles.

The problem of the nature of massive neutrinos (Dirac or Majorana?) is one of the most fundamental problem of neutrino physics. The solution of this problem will have an enormous impact on our the understanding of the origin of neutrino masses and mixing.

Neutrino oscillations is an interference phenomenon sensitive to very small values of neutrino mass-squared differences. However, by the investigation of neutrino oscillations it is impossible to decide on the nature of neutrinos ν_i: are they Dirac or Majorana particles. In order to reveal the nature of neutrinos with definite masses it is necessary to study processes in which the total lepton number L is violated.

The neutrinoless double β-decay ($0\nu\beta\beta$-decay)

$$(A, Z) \rightarrow (A, Z + 2) + e^- + e^- \tag{9.1}$$

of ^{76}Ge, ^{130}Te, ^{136}Xe and other even-even nuclei is allowed if ν_i are Majorana neutrinos. The study of the $0\nu\beta\beta$-decay is the most sensitive method of the investigation of the nature of neutrinos with definite masses. However, the probability of the process is very strongly suppressed because, first, the $0\nu\beta\beta$-decay is a process of the

© Springer International Publishing AG, part of Springer Nature 2018
S. Bilenky, *Introduction to the Physics of Massive and Mixed Neutrinos*,
Lecture Notes in Physics 947, https://doi.org/10.1007/978-3-319-74802-3_9

second order of the perturbation theory in the Fermi constant G_F and, second, this process is due to neutrino helicity-flip. As a result, the matrix element of the $0\nu\beta\beta$-decay is proportional to the effective Majorana mass $m_{\beta\beta} = \sum_i U_{ei}^2 m_i$ (m_i is the mass of the neutrino ν_i, U_{ei} is the element of the neutrino mixing matrix). Smallness of the neutrino masses is an additional reason for smallness of the probability of the $0\nu\beta\beta$-decay.

Up to now the neutrinoless double β-decay was not observed. From performed experiments impressive lower bounds for life-time of the $0\nu\beta\beta$-decay of some nuclei were obtained ($T_{1/2}^{0\nu}(^{136}\text{Xe}) > 1.1 \cdot 10^{26}$ years, $T_{1/2}^{0\nu}(^{76}\text{Ge}) > 5.2 \cdot 10^{25}$ years etc.). However, in order to reach half-lives of the $0\nu\beta\beta$-decay suggested by the neutrino oscillation data significant improvement is required. Several new experiments on the search for the $0\nu\beta\beta$-decay which could probe the region of the inverted neutrino mass hierarchy are at preparation at present.

Let us compare $0\nu\beta\beta$-decay with the following lepton number violating processes

$$K^+ \to \pi^- + \mu^+ + \mu^+ \tag{9.2}$$

and

$$\mu^- + \text{Ti} \to e^+ + \text{Ca}. \tag{9.3}$$

If ν_i are Majorana particles these processes are allowed and their matrix elements are proportional to

$$|m_{\mu\mu}| = |\sum_i U_{\mu i}^2 \, m_i| \text{ and } |m_{\mu e}| = |\sum_i U_{\mu i} \, U_{ei} \, m_i|, \tag{9.4}$$

correspondingly. Taking into account the Cauchy-Schwarz inequality and the unitarity of the mixing matrix, we have

$$|m_{ll'}| \leq \sqrt{\sum_i |U_{li}|^2 m_i^2} \sqrt{\sum_i |U_{l'i}|^2} \leq m_{max}, \tag{9.5}$$

where m_{max} is the mass of the heaviest neutrino. From the data of tritium experiments on the measurement of the neutrino mass (see the next chapter) it follows that $m_{max} \leq 2.2$ eV . Thus, we have

$$|m_{ll'}| \leq 2.2 \text{ eV} . \tag{9.6}$$

The sensitivities to the parameter $|m_{\mu\mu}|$ and $|m_{\mu e}|$ of experiments on the search for the processes (9.2) and (9.3), correspondingly, are much worse than the upper bound (9.6).

From exiting data it was found

$$\frac{\Gamma(K^+ \to \pi^- \mu^+ \mu^+)}{\Gamma(K^+ \to \text{all})} \leq 3 \cdot 10^{-9} \tag{9.7}$$

and

$$\frac{\Gamma(\mu^- \text{Ti} \to e^+ \text{Ca})}{\Gamma(\mu^- \text{Ti} \to \text{all})} \leq 1.7 \cdot 10^{-12} . \tag{9.8}$$

From these results the following bounds were obtained

$$|m_{\mu\mu}| \leq 4 \cdot 10^4 \text{ MeV}, \quad |m_{\mu e}| \leq 82 \text{ MeV} . \tag{9.9}$$

Similar bounds can be inferred from the data on the search for other lepton number violating processes. On the other side from existing experiments on the search for the 0νββ-decay it was found

$$m_{ee} \equiv m_{\beta\beta} \leq (1.4\text{–}4.5) \cdot 10^{-1} \text{ eV} . \tag{9.10}$$

The possibilities to use large targets (usually enriched in isotope which could exhibit ββ-decay), to reach small background and high energy resolution make *experiments on the search for the 0νββ-decay a unique source of information about the nature of massive neutrinos ν_i*. In the next sections we will consider this process in some details.

9.2 Basic Elements of the Theory of 0νββ-Decay

We will consider here the basic theory of neutrinoless double β-decay. Assume that an even-even nucleus (A, Z) has a mass $M_{A,Z}$ and the mass of odd-odd nucleus with the same atomic number $(A, Z + 1)$ is larger than $M_{A,Z}$. In such a case the usual β-decay $(A, Z) \to (A, Z + 1) + e^- + \bar{\nu}_e$ is forbidden. If, however, exist even-even nucleus $(A, Z+2)$ with mass smaller than $M_{A,Z}$, the nucleus (A, Z) can decay into $(A, Z + 2)$ with emission of two electrons via $(A, Z) \to (A, Z + 2) + e^- + e^- + \bar{\nu}_e + \bar{\nu}_e$ or $(A, Z) \to (A, Z + 2) + e^- + e^-$.

Let us consider the even-even nucleus ^{76}Ge. The decay ^{76}Ge $\to$ ^{76}As $+ e^- + \bar{\nu}_e$ is forbidden (the ^{76}As nucleus is heavier than ^{76}Ge). However, the transition of ^{76}Ge into lighter even-even nucleus ^{76}Se and two electrons is allowed.

Table 9.1 $\beta\beta$ candidate
nuclei

Transition	$T_0 = Q_{\beta\beta}$ (keV)	Abundance (%)
^{76}Ge $\rightarrow$ ^{76}Se	2039.6 ± 0.9	7.8
^{100}Mo $\rightarrow$ ^{100}Ru	3934 ± 6	9.6
^{130}Te $\rightarrow$ ^{130}Xe	2533 ± 4	34.5
^{136}Xe $\rightarrow$ ^{136}Ba	2479 ± 8	8.9
^{150}Nd $\rightarrow$ ^{150}Sm	3367.1 ± 2.2	5.6
^{82}Se $\rightarrow$ ^{82}Kr	2995 ± 6	9.2
^{48}Ca $\rightarrow$ ^{48}Ti	4271 ± 4	0.187

In the first column nuclei transitions are indicated; in the
second column Q-values are shown; in the third column
abundances of $\beta\beta$ candidates are presented

Two types of $\beta\beta$-decays are possible

1. Two-neutrino double β-decay ($2\nu\beta\beta$-decay)

$$(A, Z) \rightarrow (A, Z + 2) + e^- + e^- + \bar{\nu}_e + \bar{\nu}_e \; . \tag{9.11}$$

2. Neutrinoless double β-decay ($0\nu\beta\beta$-decay)

$$(A, Z) \rightarrow (A, Z + 2) + e^- + e^- \; . \tag{9.12}$$

In Table 9.1 a list of several even-even nuclei, which can have the $\beta\beta$-decay, is
presented.

The $2\nu\beta\beta$-decay is allowed, second order in the Fermi constant G_F, very rare
process. This decay was observed in the case of more than ten different nuclei with
half-lives in the range $(10^{18}$–$10^{20})$ years.

The $0\nu\beta\beta$-decay is allowed only in the case if the total lepton number is not
conserved and neutrinos with definite masses are Majorana particles. We will
consider this process assuming that neutrinos have the standard charged current
interaction

$$\mathcal{H}_I(x) = \frac{G_F}{\sqrt{2}} 2 \, \bar{e}_L(x) \gamma^\alpha \, \nu_{eL}(x) \, j_\alpha^{CC}(x) + \text{h.c.} \tag{9.13}$$

Here G_F is the Fermi constant and

$$\nu_{eL}(x) = \sum_{i=1}^{3} U_{ei} \, \nu_{iL}(x) \; , \tag{9.14}$$

where $\nu_i(x)$ is the field of the Majorana neutrino with the mass m_i ($\nu_i^c(x) = C\bar{\nu}_i^T(x) = \nu_i(x)$)

$$j_\alpha^{CC}(x) = \cos\theta_C \, j_\alpha(x), \tag{9.15}$$

Fig. 9.1 Feynman diagram
of the neutrinoless double
β-decay

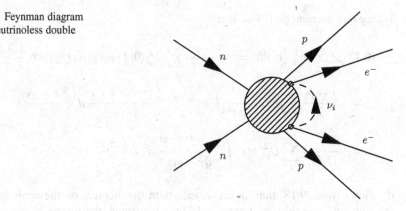

Fig. 9.1 Feynman diagram of the neutrinoless double β-decay

where $j_\alpha(x)$ is the $\Delta S = 0$ hadronic weak charged current and θ_C is the Cabibbo angle.

The neutrinoless double β-decay is the second order in G_F process with the virtual neutrinos. The Feynman diagram of the transition $n + n \to p + p + e^- + e^-$ is presented in Fig. 9.1. The matrix element of the $0\nu\beta\beta$-decay is given by the following expression

$$\langle f|S^{(2)}|i\rangle = 4\frac{(-i)^2}{2!}\left(\frac{G_F \cos\theta_C}{\sqrt{2}}\right)^2$$

$$\times N_{p_1} N_{p_2} \int \bar{u}_L(p_1)e^{ip_1x_1}\gamma^\alpha \langle 0|T(\nu_{eL}(x_1)\, \nu_{eL}^T(x_2))|0\rangle$$

$$\times \gamma^{\beta T} \bar{u}_L^T(p_2)e^{ip_2x_2}\langle N_f|T(J_\alpha(x_1)J_\beta(x_2))|N_i\rangle\, d^4x_1 d^4x_2 - (p_1 \rightleftarrows p_2). \tag{9.16}$$

Here p_1 and p_2 are electron momenta, $J_\alpha(x)$ is the weak charged current in the Heisenberg representation, N_i and N_f are the states of the initial and the final nuclei with 4-momenta $P_i = (E_i, \mathbf{p_i})$ and $P_f = (E_f, \mathbf{p_f})$ and $N_p = \frac{1}{(2\pi)^{3/2}\sqrt{2p^0}}$ is the standard normalization factor. Let us consider the neutrino propagator. From the Majorana condition we have

$$\nu_i^T(x) = -\bar{\nu}_i\, C. \tag{9.17}$$

Taking into account (9.17) we find[1]

$$\langle 0|T(\nu_{eL}(x_1)\nu_{eL}^T(x_2)|0\rangle = -\frac{1-\gamma_5}{2}\sum_k U_{ek}^2 \langle 0|T(\nu_k(x_1)\bar{\nu}_k(x_2))|0\rangle \frac{1-\gamma_5}{2} C$$

$$= -\frac{1-\gamma_5}{2}\sum_k U_{ek}^2 \frac{i}{(2\pi)^4}\int e^{-i\bar{q}\,(x_1-x_2)}\frac{\gamma\cdot\bar{q}+m_k}{\bar{q}^2-m_k^2}d^4\bar{q}\,\frac{1-\gamma_5}{2}C$$

$$= -\frac{i}{(2\pi)^4}\sum_k U_{ek}^2\,m_k\int\frac{e^{-i\bar{q}\,(x_1-x_2)}}{\bar{q}^2-m_k^2}d^4\bar{q}\,\frac{1-\gamma_5}{2}C\,. \qquad (9.18)$$

It follows from (9.18) that, in accordance with the theorem on the equivalence of theories with massless Majorana and Dirac neutrinos, the matrix element of the $0\nu\beta\beta$-decay is equal to zero in the case of massless neutrinos. This is connected with the fact that only left-handed neutrino fields enter into the Hamiltonian of the weak interaction.

Let us consider the second term of the matrix element (9.16). We have

$$\bar{u}_L(p_1)\gamma^\alpha(1-\gamma_5)\gamma^\beta C\bar{u}_L^T(p_2) = \bar{u}_L(p_2)C^T\gamma_\beta^T(1-\gamma_5^T)\gamma^{\alpha T}\bar{u}_L^T(p_1)$$

$$= -\bar{u}_L(p_2)\gamma^\beta(1-\gamma_5)\gamma^\alpha C\bar{u}_L^T(p_1)\,. \quad (9.19)$$

Taking into account (9.19) and the relation

$$T(J_\beta(x_2)J_\alpha(x_1)) = T(J_\alpha(x_1)J_\beta(x_2)) \qquad (9.20)$$

we find that the second term of the matrix element (9.16) is equal to the first one. For the matrix element of the $0\nu\beta\beta$-decay we obtain the following expression

$$\langle f|S^2|i\rangle = -4\left(\frac{G_F\cos\theta_C}{\sqrt{2}}\right)^2$$

$$\times N_{p_1}N_{p_2}\int \bar{u}_L(p_1)e^{ip_1x_1}\gamma^\alpha\frac{i}{(2\pi)^4}\sum_k U_{ek}^2 m_k\int\frac{e^{-i\bar{q}\,(x_1-x_2)}}{\bar{q}^2-m_k^2}d^4\bar{q}$$

$$\times\frac{1-\gamma_5}{2}\gamma^\beta C\,\bar{u}_L^T(p_2)e^{ip_2x_2}\,\langle N_f|T(J_\alpha(x_1)J_\beta(x_2))|N_i\rangle\,d^4x_1d^4x_2\,.$$

$$(9.21)$$

---

[1] Notice that in the case of the Dirac neutrinos $\langle 0|\nu_{eL}(x_1)\nu_{eL}^T(x_2)|0\rangle =$ $\frac{1-\gamma_5}{2}\sum_k U_{ek}^2\langle 0|\nu_k(x_1)\nu_k^T(x_2)|0\rangle\frac{1-\gamma_5^T}{2} = 0$. The neutrinoless double β-decay is obviously forbidden in the Dirac case.

The calculation of the nuclear part of the matrix element of the $0\nu\beta\beta$-decay is a complicated nuclear problem. There exist several approximate methods which are used in such calculations. We will present now the matrix element of the $0\nu\beta\beta$-decay in a form which is usually used in calculations of the nuclear part of the matrix element.

Let us perform in (9.21) the integration over the time variables x_2^0 and x_1^0. The integral over x_2^0 can be presented in the form

$$\int_{-\infty}^{\infty} ...dx_2^0 = \int_{-\infty}^{x_1^0} ...dx_2^0 + \int_{x_1^0}^{\infty} ...dx_2^0 . \qquad (9.22)$$

In the first integral we have $x_1^0 > x_2^0$. In this region we can perform integration over $\bar{q}^0$ in the expression (9.21). We find[2]

$$\frac{i}{(2\pi)^4} \int \frac{e^{-i\bar{q}(x_1-x_2)}}{\bar{q}^2 - m_k^2} d^4\bar{q} = \frac{1}{(2\pi)^3} \int \frac{e^{-i\bar{q}_k^0(x_1^0-x_2^0)+i\mathbf{q}(\mathbf{x}_1-\mathbf{x}_2)}}{2\,\bar{q}_k^0} d^3q , \qquad (9.23)$$

where

$$\bar{q}_k^0 = \sqrt{(\mathbf{q})^2 + m_k^2} . \qquad (9.24)$$

In the second integral of the expression (9.22) we have $x_1^0 < x_2^0$. In this region we find

$$\frac{i}{(2\pi)^4} \int \frac{e^{-i\bar{q}(x_1-x_2)}}{\bar{q}^2 - m_k^2} d^4\bar{q} = \frac{1}{(2\pi)^3} \int \frac{e^{-i\bar{q}_k^0(x_2^0-x_1^0)+i\mathbf{q}(\mathbf{x}_2-\mathbf{x}_1)}}{2\,\bar{q}_k^0} d^3q. \qquad (9.25)$$

Let us consider the matrix element $\langle N_f | T(J_\alpha(x_1) J_\beta(x_2)) | N_i \rangle$. From the invariance under the translations we have

$$J_\alpha(x) = e^{iHx^0} J_\alpha(\mathbf{x}) e^{-iHx^0} \quad (J_\alpha(0, \mathbf{x}) \equiv J_\alpha(\mathbf{x})), \qquad (9.26)$$

where H is the total Hamiltonian. Using this relation at $x_1^0 > x_2^0$ we find

$$\langle N_f | T(J_\alpha(x_1) J_\beta(x_2)) | N_i \rangle = \langle N_f | J_\alpha(x_1) J_\beta(x_2) | N_i \rangle$$
$$= \sum_n e^{i(E_f-E_n)x_1^0} e^{i(E_n-E_i)x_2^0} \langle N_f | J_\alpha(\mathbf{x}_1) | N_n \rangle \langle N_n | J_\beta(\mathbf{x}_2) | N_i \rangle. \qquad (9.27)$$

Here $|N_n\rangle$ is the vector of the state of the intermediate nucleus with four-momentum $P_n = (E_n, \mathbf{p_n})$ and the sum over the total system of states $|N_n\rangle$ is performed. In the

---

[2]We took into account that in the propagator $m_k^2 \to m_k^2 - i\epsilon$.

region $x_1^0 < x_2^0$ we have

$$\langle N_f | T(J_\alpha(x_1) J_\beta(x_2)) | N_i \rangle = \langle N_f | J_\alpha(x_2) J_\beta(x_1) | N_i \rangle$$

$$= \sum_n e^{i(E_f - E_n)x_2^0} e^{i(E_n - E_i)x_1^0} \langle N_f | J_\alpha(\mathbf{x}_2) | N_n \rangle \langle N_n | J_\beta(\mathbf{x}_1) | N_i \rangle. \quad (9.28)$$

From Eqs. (9.23) and (9.27) we find

$$\int_{-\infty}^{\infty} dx_1^0 \int_{-\infty}^{x_1^0} dx_2^0 \langle N_f | J_\alpha(\mathbf{x}_1) J_\beta(\mathbf{x}_2) | N_i \rangle e^{i(p_1^0 x_1^0 + p_2^0 x_2^0)} e^{i \bar{q}_k^0 (x_2^0 - x_1^0)}$$

$$= -i \sum_n \frac{\langle N_f | J_\alpha(\mathbf{x}_1) | N_n \rangle \langle N_n | J_\beta(\mathbf{x}_2) | N_i}{E_n + p_2^0 + \bar{q}_k^0 - E_i - i\epsilon} 2\pi\delta(E_f + p_1^0 + p_2^0 - E_i).$$

$$(9.29)$$

Analogously, from (9.25) and (9.28) we obtain the following relation

$$\int_{-\infty}^{\infty} dx_1^0 \int_{x_1^0}^{\infty} dx_2^0 \langle N_f | J_\alpha(\mathbf{x}_2) J_\beta(\mathbf{x}_1) | N_i \rangle e^{i(p_1^0 x_1^0 + p_2^0 x_2^0)} e^{i \bar{q}_k^0 (x_2^0 - x_1^0)}$$

$$= -i \sum_n \frac{\langle N_f | J_\beta(\mathbf{x}_2) | N_n \rangle \langle N_n | J_\alpha(\mathbf{x}_1) | N_i}{E_n + p_1^0 + \bar{q}_k^0 - E_i - i\epsilon} 2\pi\delta(E_f + p_1^0 + p_2^0 - E_i).$$

$$(9.30)$$

In Eqs. (9.29) and (9.30) we used the relations

$$\int_{-\infty}^{0} e^{i a x_2^0} dx_2^0 \rightarrow \int_{-\infty}^{0} e^{i(a - i\epsilon)x_2^0} dx_2^0 = \lim_{\epsilon \to 0} \frac{-i}{a - i\epsilon} \quad (9.31)$$

and

$$\int_{0}^{-\infty} e^{i a x_2^0} dx_2^0 \rightarrow \int_{0}^{\infty} e^{i(a + i\epsilon)x_2^0} dx_2^0 = \lim_{\epsilon \to 0} \frac{i}{a + i\epsilon}. \quad (9.32)$$

which are based on the standard assumption that interaction is turned off at $\pm\infty$.

Taking into account all these relations, for the matrix element of the neutrinoless double β-decay we find the following expression

$$\langle f | S^{(2)} | i \rangle = i \left(\frac{G_F \cos\theta_C}{\sqrt{2}} \right)^2$$

$$\times N_{p_1} N_{p_2} \bar{u}(p_1) \gamma^\alpha \gamma^\beta (1 + \gamma_5) C \bar{u}^T(p_2) \int d^3x_1 d^3x_1 e^{-i\mathbf{p}_1\mathbf{x}_1 - i\mathbf{p}_2\mathbf{x}_2}$$

$$\times \sum_k U_{ek}^2 m_k \frac{1}{(2\pi)^3} \int \frac{e^{i\mathbf{q}\,(\mathbf{x_1}-\mathbf{x_2})}}{\bar{q}_k^0} d^3q \Big[\sum_n \frac{\langle N_f|\, J_\alpha(\mathbf{x_1})|N_n\rangle\langle N_n|\, J_\beta(\mathbf{x_2})|N_i\rangle}{E_n + p_2^0 + \bar{q}_k^0 - E_i - i\epsilon}$$

$$+ \sum_n \frac{\langle N_f|\, J_\beta(\mathbf{x_2})|N_n\rangle\langle N_n|\, J_\alpha(\mathbf{x_1}))|N_i\rangle}{E_n + p_1^0 + \bar{q}_k^0 - E_i - i\epsilon} \Big] 2\pi\delta(E_f + p_1^0 + p_2^0 - E_i). \quad (9.33)$$

Equation (9.33) is the exact expression for the matrix element of the $0\nu\beta\beta$-decay in the second order of the perturbation theory. We will consider major $0^+ \to 0^+$ transitions of even-even nuclei. For these transitions the following approximations are standard ones.

1. Small neutrino masses can be safely neglected in the expression for the neutrino energy $\bar{q}_k^0 = \sqrt{q^2 + m_k^2}$, $q \equiv |\mathbf{q}|$. In fact, from the uncertainty relation for the average neutrino momentum we have $q \simeq 1/r$, where r is the average distance between two nucleons in a nucleus. Taking into account that $r \simeq 10^{-13}$ cm, we find that $q \simeq 100\,\mathrm{MeV}$. For the neutrino masses we have the upper bound $m_k \lesssim 1\,\mathrm{eV}$. Thus, we have $\bar{q}_k^0 \simeq q$.

2. Long-wave approximation. We have $|\mathbf{p}_k \cdot \mathbf{x}_k| \le |\mathbf{p}_k| R \simeq A^{1/3} \frac{p_k}{100\,\mathrm{MeV}}$ ($k = 1, 2$; $R \simeq 1.2 \cdot 10^{-13} A^{1/3}$ cm is the radius of the nucleus). Taking into account that $p_k \lesssim 1\,\mathrm{MeV}$, we obtain $e^{-i\mathbf{p}_k \cdot \mathbf{x}_k} \simeq 1$. Thus, in the $0\nu\beta\beta$-decay the two electrons are produced predominantly in the S-states.

3. Closure approximation.

   The energy of the virtual neutrino $q \simeq 100\,\mathrm{MeV}$ is much larger than the excitation energy of the intermediate states of a nucleus $(E_n - E_i)$. Taking this into account we can replace the energy of intermediate states E_n by average energy $\overline{E}$. In this approximation (which is called *the closure approximation*) we can perform the sum over the total system of intermediate states $|N_n\rangle$ in the matrix element (9.33).

   Let us consider energy denominators in (9.33). In the laboratory frame we have

$$\overline{E} + q + p_{1,2}^0 - M_i = \overline{E} + \Big(\frac{p_1^0 + p_2^0}{2}\Big) \pm \Big(\frac{p_1^0 - p_2^0}{2}\Big) + q - M_i \simeq q + \overline{E} - \frac{M_i + M_f}{2},$$
$$(9.34)$$

where we take into account that $\frac{|p_1^0 - p_2^0|}{2} \ll q$ and neglect nuclear recoil.

We have

$$\sum_n \frac{\langle N_f|\, J_\alpha(\mathbf{x_1})|N_n\rangle\langle N_n|\, J_\beta(\mathbf{x_2})|N_i\rangle}{E_n + p_2^0 + q_k^0 - E_i - i\epsilon} + \sum_n \frac{\langle N_f|\, J_\beta(\mathbf{x_2})|N_n\rangle\langle N_n|\, J_\alpha(\mathbf{x_1})|N_i\rangle}{E_n + p_1^0 + q_k^0 - E_i - i\epsilon}$$

$$= \frac{1}{\Big(q + \overline{E} - \frac{M_i + M_f}{2}\Big)} \langle N_f|\, \big(J_\alpha(\mathbf{x_1}) J_\beta(\mathbf{x_2}) + J_\beta(\mathbf{x_2}) J_\alpha(\mathbf{x_1})\big)\,|N_i\rangle. \quad (9.35)$$

4. The nonrelativistic impulse approximation for the hadronic charged current $J_\alpha(\mathbf{x})$:

$$J_\alpha(\mathbf{x}) = \sum_n \delta(\mathbf{x} - \mathbf{r}_n)\, \tau_+^n g_{\alpha\beta} J_n^\beta(q^2), \tag{9.36}$$

where $g_{\alpha\beta}$ is metric tensor ($g_{00} = 1,\, g_{ii} = -1,\, g_{\alpha\beta} = 0,\, \alpha \neq \beta$) and

$$J_n^0(q^2) = g_V(q^2), \quad \mathbf{J}_n(q^2) = g_A(q^2)\boldsymbol{\sigma}_n + ig_M(q^2)\frac{\boldsymbol{\sigma}_n \times \mathbf{q}}{2M} - g_P(q^2)\frac{\boldsymbol{\sigma}_n \cdot \mathbf{q}}{2M}\mathbf{q}. \tag{9.37}$$

Here $g_V(q^2)$, $g_A(q^2)$, $g_M(q^2)$ and $g_P(q^2)$ are CC vector, axial, magnetic and pseudoscalar formfactors of the nucleon, σ_i and τ_i are Pauli matrices, $\tau_+ = \frac{1}{2}(\tau_1 + i\tau_2)$ and index n runs over all nucleons in a nucleus. We have $g_V(0) = 1$, $g_A(0) = g_A \simeq 1.27$ and $g_M(0) = \mu_p - \mu_n$ (μ_p and μ_n are the anomalous magnetic moments of the proton and the neutron). From PCAC it follows that the pseudoscalar formfactor is given by the expression $g_P(q^2) = 2Mg_A/(q^2 + m_\pi^2)$.

The approximate matrix element of the $0\nu\beta\beta$-decay has the form

$$\langle f|S^{(2)}|i\rangle = im_{\beta\beta}\left(\frac{G_F \cos\vartheta_C}{\sqrt{2}}\right)^2 N_{p_1} N_{p_2}\bar{u}(p_1)\gamma^\alpha\gamma^\beta(1 + \gamma_5)C\bar{u}^T(p_2)$$

$$\times \int d^3x_1 d^3x_2 \frac{1}{(2\pi)^3}\int d^3q \frac{e^{i\mathbf{q}\cdot(\mathbf{x}_1 - \mathbf{x}_2)}}{q\left(q + \overline{E} - \frac{M_i + M_f}{2}\right)}\Big[\langle N_f|(J_\alpha(\mathbf{x}_1)J_\beta(\mathbf{x}_2)$$

$$+ J_\beta(\mathbf{x}_2)J_\alpha(\mathbf{x}_1))|N_i\rangle\Big]\, 2\pi\,\delta(M_f + p_1^0 + p_2^0 - M_i), \tag{9.38}$$

where

$$m_{\beta\beta} = \sum_k U_{ek}^2 m_k \tag{9.39}$$

is *the effective Majorana mass.*

Let us stress that the matrix element of the $0\nu\beta\beta$-decay is proportional to the effective Majorana neutrino mass. This is a general consequence of the neutrino mixing, of the smallness of neutrino mass with respect to the neutrino momentum and of the fact that fields of neutrinos with definite masses are left-handed fields.

It is obvious that $\tau_+^n \tau_+^n = 0$. Thus, in the impulse approximation we have

$$J_\alpha(\mathbf{x}_1)\, J_\beta(\mathbf{x}_2) = J_\beta(\mathbf{x}_2)\, J_\alpha(\mathbf{x}_1). \tag{9.40}$$

Further, the matrix $\gamma^\alpha \gamma^\beta$ in the leptonic part of the matrix element (9.38) can be presented in the form

$$\gamma^\alpha \gamma^\beta = g^{\alpha\beta} + \frac{1}{2}(\gamma^\alpha \gamma^\beta - \gamma^\beta \gamma^\alpha). \qquad (9.41)$$

It follows from (9.40) that the second term of this relation does not give contribution to the matrix element. Further from (9.36) we find

$$J_\alpha(\mathbf{x}_1) J_\beta(\mathbf{x}_2) g^{\alpha\beta} = \sum_{n,m} \delta(\mathbf{x}_1 - \mathbf{r}_n)\delta(\mathbf{x}_2 - \mathbf{r}_m)\tau_+^n \tau_+^m (J_n^0 J_m^0 - \mathbf{J_n} \cdot \mathbf{J_m}) \qquad (9.42)$$

Taking into account all these relations after the integration over $\mathbf{x}_1$ and $\mathbf{x}_2$ from (9.38) for the matrix element of the $0\nu\beta\beta$-decay we find the following expression

$$\langle f|S^2|i\rangle = im_{\beta\beta}\left(\frac{G_F \cos \vartheta_C}{\sqrt{2}}\right)^2 \frac{1}{(2\pi)^3 \sqrt{p_1^0 p_2^0}} \bar{u}(p_1)(1+\gamma_5)C\bar{u}^T(p_2)$$

$$\times \langle N_f|\left(\sum_{n,m} \frac{1}{(2\pi)^3} \int d^3q \, \frac{e^{i\mathbf{q}\cdot\mathbf{r}_{nm}}}{q(q+\overline{E} - \frac{M_i+M_f}{2})}\tau_+^n \tau_+^m (J_n^0 J_m^0 - \mathbf{J_n}\mathbf{J_m})\right)|N_i\rangle$$

$$\times 2\pi\delta(M_f + p_1^0 + p_2^0 - M_i), \qquad (9.43)$$

where $\mathbf{r}_{nm} = \mathbf{r}_n - \mathbf{r}_m$. We have $\mathbf{q}\cdot\mathbf{r}_{nm} = q\,r_{nm}\cos\theta$. After the integration over the angle θ we find

$$\frac{1}{(2\pi)^3}\int \frac{e^{i\mathbf{q}\cdot\mathbf{r}_{nm}}d^3q}{q(q+\overline{E}-\frac{1}{2}(M_i+M_f))} = \frac{1}{2\pi^2 r_{nm}}\int_0^\infty \frac{\sin(qr_{nm})\,dq}{q+\overline{E}-\frac{1}{2}(M_i+M_f)}. \qquad (9.44)$$

The matrix element of the $0\nu\beta\beta$-decay takes the form

$$\langle f|S^2|i = -im_{\beta\beta}\left(\frac{G_F \cos \vartheta_C}{\sqrt{2}}\right)^2 \frac{1}{2(2\pi)^6 \sqrt{p_1^0 p_2^0}} \frac{1}{R}\bar{u}(p_1)(1+\gamma_5)C\bar{u}^T(p_2)$$

$$\times M^{0\nu} \, \delta(p_1^0 + p_2^0 + M_f - M_i), \qquad (9.45)$$

where R is the radius of the nucleus and the nuclear matrix element $M^{0\nu}$ is given by the expression

$$M^{0\nu} = \langle \Psi_f| \sum_{n,m} H(r_{nm}, \overline{E})\tau_+^n \tau_+^m \, (\mathbf{J_n}\mathbf{J_m} - J_n^0 J_m^0)|\Psi_i\rangle. \qquad (9.46)$$

Here

$$H(r_{nm}, \overline{E}) = \frac{2R}{\pi \, r_{nm}} \int_0^\infty \frac{\sin(q \, r_{nm}) \, dq}{q + \overline{E} - \frac{1}{2}(M_i + M_f)}. \tag{9.47}$$

and $|\Psi_i\rangle$ and $|\Psi_f\rangle$ are wave functions of the initial and final nuclei.

The major contribution to the nuclear matrix element of the $0\nu\beta\beta$-decay give vector and axial terms in (9.37). Taking into account only these terms we have

$$M^{0\nu} = g_A^2 \left(M_{GT}^{0\nu} - \frac{1}{g_A^2} M_F^{0\nu} \right). \tag{9.48}$$

Here

$$M_{GT}^{0\nu} = \langle \Psi_f | \sum_{n,m} H(r_{nm}, \overline{E}) \, \tau_+^n \tau_+^m \, \sigma^n \sigma^m |\Psi_i\rangle \tag{9.49}$$

is the Gamov-Teller matrix element and

$$M_F^{0\nu} = \langle \Psi_f | \sum_{n,m} H(r_{nm}, \overline{E}) \, \tau_+^n \tau_+^m |\Psi_i\rangle \tag{9.50}$$

is the Fermi matrix element.

The function $H(r_{nm}, \overline{E})$ is called a neutrino potential. Taking into account that $q \gg \overline{E} - (M_i + M_f)/2$, for the neutrino potential we obtain the following approximate expression

$$H(r) \simeq \frac{2R}{\pi r} \int_0^\infty \frac{\sin(q \, r)}{q} \, dq = \frac{2R}{\pi r} \left(\frac{\pi}{2} \right) = \frac{R}{r}. \tag{9.51}$$

Let us calculate now the probability of the $0\nu\beta\beta$-decay. We have

$$\sum_{r_1, r_2} \left| \bar{u}^{r_1}(p_1)(1 + \gamma_5) C (\bar{u}^{r_2}(p_2))^T \right|^2$$

$$= \text{Tr} \left[(1 + \gamma_5)(\gamma \cdot p_2 - m_e)(1 - \gamma_5)(\gamma \cdot p_1 + m_e) \right]$$

$$= 8 \, p_1 \cdot p_2. \tag{9.52}$$

From Eqs. (9.45) and (9.52) for the decay rate of the $0\nu\beta\beta$-decay we find the following expression

$$d\Gamma^{0\nu} = |m_{\beta\beta}|^2 \, |M^{0\nu}|^2 \, \frac{(G_F \cos\theta_C)^4}{(2\pi)^5 R^2} \, (E_1 E_2 - |\mathbf{p}_1||\mathbf{p}_2| \cos\theta)$$

$$\times F(E_1, (Z+2)) F(E_2, (Z+2)) |\mathbf{p}_1||\mathbf{p}_2| \sin\theta \, d\theta \, dE_1. \tag{9.53}$$

Here $E_{1,2} = p^0_{1,2}$ are the energies of the emitted electrons ($E_2 = M_i - M_f - E_1$) and θ is the angle between the electron momenta $\mathbf{p}_1$ and $\mathbf{p}_2$. The function $F(E, Z)$ describes final state electromagnetic interaction of the electron and the nucleus. For a point-like nucleus it is given by the Fermi function

$$F(E, Z) = \frac{2\pi\eta}{1 - e^{-2\pi\eta}}, \quad \eta = Z\alpha \frac{m_e}{p}. \tag{9.54}$$

From (9.53) follows that for the ultra relativistic electrons θ-dependence of the decay rate is given by the factor $(1 - \cos\theta)$. Thus, ultra relativistic electrons can not be emitted in the same direction. This is connected with the fact the high-energy electrons produced due to CC weak interaction have negative helicity. If the two electrons are emitted in the same direction, the projection of their total angular momentum on the direction of the momentum is equal to -1. In $0^+ \to 0^+$ nuclear transitions this configuration is forbidden by angular momentum conservation.

From Eq. (9.53) for the half-life of the neutrinoless double β-decay $(A, Z) \to (A, Z + 2) + e^- + e^-$ we find the following expression

$$(T^{0\nu}_{1/2})^{-1} = \frac{\Gamma^{0\nu}}{\ln 2} = |m_{\beta\beta}|^2 |M^{0\nu}|^2 G^{0\nu}(Q, Z), \tag{9.55}$$

where the phase-space factor is given by[3]

$$G^{0\nu}(Q, Z) = \frac{1}{2\ln 2(2\pi)^5} (G_F \cos\theta_C)^4$$

$$\times \frac{1}{R^2} \int_0^Q dT_1 \int_0^\pi \sin\theta d\theta \, |\mathbf{p}_1||\mathbf{p}_1| \, (E_1 E_2 - |\mathbf{p}_1||\mathbf{p}_2| \cos\theta)$$

$$\times F(E_1, (Z + 2)) \, F(E_2, (Z + 2)). \tag{9.56}$$

Here $T_1 = E_1 - m_e$ is the kinetic energy the electron and $Q = M_i - M_f - 2m_e$ is the total released kinetic energy.

Thus for small Majorana neutrino masses *the total rate of the $0\nu\beta\beta$-decay is the product of three factors*:

1. The modulus squared of the effective Majorana mass.
2. Square of nuclear matrix element.
3. The known factor $G^{0\nu}(Q, Z)$.

The values of the factor $G^{0\nu}(Q, Z)$ for $0\nu\beta\beta$-decays, which are searching for in different experiments, are presented in Table 9.2. The problem of calculation of nuclear matrix elements we will discuss later in this chapter. The effective Majorana mass is given by the relation $m_{\beta\beta} = \sum_k U^2_{ek} m_k$. Neutrino mixing angles and

---

[3] An additional factor 1/2 is due to the fact that in the final state there are two identical electrons.

Table 9.2 The phase factor $G^{0\nu}(Q, Z)$ for some $0\nu\beta\beta$ transitions

$0\nu\beta\beta$ transition	$G^{0\nu}(Q, Z)(10^{-26}$ years^{-1} eV$^{-2})$
^{76}Ge $\rightarrow$ ^{76}Se	0.9049
^{100}Mo $\rightarrow$ ^{100}Ru	6.097
^{130}Te $\rightarrow$ ^{130}Xe	5.446
^{136}Xe $\rightarrow$ ^{136}Ba	5.584
^{150}Nd $\rightarrow$ ^{150}Sm	24.14
^{82}Se $\rightarrow$ ^{82}Kr	3.891
^{48}Ca $\rightarrow$ ^{48}Ti	9.501

neutrino mass-squared differences were determined from the data of the neutrino oscillation experiments. Taking into account existing data, we will consider now expected values of the effective Majorana mass.

9.3   Effective Majorana Mass

In order to determine the effective Majorana mass

$$|m_{\beta\beta}| = |\sum_{k=1}^{3} U_{ek}^2 m_k| \tag{9.57}$$

we need to know absolute values of neutrino masses and elements of the Majorana neutrino mixing matrix which in the standard parametrization have the form

$$U_{e1} = \cos\theta_{13}\cos\theta_{12}e^{i\bar{\alpha}_1}, \ U_{e2} = \cos\theta_{13}\sin\theta_{12}e^{i\bar{\alpha}_2}, \ U_{e3} = \sin\theta_{13}e^{i\bar{\alpha}_3} \ (\bar{\alpha}_3 \equiv -\delta), \tag{9.58}$$

where $\bar{\alpha}_i$ are Majorana phases.

From neutrino oscillation data we know the values of the neutrino mixing angles and two mass-squared differences (solar Δm_S^2 and atmospheric Δm_A^2). The value of the smallest neutrino mass m_{min} and Majorana phases are unknown parameters. From the data analysis it was established that the solar mass-squared difference is much smaller than the atmospheric one:

$$\frac{\Delta m_S^2}{\Delta m_A^2} \simeq \frac{1}{30}. \tag{9.59}$$

For three massive neutrinos two types of neutrino mass spectra are possible. Usually, neutrino masses are labeled in such a way that $\Delta m_{12}^2 = m_2^2 - m_1^2 > 0$ and $\Delta m_{12}^2 = \Delta m_S^2$. For the third mass m_3 there are two possibilities. Correspondingly, there are two possible neutrino mass spectra:

1. Normal ordering (NO)

$$m_1 < m_2 < m_3, \quad \Delta m_{23}^2 = \Delta m_A^2. \tag{9.60}$$

For the neutrino masses we have in this case

$$m_1 = m_{\min}, \ m_2 = \sqrt{\Delta m_S^2 + m_{\min}^2}, \ m_3 = \sqrt{\Delta m_A^2 + \Delta m_S^2 + m_{\min}^2}.$$

2. Inverted ordering (IO)

$$m_3 < m_1 < m_2, \quad |\Delta m_{13}^2| = \Delta m_A^2 \tag{9.61}$$

For the neutrino masses in the IO case we have

$$m_3 = m_{\min}, \ m_1 = \sqrt{\Delta m_A^2 + m_{\min}^2}, \ m_2 = \sqrt{\Delta m_A^2 + \Delta m_S^2 + m_{\min}^2}.$$

Three neutrino mass spectra are of a special interest.

9.3.1 Hierarchy of the Neutrino Masses

$$m_1 \ll m_2 \ll m_3. \tag{9.62}$$

In this case we have

$$m_1 \ll \sqrt{\Delta m_S^2} \simeq 8.7 \cdot 10^{-3} \ \text{eV}^2, \quad m_2 \simeq \sqrt{\Delta m_S^2}, \quad m_3 \simeq \sqrt{\Delta m_A^2}. \tag{9.63}$$

Neglecting the contribution of m_1 and using the standard parametrization of the neutrino mixing matrix we find

$$|m_{\beta\beta}| \simeq \left| \cos^2 \theta_{13} \sin^2 \theta_{12} \sqrt{\Delta m_S^2} + e^{2i\alpha} \sin^2 \theta_{13} \sqrt{\Delta m_A^2} \right|. \tag{9.64}$$

Here $\alpha = \bar{\alpha}_3 - \bar{\alpha}_2$ is (unknown) Majorana phase difference.

The first term in Eq. (9.64) is small because of the smallness of Δm_S^2. The contribution of the "large" Δm_A^2 to $|m_{\beta\beta}|$ is suppressed by the small factor $\sin^2 \theta_{13}$. Using the best fit values of the parameters we have

$$\cos^2 \vartheta_{13} \sin^2 \vartheta_{12} \sqrt{\Delta m_S^2} \simeq 3 \cdot 10^{-3} \ \text{eV}, \quad \sin^2 \vartheta_{13} \sqrt{\Delta m_A^2} \simeq 1 \cdot 10^{-3} \ \text{eV}. \tag{9.65}$$

From (9.64) and (9.65) we find the following expected range for the effective Majorana mass in the case of the normal hierarchy of the neutrino masses

$$2 \times 10^{-3} \ \text{eV} \lesssim |m_{\beta\beta}| \lesssim 4 \times 10^{-3} \ \text{eV}. \tag{9.66}$$

Thus, expected values of the effective Majorana mass $|m_{\beta\beta}|$ in the case of the normal hierarchy is significantly smaller than the expected sensitivity of the future experiments on the search for $0\nu\beta\beta$-decay (see later). Observation of the $0\nu\beta\beta$-decay in this case will be a real challenge.

9.3.2 Inverted Hierarchy of the Neutrino Masses

For the neutrino masses we have in this case

$$m_3 \ll \sqrt{\Delta m_A^2}, \quad m_1 \simeq \sqrt{\Delta m_A^2}, \quad m_2 \simeq \sqrt{\Delta m_A^2}\left(1 + \frac{\Delta m_S^2}{2\,\Delta m_A^2}\right). \tag{9.67}$$

In the expression for the effective Majorana mass $|m_{\beta\beta}|$ the lightest mass m_3 is multiplied by the small parameter $\sin^2\theta_{13}$. Neglecting the contribution of this term and also neglecting the small term $\frac{\Delta m_S^2}{2\Delta m_A^2}$ in (9.67) we find

$$|m_{\beta\beta}| \simeq \cos^2\theta_{13}\sqrt{\Delta m_A^2}\,(1 - \sin^2 2\theta_{12}\sin^2\alpha)^{\frac{1}{2}}, \tag{9.68}$$

where $\alpha = \bar{\alpha}_2 - \bar{\alpha}_1$ is the only unknown parameter in this expression.

From (9.68) we find

$$\cos^2\theta_{13}\cos 2\theta_{12}\sqrt{\Delta m_A^2} \leq |m_{\beta\beta}| \leq \cos^2\theta_{13}\sqrt{\Delta m_A^2}, \tag{9.69}$$

where upper and lower bounds corresponds, respectively, to equal and opposite CP parities of ν_1 and ν_2 in the case of CP conservation in the lepton sector.[4] Using the values of the parameters in Eq. (9.69) we have

$$2 \times 10^{-2} \lesssim |m_{\beta\beta}| \lesssim 5 \times 10^{-2}\,\text{eV}. \tag{9.70}$$

The anticipated sensitivities to the effective Majorana mass of the next generation of the experiments on the search for the $0\nu\beta\beta$-decay are in the range (9.70). Thus, the future $0\nu\beta\beta$-decay experiments will probe the inverted hierarchy of the neutrino masses.

---

[4] In fact, from CP invariance follows that $U_{ei} = U_{ei}^*\eta_i$, where $\eta_i = \pm i$ is the CP parity of the Majorana neutrino with mass m_i. From this condition we find $e^{2i\bar{\alpha}_i} = \eta_i$. Thus, we have $e^{2i(\bar{\alpha}_2-\bar{\alpha}_1)} = e^{2i\alpha} = \eta_2\eta_1^*$. If $\eta_2 = \eta_1$ we have $\alpha = 0, \pi$ (the upper bound in Eq. (9.69)), and if $\eta_2 = -\eta_1$ we have $\alpha = \pm\pi/2$ (the lower bound in Eq. (9.69)).

9.3.3   Quasi-Degenerate Neutrino Mass Spectrum

If $m_{min} \gg \sqrt{\Delta m_A^2} \simeq 5 \cdot 10^{-2}$ eV the spectrum of neutrino masses is quasi-degenerate

$$m_1 \simeq m_2 \simeq m_3 \simeq m_{min}. \tag{9.71}$$

For the Majorana neutrino mass we have in this case

$$|m_{\beta\beta}| \simeq a\, m_{min}, \tag{9.72}$$

where $a = |\sum_{i=1}^{3} U_{ei}^2|$. Neglecting the contribution of the small angle θ_{13} we have

$$a = (1 - \sin^2 2\theta_{12} \sin^2 \alpha)^{\frac{1}{2}}, \quad \cos 2\theta_{12} \leq a \leq 1. \tag{9.73}$$

Notice that in the case of the quasi-degenerate neutrino mass spectrum the value of m_{min} can be determined from experiments on the measurement of the effective neutrino mass m_β via investigation of the end-point part of the tritium β-spectrum (see the next chapter). We have

$$m_{min} = m_\beta. \tag{9.74}$$

We have considered three neutrino mass spectra with the values of m_{min} of a special interest (very small or relatively large). In Fig.9.2 the effective Majorana mass for the case of normal and inverted neutrino mass ordering as a function of m_{min} is presented. Uncertainties of the parameters Δm_S^2, Δm_A^2 and $\sin^2 \theta_{12}$ and possible values of the Majorana phase differences are taken into account in Fig.9.2. Notice that for the normal neutrino mass ordering in the region $(2 \cdot 10^{-3} \lesssim m_{min} \lesssim 7 \cdot \times 10^{-3})$ eV the Majorana neutrino mass is very small.

9.4   On the Nuclear Matrix Elements of the $0\nu\beta\beta$-Decay

The effective Majorana mass $|m_{\beta\beta}|$ is not a directly measurable quantity. From the measurement of the half-life of the $0\nu\beta\beta$-decay only *the product of the effective Majorana mass and nuclear matrix element* can be obtained (see relation (9.55)). If the $0\nu\beta\beta$-decay will be observed, it will be proved that neutrinos with definite masses are Majorana particles and total lepton number is violated. However, in order to determine very important quantity $m_{\beta\beta}$ we need to know nuclear matrix elements (NME) which must be calculated.

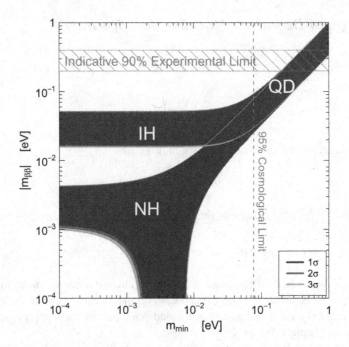

Fig. 9.2 Effective Majorana mass for the normal, inverted and quasi-degenerate neutrino mass spectra as a function of m_{min} (arXiv:1411.4791)

The calculation of NME is a complicated nuclear problem. Five different many-body approximate methods have been used for the calculation of NME of neutrinoless double-β decays of different even-even nuclei:

- Nuclear Shell Model (NSM).
- Quasi-Particle Random Phase Approximation (QRPA).
- Interacting Boson Model (IBM).
- Energy Density Functional Method (EDF).
- Projected Hartree-Fock-Bogoliubov Method (PHFB).

The consideration of these many-body methods is out of the scope of this book. In Table 9.3 we present ranges of calculated values of NME of the $0\nu\beta\beta$ decays of some nuclei of experimental interest. In the third column of Table 9.3 ratios of maximal and minimal values of the nuclear matrix elements and in the fourth column ranges of expected halve-lives of the $0\nu\beta\beta$ decays under the assumption that $|m_{\beta\beta}| = 0.1$ eV are presented.

As it is seen from Table 9.3 there is about one order of magnitude difference between different model calculations of halve-lives of the $0\nu\beta\beta$-decay. Thus, the present situation with the calculation of NMEs is far from satisfactory. Further progress in mandatory. Taking into account a complexity of the problem of reliable

Table 9.3 Calculated values of $|M^{0\nu}|$, ratio $\frac{|M^{0\nu}|_{max}}{|M^{0\nu}|_{min}}$ and half-lives of the $0\nu\beta\beta$-decay of some nuclei of experimental interest, calculated under the assumption that $|m_{\beta\beta}| = 0.1$ eV

$\beta\beta$ transition	$|M^{0\nu}|$	$\frac{|M^{0\nu}|_{max}}{|M^{0\nu}|_{min}}$	$T_{1/2}^{0\nu} (m_{\beta\beta}=0.1\,\text{eV})$ [10^{26} year]
^{76}Ge $\rightarrow$ ^{76}Se	3.59–10.39	2.9	1.0–8.6
^{100}Mo $\rightarrow$ ^{100}Ru	4.39–12.13	2.8	0.1–0.8
^{130}Te $\rightarrow$ ^{130}Xe	2.06–8.00	3.9	0.3–4.3
^{136}Xe $\rightarrow$ ^{136}Ba	1.85–6.38	3.4	0.4–5.2
^{150}Nd $\rightarrow$ ^{150}Sm	1.48–5.80	3.9	0.1–1.9
^{82}Se $\rightarrow$ ^{82}Kr	3.41–8.84	2.6	0.3–2.2
^{48}Ca $\rightarrow$ ^{48}Ti	0.89–4.14	4.6	0.6–13.3

treatment of many-body nuclear system it will be also important to find a way to check different calculations of NMEs of the $0\nu\beta\beta$-decay.

Let us notice that NME for ^{76}Ge and heavier nuclei, calculated in the framework of the Nuclear Shell Model, practically do not depend on a nucleus. It follows from (9.55) that in this case the ratio of halve-lives of different nuclei is determined by the ratio of known phase space factors:

$$\frac{T_{1/2}^{0\nu}(A, Z)}{T_{1/2}^{0\nu}(A', Z')} = \frac{G^{0\nu}(Q', Z')}{G^{0\nu}(Q, Z)}. \tag{9.75}$$

Thus, observation of the $0\nu\beta\beta$-decay of two and more nuclei would allow to check NSM. If the relation (9.75) would be in agreement with experiment this would allow to determine $|m_{\beta\beta}|$ and to check the three-neutrino mechanism of the $0\nu\beta\beta$-decay.

9.5 Experiments on the Search for $0\nu\beta\beta$-Decay

If neutrinos with definite masses are Majorana particles, neutrinoless double β-decay of some even-even nuclei is allowed but, as we discussed before, the probability of the decay is extremely small. Experiments on the observation of this process are very challenging.[5] Main signature of the process is a monochromatic peak in distribution of the sum of energies of two electrons (which in the case of the $0\nu\beta\beta$-decay is equal to the energy release in the nuclear transition Q). There are, however, different sources of background such as $2\nu\beta\beta$-decay, natural radioactivity, cosmic rays etc.

---

[5]If $|m_{\beta\beta}| \simeq \sqrt{\Delta m_A^2} \simeq 5 \cdot 10^{-2}$ eV (inverted hierarchy) in one ton of isotopically enriched detector about one $0\nu\beta\beta$-decay event per year is expected.

In experiments on the search for the $0\nu\beta\beta$-decay

- very high energy resolution is required;
- background level must be very low;
- high detection efficiency must be reached;
- large mass of isotopes, for which $0\nu\beta\beta$-decay is allowed, must be used.

In order to reach low background, $0\nu\beta\beta$ experiments must be performed in underground laboratories, low-radioactivity materials for detectors have to be used and effective shielding against external radioactivity must be provided.

At present there exist data of many experiments on the search for $0\nu\beta\beta$-decay of different nuclei. Up to now no evidence for the $0\nu\beta\beta$-decay was found. We will briefly discuss only recent experiments in which the best sensitivity was reached.

1. In the **GERDA** experiment (Gran Sasso underground laboratory) the $0\nu\beta\beta$-decay of ^{76}Ge (^{76}Ge $\to$ ^{76}Se $+ e^- + e^-$) is searched for. Germanium detectors with ^{76}Ge fraction, enriched from 7.8% (natural abundance) to 87%, are employed in the experiment. Detectors operate in radio-pure liquid argon LAr used for cooling and for shielding against external background.

    In the Phase-I of the experiment the lower bound $T_{1/2}^{0\nu} > 2.1 \cdot 10^{25}$ year (90%CL) was obtained with an exposure 21.6 kg year and background $B = 0.01 \frac{\text{counts}}{\text{keV kg year}}$. Ten germanium detectors with total mass 17.6 kg were deployed in the Phase I.

    In the Phase-II, started in December 2015, 37 enriched germanium detectors with total mass 35.6 kg were used and the level of the background $(0.7^{+1.1}_{-0.5}) \cdot 10^{-3} \frac{\text{counts}}{\text{counts}}$ keV kg year at $Q = 2039 \pm 0.007$ keV was reached. No $0\nu\beta\beta$ signal was observed. Combing data of Phase I and Phase II it was found that $T_{1/2}^{0\nu} > 5.3 \cdot 10^{25}$ year (90%CL). Taking into account existing uncertainties of NME for the effective Majorana mass the following upper bound $|m_{\beta\beta}| < (0.15–0.33)$ eV was obtained.

2. The $0\nu\beta\beta$-decay of ^{136}Xe (^{136}Xe $\to$ ^{136}Ba $+ e^- + e^-$) is searched for in the **KamLAND-Zen** experiment (Kamioka underground Observatory). In this experiment a balloon filled with 13 ton Xe-loaded liquid scintillator is imbedded in the center of the KamLAND detector (1 kton of liquid scintillator). In the Phase-I of the experiment 320 kg of enriched xenon gas (90.6% of ^{136}Xe) was dissolved in the liquid scintillator. After 89.5 kg year of exposure it was found the following bound $T_{1/2}^{0\nu} > 1.9 \cdot 10^{25}$ year (90%CL).

    In the Phase-II of the experiment after purification of the Xe-loaded liquid scintillator the significant reduction of the background was achieved. Combining Phase-I and Phase-II results, the limit $T_{1/2}^{0\nu} > 1.07 \cdot 10^{26}$ year (90%CL) was obtained. From this limit for the effective Majorana mass it was found $|m_{\beta\beta}| < (0.06–0.16)$ eV.

3. In the **EXO-200** experiment (WIIP underground site in New Mexico) the $0\nu\beta\beta$-decay of ^{136}Xe is investigated. In this experiment Time Projection Chambers are filled with 200 kg of enriched liquid xenon (80% of ^{136}Xe). The background level

$B = (1.7 \pm 0.2) \cdot 10^{-3} \frac{\text{counts}}{\text{keV kg year}}$ and energy resolution 3.5% at $Q = 2.458$ MeV are reached. The lower bound $T^{0\nu}_{1/2} > 1.1 \cdot 10^{25}$ year (90%CL) was obtained in the EXO-200 experiment. From this bound for the effective Majorana mass the following upper bound $|m_{\beta\beta}| < (0.19\text{--}0.45)$ eV was inferred.

4. In the cryogenic **CUORE-0** experiment (Gran Sasso underground laboratory) the neutrinoless double β-decay of ^{130}Te (^{130}Te $\rightarrow$ ^{130}Xe $+ e^- + e^-$) was searched for. In this experiment TeO$_2$ crystals are arranged in a tower (total mass 39 kg and ^{130}Te mass, due to natural abundance, is 10.9 kg). The energy resolution (5.1 ± 0.3) keV and the background $B = 0.058 \frac{\text{counts}}{\text{counts}}$ keV kg year were reached in the CUORE-0 experiment. After 9.8 kg year of exposure the lower bound $T^{0\nu}_{1/2} > 2.7 \cdot 10^{24}$ year (90%CL) was found. Combining data of the CUORE-0 experiment and of the previous experiment CUORECINO it was obtained $T^{0\nu}_{1/2} > 4.0 \cdot 10^{24}$ year (90%CL). From this result the following upper bound $|m_{\beta\beta}| < (0.27\text{--}0.76)$ eV was inferred.

5. In the **NEMO-3** experiment (Frejus Underground Laboratory) the $0\nu\beta\beta$-decay of the different nuclei (^{100}Mo, ^{82}Se, ^{96}Zr, ^{48}Ca, ^{150}Nd) was investigated. Electrons in this experiment were identified by the curvature in the magnetic field and their energy was measured in the calorimeter. The most stringent bound was obtained for ^{100}Mo (7 kg of enriched Mo was utilized). For half-live it was found the following lower bound $T^{0\nu}_{1/2} > 1.1 \cdot 10^{24}$ year (90%CL). From this result it follows that $|m_{\beta\beta}| < (0.3\text{--}0.9)$ eV.

Future experiments on the search for neutrinoless double β-decay are aimed to probe inverted hierarchy region ($|m_{\beta\beta}| \simeq$ (a few) 10^{-2} eV).

The GERDA cryostat will host a 200 kg germanium detector. The sensitivity of this detector will reach $T^{0\nu}_{1/2} \simeq 1 \cdot 10^{27}$ year. The 200 kg experiment could be a first step for a 1 ton experiment with sensitivity $T^{0\nu}_{1/2} \simeq 1 \cdot 10^{28}$ year which corresponds to $|m_{\beta\beta}| \simeq (1\text{--}2) \cdot 10^{-2}$ eV.

The KamLAND-Zen collaboration plan to increase the volume of Xe-loaded liquid scintillator which will be loaded with 800 kg of enriched Xe. This will allow to probe $|m_{\beta\beta}| \simeq 5 \cdot 10^{-2}$ eV.

The next step of the EXO experiment will the nEXO experiment with 5 ton of enriched Xe and sensitivity $T^{0\nu}_{1/2} \simeq 6 \cdot 10^{27}$ year. After 5 years of running the collaboration plan to reach $|m_{\beta\beta}| \simeq (1.5\text{--}2.5) \cdot 10^{-2}$ eV.

In the cryogenic CUORE experiment (741 kg of TeO$_2$ with 206 kg of $\beta\beta$ nuclei ^{130}Te) the sensitivity $T^{0\nu}_{1/2} \simeq 9.5 \cdot 10^{25}$ year is planned to be reached. Such sensitivity correspond to $|m_{\beta\beta}| \simeq (5\text{--}15) \cdot 10^{-2}$ eV.

Several other high-sensitivity experiments on the search for the $0\nu\beta\beta$-decay are in preparation: Majorana (^{76}Ge), SNO+ (^{130}Te), COBRA (^{116}Cd), SuperNEMO (^{82}Se), CANDELS (^{48}Ca), NEXT (^{136}Xe) and others.

Chapter 10
On Absolute Values of Neutrino Masses

10.1 Masses of Muon and Tau Neutrinos

We have seen in the previous sections that neutrino oscillation experiments allow us to obtain the values of the neutrino mass-squared differences. Information about the absolute values of the neutrino masses can be inferred from experiments on the precise measurement of the kinematics of decays of different particles, from cosmological data and from experiments on the measurement of the neutrinoless double β-decay.

In this section we will briefly discuss an information on "masses" of muon and tau neutrinos.[1]

The most precise upper bound on the mass of the muon neutrino was obtained from the measurement of the muon momentum in the decay

$$\pi^+ \rightarrow \mu^+ + \nu_\mu. \tag{10.1}$$

From the energy-momentum conservation for the mass of neutrino, produced in the decay $\pi^+ \rightarrow \mu^+ \nu_\mu$, we find the following expression

$$m_{\nu_\mu}^2 = m_\pi^2 + m_\mu^2 - 2m_\pi\sqrt{m_\mu^2 + (p_\mu)^2}, \tag{10.2}$$

where m_π and m_μ are masses of the pion and muon and p_μ is the momentum of the muon (in the pion rest frame).

---

[1] Let us notice that in the case of the neutrino mixing the states of the flavor neutrinos ν_e, ν_μ, ν_τ are not states with definite masses. However, if the spectrum of the neutrino masses is degenerate ($m_i \gg \sqrt{\Delta m_A^2} \simeq 2.5 \cdot 10^{-2}$ eV^2) only common (minimal) neutrino mass $m_{\min}$ can be determined from experiments on the investigation of different decays.

© Springer International Publishing AG, part of Springer Nature 2018
S. Bilenky, *Introduction to the Physics of Massive and Mixed Neutrinos*,
Lecture Notes in Physics 947, https://doi.org/10.1007/978-3-319-74802-3_10

In the most precise PSI experiment for the muon momentum the value

$$p_\mu = (29.79200 \pm 0.00011) \text{ MeV} \tag{10.3}$$

was found. From (10.2) and (10.3) it follows that

$$m_{\nu_\mu}^2 = (-0.016 \pm 0.023) \text{ MeV}. \tag{10.4}$$

This result implies the following upper bound for the muon neutrino mass

$$m_{\nu_\mu} < 190 \text{ keV}. \tag{10.5}$$

The upper bound for the mass of the tau neutrino was obtained from a study of the decays

$$\tau^- \to 2\pi^- + \pi^+ + \nu_\tau, \quad \tau^- \to 3\pi^- + 2\pi^+ + (\pi^0) + \nu_\tau \tag{10.6}$$

in the ALEPH experiment (CERN). From this experiment was found the bound

$$m_{\nu_\tau} < 18.2 \text{ MeV}. \tag{10.7}$$

10.2 Effective Neutrino Mass from the Measurement of the High-Energy Part of the β-Spectrum of Tritium

The most stringent upper bound on the absolute value of the neutrino mass was obtained from the precise measurement of the high-energy part of the β-spectrum of tritium

$$^3H \to\, ^3He + e^- + \bar{\nu}_e. \tag{10.8}$$

The effective Hamiltonian of the β-decay is given by the expression

$$\mathcal{H}_I = \frac{G_F}{\sqrt{2}} 2\bar{e}_L \gamma_\alpha \nu_{eL} \, j^\alpha + \text{h.c.}, \tag{10.9}$$

where j^α is the hadronic charged current and

$$\nu_{eL} = \sum_i U_{li} \nu_{iL} \tag{10.10}$$

is the mixed field of the electron neutrino.

The β-spectrum of the decay (10.8) is given by the following expression

$$\frac{d\Gamma}{dE} = \sum_i |U_{ei}|^2 \frac{d\Gamma_i}{dE}, \tag{10.11}$$

where

$$\frac{d\Gamma_i}{dE} = Cp(E+m_e)(E_0-E)\sqrt{(E_0-E)^2-m_i^2}\, F(E)\,\theta(E_0-E-m_i). \tag{10.12}$$

Here p and E are the momentum and kinetic energy of the electron, $E_0 \simeq 18.6\,\text{keV}$ is the energy release and $F(E)$ is the Fermi function, which describes the Coulomb interaction of the final electron and ^3He nucleus. The constant C is given by the expression

$$C = \frac{G_F^2 m_e^5}{2\pi^3}\cos^2\theta_C |M|^2, \tag{10.13}$$

where m_e is the electron mass, θ_C is the Cabibbo angle and M is the nuclear matrix element. Let us stress that the nuclear matrix element is a constant and the shape of the electron spectrum of allowed tritium decay (10.8) is determined by the phase space factor. The neutrino mass m_i enters in (10.12) only through the neutrino momentum $p_i = \sqrt{(E_0-E)^2-m_i^2}$.

As is seen from (10.12), the largest distortion of the electron spectrum can be observed in the region $E_{\max} - E \simeq m_i$, where $E_{\max} = E_0 - m_i$ is the maximal electron energy. However, for $m_i \simeq 1\,\text{eV}$ only a very small fraction (about $2\cdot 10^{-13}$) of the tritium decays give a contribution to this region. In order to increase the luminosity of the tritium experiments a much larger part of the β-spectrum is used for the analysis of the effect of the neutrino mass. Taking into account experimental conditions we have

$$\sum_i \sqrt{(E_0-E)^2-m_i^2}\,|U_{ei}|^2 \simeq (E_0-E)\left(1+\frac{m_\beta^2}{2(E_0-E)^2}\right) \simeq \sqrt{(E_0-E)^2-m_\beta^2}, \tag{10.14}$$

where

$$m_\beta = \sqrt{\sum_i |U_{ei}|^2 m_i^2}. \tag{10.15}$$

is the effective neutrino mass. The tritium electron spectrum takes the form

$$\frac{d\Gamma}{dE} \simeq Cp(E+m_e)(E_0-E)\sqrt{(E_0-E)^2-m_\beta^2}\, F(E). \tag{10.16}$$

In order to measure the distortion of electron spectrum due to small neutrino mass any experiment must have

- a high energy resolution,
- a well-known spectrometer resolution,
- intense tritium source and ability to detect a large number of decays.

The best upper bounds on the effective neutrino mass m_β were obtained in the Mainz and Troitsk tritium experiments. In the **Mainz** experiment, frozen molecular tritium condensed on the graphite substrate was used as a tritium source. The electron spectrum was measured by an integral spectrometer with a retarding electrostatic filter which combine high luminosity with high resolution (4.8 eV). In analysis of the experimental data four free parameters were used: the normalization C, the background B, the released energy E_0 and the effective neutrino mass-squared m_β^2. From the fit of the data it was found that $E_0 = 18.575$ eV.

For the determination of the effective neutrino mass the last 70 eV of the spectrum was used in the experiment. From the combined analysis of all data it was found

$$m_\beta^2 = (-1.2 \pm 2.2 \pm 2.1) \, \text{eV}^2. \tag{10.17}$$

From (10.17) the following upper bound for the effective neutrino mass was obtained

$$m_\beta < 2.3 \, \text{eV} \quad (95\% \, CL). \tag{10.18}$$

In the **Troitsk** tritium experiment a windowless gaseous molecular source was used. The electron spectrum was measured by an integral electrostatic spectrometer of the same type as in the Mainz experiment. The resolution of the spectrometer was (3–4) eV.

In the fit of the Troitsk data the same four free parameters, as in the Mainz experiment, were used. From the analysis of the data, for the parameter m_β^2 it was found

$$m_\beta^2 = (0.67 \pm 2.53) \, \text{eV}^2. \tag{10.19}$$

From (10.19) the following upper bound for the effective neutrino mass was obtained

$$m_\beta < 2.12 \, \text{eV (Bayesian statistics)}, \quad m_\beta < 2.05 \, \text{eV (Feldman–Cousins)}. \tag{10.20}$$

The experiment of the next generation on the measurement of the neutrino mass will be the Karlsruhe Tritium Neutrino Experiment (**KATRIN**). In this experiment two tritium sources will be used: a gaseous T_2 source, as in the Troitsk experiment, and a frozen tritium source, as in the Mainz experiment. The integral spectrometer with a retarding electrostatic filter will have two parts: the pre-spectrometer, which will select electrons in the last $\sim 100\,\text{eV}$ part of the spectrum, and the large main spectrometer. It will have high luminosity, low background and high energy resolution (1 eV).

It is anticipated that after 5 years of running a sensitivity to the effective neutrino mass 0.2 eV at 95% CL will be reached and the mass $m_\beta = 0.35\,\text{eV}$ ($m_\beta = 0.30\,\text{eV}$) can be measured with $5\,\sigma$ ($3\,\sigma$) significance.

An ambitious future **Project 8** tritium experiment will be based on a new technology of the measurement of the endpoint tritium electron spectrum by the Cyclotron Radiation Emission Spectroscopy technique. In this experiment gaseous *atomic* tritium will be stored and decay in an uniform magnetic field. Magnetically-trapped electrons execute cyclotron motion and produce microwave radiation. The energy of the electron determines the frequency of the radiation. The precise measurement of the frequency allows to obtain excellent energy resolution. At the kinetic energy of electron 18.6 keV (tritium end-point energy) the energy resolution 1 eV can be achieved if electron is trapped and observed during several microseconds. At the final phase of the Project 8 experiment the sensitivity $m_\beta \simeq 4 \cdot 10^{-2}$ eV is planned to be achieved. In case of the inverted hierarchy of neutrino masses $m_\beta \simeq \sqrt{\Delta m_A^2} \simeq 5 \cdot 10^{-2}$ eV. Thus the Project 8 experiment will probe IH neutrino mass spectrum.

Chapter 11
Neutrino Oscillation Experiments

11.1 Introduction

A long period of the searching for neutrino oscillations started in 1970 with the Homestake solar neutrino radiochemical experiment by Davis et al. In this experiment, the observed rate of solar ν_e was found to be two to three times smaller that the rate, predicted by the Standard Solar Model (SSM). This discrepancy was called *the solar neutrino problem*.

Before the Homestake experiment started, B. Pontecorvo suggested that because of neutrino oscillations the observed flux of the solar neutrinos might be two times smaller than the predicted flux.[1] After the Davis results were obtained the idea of neutrino oscillations as a possible reason for the solar neutrino deficit became more and more popular.

In the eighties, the second solar neutrino experiment Kamiokande was performed. In this direct-counting experiment a large water-Cherenkov detector was used. The solar neutrino rate measured by the Kamiokande experiment was also smaller than the rate predicted by the SSM.

In the Homestake and Kamiokande experiments high-energy solar neutrinos, produced mainly in the decay of ^{8}B, were detected. The flux of these neutrinos is about 10^{-4} of the total solar neutrino flux and the predicted value of the flux depends on the model.

In the nineties new radiochemical solar neutrino experiments SAGE and GALLEX were performed. In these experiments neutrinos from all reactions of the solar pp and CNO cycles, including low-energy neutrinos from the reaction $pp \rightarrow de^+\nu_e$, were detected. This reaction gives the largest contribution to the flux of the solar neutrinos. The flux of the pp neutrinos can be predicted in a model independent way. The event rates measured in the SAGE and GALLEX

---

[1] At that time only ν_e and ν_μ were known.

© Springer International Publishing AG, part of Springer Nature 2018
S. Bilenky, *Introduction to the Physics of Massive and Mixed Neutrinos*,
Lecture Notes in Physics 947, https://doi.org/10.1007/978-3-319-74802-3_11

experiments were approximately two times smaller than the predicted rates. Thus, in these experiments important evidence was obtained in favor of the disappearance of solar ν_e on the way from the central region of the sun, where solar neutrinos are produced, to the earth.

Another indications in favor of neutrino oscillations were obtained in the nineties in the Kamiokande and IMB neutrino experiments in which atmospheric muon and electron neutrinos were detected. These neutrinos are produced in decays of pions and kaons, created in interactions of cosmic rays with nuclei in the atmosphere, and in decays of muons, which are produced in the decays of pions and kaons. It was found in these experiments that the ratio of the numbers of ν_μ and ν_e events is significantly smaller than the predicted (practically model independent) ratio.

On the other side, no indications in favor of neutrino oscillations were found in the eighties and nineties in numerous reactor and accelerator short baseline experiments.[2]

A first model independent evidence in favor of neutrino oscillations was obtained in 1998 in the water-Cherenkov Super-Kamiokande experiment. In this experiment a significant up-down asymmetry of the high-energy atmospheric neutrino muon events was observed. It was discovered that the number of up-going high-energy muon neutrinos, passing through the earth, is about two times smaller than the number of the down-going muon neutrinos coming directly from the atmosphere.

In 2002 in the SNO solar neutrino experiment evidence in the favor of the disappearance of solar ν_e was obtained. In this experiment solar neutrinos were detected through the observation of CC and NC reactions. *A Model independent evidence of the disappearance of solar ν_e* was obtained. It was shown that the flux of the solar ν_e is approximately three times smaller than the flux of ν_e, ν_μ and ν_τ.

In 2002 in the KamLAND reactor neutrino experiment a model independent evidence in favor of oscillations of reactor antineutrinos was obtained. In this experiment was found that the number of reactor $\bar{\nu}_e$ events at the average distance of ~ 180 km from the reactors is about 0.6 of the number of the expected events. In 2004 a significant distortion of the $\bar{\nu}_e$ spectrum was observed in the KamLAND experiment.

All these experiments complete the first period of the brilliant discovery of neutrino oscillations. It was proven that neutrinos have small masses and that the flavor neutrinos ν_e, ν_μ, ν_τ are "mixed particles". All observed data can be described if we assume the three-neutrino mixing. The values of four neutrino oscillation parameters (two-mass squared differences and two mixing angles) were determined.

The muon neutrino disappearance were observed in the accelerator long-baseline K2K, MINOS and later T2K and NOvA experiments. These experiments confirm the results obtained in the pioneer atmospheric Super-Kamiokande experiment.

Oscillations of atmospheric and accelerator neutrinos (solar and reactor Kam-LAND neutrinos) are determined mainly by the large mass-squared difference Δm_A^2

---

[2]Notice that the spectrum of the reactor $\bar{\nu}_e$ was recently recalculated. As a result, old reactor data are considered at present as an indication in favor of active-sterile neutrino oscillations (see later).

and large mixing angle θ_{23} (small mass-squared difference Δm_S^2 and large mixing angle θ_{12}). During many years from the reactor CHOOZ experiment only upper bound for the small mixing angle θ_{13} was known.

The angle θ_{13} determines subdominant $\nu_\mu \leftrightarrows \nu_e$ oscillations of accelerator and atmospheric neutrinos and disappearance of reactor antineutrinos driven by the atmospheric mass-squared difference. First evidence in favor of $\nu_\mu \leftrightarrows \nu_e$ oscillations was obtained in the accelerator T2K experiment. In 2012–2016 the angle θ_{13} was measured with high precision in the reactor Daya Bay, RENO and Double Chooz experiments. This was very important development in the investigation of a new phenomenon, neutrino oscillations. The way to the determination of the character of the neutrino mass spectrum (normal or inverted mass ordering?) and to the measurement of the CP phase δ was open.

In this chapter we will briefly discuss the major neutrino oscillation experiments.

11.2 Solar Neutrino Experiments

11.2.1 Introduction

Solar ν_e's are produced in reactions of the thermonuclear pp and CNO cycles in which the energy of the sun is generated. The thermonuclear reactions are going on in the central, most hot region of the sun. In this region the temperature is about $15 \cdot 10^6$ K. At such a temperature the major contribution to the energy production is given by the pp cycle. The estimated contribution of the CNO cycle to the sun energy production is about 1%.[3]

The pp cycle starts with the pp and pep reactions

$$p + p \rightarrow d + e^+ + \nu_e \quad \text{and} \quad p + e^- + p \rightarrow d + \nu_e. \tag{11.1}$$

The pp reaction gives the dominant contribution to the deuterium production (99.77%). The contribution of the pep reaction is 0.23%.

Deuterium and proton produce ^{3}He in the reaction

$$p + d \rightarrow\ ^3\text{He} + \gamma. \tag{11.2}$$

Nuclei ^{3}He disappear due to the following three reactions

$$^3\text{He} +\ ^3\text{He} \rightarrow\ ^4\text{He} + p + p \quad (84.92\%). \tag{11.3}$$

$$^3\text{He} + p \rightarrow\ ^4\text{He} + e^+ + \nu_e \quad (\text{about } 10^{-5}\%) \tag{11.4}$$

$$^3\text{He} +\ ^4\text{He} \rightarrow\ ^7\text{Be} + \gamma \quad (15.08\%) \tag{11.5}$$

---

[3]In stars significantly heavier than the sun the central temperatures are higher and the CNO cycle gives important contribution to the energy production.

In the first two reactions ^{4}He is produced. Nuclei ^{7}Be, produced in the third reaction, take part in two chains of reactions terminated by the production of ^{4}He nuclei

$$^7\text{Be} + e^- \rightarrow {}^7\text{Li} + \nu_e, \quad {}^7\text{Li} + p \rightarrow {}^4\text{He} + {}^4\text{He}. \tag{11.6}$$

and

$$p + {}^7\text{Be} \rightarrow {}^8\text{B} + \gamma, \quad {}^8\text{B} \rightarrow {}^8\text{Be}^* + e^+ + \nu_e, \quad {}^8\text{Be}^* \rightarrow {}^4\text{He} + {}^4\text{He}. \tag{11.7}$$

Positrons annihilate with electrons and produce photons. Thus, the energy of the sun is generated in the transition[4]

$$4p + 2e^- \rightarrow {}^4\text{He} + 2\nu_e + Q, \tag{11.8}$$

where

$$Q = 4m_p + 2m_e - m_{^4\text{He}} \simeq 26.73 \,\text{MeV} \tag{11.9}$$

is the energy produced in the transition (11.8).[5] From (11.8) follows that the production of $\frac{1}{2}Q \simeq 13.36\,\text{MeV}$ is accompanied by the emission of one neutrino. Let us consider a neutrino with energy E. The production of such neutrino is accompanied by the emission of luminous energy equal to $\frac{1}{2}Q - E$. If $\phi_r(E)$ is the flux of neutrinos from the source r ($r = pp$,[7] Be,[8] B, ...) on the earth, we have the following relation

$$\sum_r \int (\frac{1}{2}Q - E)\, \phi_r(E)\, dE = \frac{\mathscr{L}_\odot}{4\pi R^2}, \tag{11.10}$$

where $\mathscr{L}_\odot$ is the luminosity of the sun and R is the sun-earth distance.

The relation (11.10) is called *luminosity relation*. It is a general constraint on the fluxes of solar neutrinos. The luminosity relation is based on the following assumptions

1. The solar energy is of thermonuclear origin.
2. The sun is in a stationary state.

---

[4] The CNO cycle is the following chain of reactions: $p + {}^{12}\text{C} \rightarrow {}^{13}\text{N} + \gamma$, $^{13}\text{N} \rightarrow {}^{13}\text{C} + e^+ + \nu_e$, $p + {}^{13}\text{C} \rightarrow {}^{14}\text{N} + \gamma$, $p + {}^{14}\text{N} \rightarrow {}^{15}\text{O} + \gamma$, $^{15}\text{O} \rightarrow {}^{15}\text{N} + e^+ + \nu_e$. There are two branches of reactions with nuclei ^{15}N, terminated with the production of ^{4}He: $p + {}^{15}\text{N} \rightarrow {}^{12}\text{C} + {}^4\text{He}$ or $p + {}^{15}\text{N} \rightarrow {}^{16}\text{O} + \gamma$, $p + {}^{16}\text{O} \rightarrow {}^{17}\text{F} + \gamma$, $^{17}\text{F} \rightarrow {}^{17}\text{O} + e^+ + \nu_e$, $p + {}^{17}\text{O} \rightarrow {}^{14}\text{N} + {}^4\text{He}$.
[5] The energy, produced by the sun, is emitted in the form of photons (about 98%) and neutrinos (about 2%).

Table 11.1 Solar neutrino-producing reactions and SSM neutrino fluxes

Abbreviation	Reaction	SSM flux (cm^{-2} s^{-1})	Neutrino energy (MeV)
pp	$p + p \to d + e^+ + \nu_e$	$5.97\,(1 \pm 0.006) \cdot 10^{10}$	≤ 0.42
pep	$p + e^- + p \to d + \nu_e$	$1.41\,(1 \pm 0.011) \cdot 10^{8}$	1.44
^{7}Be	$e^- + {}^7\text{Be} \to {}^7\text{Li} + \nu_e$	$5.07\,(1 \pm 0.06) \cdot 10^{9}$	0.86
^{8}B	$^8\text{B} \to {}^8\text{Be}^* + e^+ + \nu_e$	$5.94\,(1 \pm 0.11) \cdot 10^{6}$	$\lesssim 15$
hep	$^3\text{He} + p \to {}^4\text{He} + e^+ + \nu_e$	$7.90\,(1 \pm 0.15) \cdot 10^{3}$	≤ 18.8
^{13}N	$^{13}\text{N} \to {}^{13}\text{C} + e^+ + \nu_e$	$2.88\,(1 \pm 0.15) \cdot 10^{8}$	≤ 1.20
^{15}O	$^{15}\text{O} \to {}^{15}\text{N} + e^+ + \nu_e$	$2.15\,(1 \pm 0.17) \cdot 10^{8}$	≤ 1.73
^{17}F	$^{17}\text{F} \to {}^{17}\text{O} + e^+ + \nu_e$	$5.82\,(1 \pm 0.19) \cdot 10^{6}$	≤ 1.74

The last assumption is connected with the fact that neutrinos observed in a detector were produced about 8 min before the detection. On the other side it takes about 10^5 years for photons produced in the central region of the sun to reach its surface.

We can rewrite the luminosity relation in the form

$$\sum_r \left(\frac{Q}{2} - \overline{E}_r \right) \Phi_r = \frac{\mathcal{L}_\odot}{4\pi R^2}. \qquad (11.11)$$

Here

$$\overline{E}_r = \frac{1}{\Phi_r} \int E\, \phi_r(E)\, dE \qquad (11.12)$$

is the average neutrino energy from the source r and $\Phi_r = \int \phi_r(E)\, dE$ is the total flux of neutrinos from the source r.

The calculation of neutrino fluxes from different reactions can be done only in the framework of a solar model. Usually the results of the Standard Solar Model (SSM) calculations are used.[6] In Table 11.1 we present SSM fluxes of ν_e from different reactions. In this Table we included also SSM fluxes from the following reactions of the CNO cycle: $^{13}\text{N} \to {}^{13}\text{C} + e^+ + \nu_e$, $^{15}\text{O} \to {}^{15}\text{N} + e^+ + \nu_e$ and $^{17}\text{F} \to {}^{17}\text{O} + e^+ + \nu_e$. In the last column of Table 11.1 neutrino energies are given.

It is evident from Table 11.1 that the second term of the luminosity relation (11.10) is much smaller than the first one. If we neglect this term, we find

---

[6]The Standard Solar Model is based on the assumption that the sun is a spherically symmetric plasma sphere in hydrostatic equilibrium. The effects of rotation and of the magnetic field are neglected.

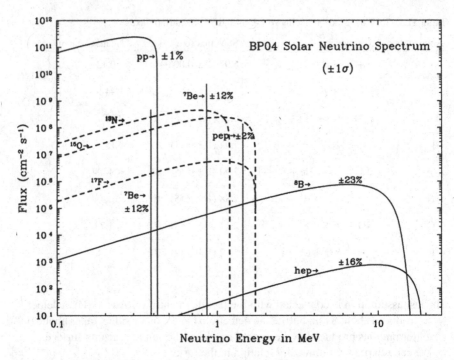

Fig. 11.1 Predicted by the Standard Solar Model spectra of solar neutrinos from different reactions

the following estimate for the total flux of neutrinos

$$\Phi = \sum_r \Phi_r \simeq \frac{\mathscr{L}_\odot}{2\pi R^2 Q}. \tag{11.13}$$

Taking into account that $\mathscr{L}_\odot = 2.40 \cdot 10^{39}\,\mathrm{MeV\,s^{-1}}$ and $R = 1.496 \cdot 10^{13}\,\mathrm{cm}$ we find

$$\Phi \simeq 6 \cdot 10^{10}\,\mathrm{cm^{-2}\,s^{-1}}. \tag{11.14}$$

In Fig. 11.1 predicted by the SSM spectra of neutrinos from different reactions are presented.

11.2.2   Homestake Chlorine Solar Neutrino Experiment

The pioneer experiment, in which solar electron neutrinos were detected, was the **Homestake** experiment by R. Davis et al.[7] The experiment continued from 1968 till 1994.

---

[7] For this experiment R. Davis was awarded with the Nobel Prize in 2002.

In the Davis experiment radiochemical chlorine-argon method, proposed by B. Pontecorvo in 1946, was used. Solar electron neutrinos were detected through the observation of the reaction

$$\nu_e + {}^{37}\text{Cl} \rightarrow e^- + {}^{37}\text{Ar} . \tag{11.15}$$

The ^{37}Ar atoms are radioactive. They decay via electron-capture with emission of Auger electrons. The half-life of the decay is 34.8 days.

A tank filled with 615 tons of liquid tetrachloroethylene (C_2Cl_4) was a part of the detector in the Davis experiment. In order to decrease the cosmic ray background, the experiment was performed in the Homestake mine (USA) at depth of about 1480 m (4100 m water equivalent). The radioactive ^{37}Ar atoms, produced by solar ν_e via the reaction (11.15) during the exposure time (about 2 months), were extracted from the tank by purging with ^{4}He gas. About 16 atoms of ^{37}Ar were extracted during one exposure run. The gas with radioactive ^{37}Ar atoms was placed into a low-background proportional counter in which the signal (Auger electrons) was detected. An important feature of the experiment was the measurement of the rise time of the signal. This allowed to suppress the background.

The energy threshold of the Cl-Ar reaction is equal to 0.814 MeV, i.e. it is larger than the maximal energy of pp neutrinos, constituting the major part of the solar neutrinos flux (see Table 11.1). At high ^{8}B energies the transition to an excited state of ^{37}Ar significantly increase the cross section of the process (11.15). As a result, the dominant contribution to the counting rate give the high energy ^{8}B neutrinos. The SSM contribution of the ^{8}B neutrinos to the event rate is approximately equal to 5.8 SNU.[8] The SSM contribution to the event rate of ^{7}Be neutrinos is approximately equal to 1.2 SNU. Other much smaller contributions come from pep and CNO neutrinos.

The averaged over 108 runs (between 1970 and 1994) event rate, measured in the Homestake experiment, is equal to

$$R_{\text{Cl}} = (2.56 \pm 0.16 \pm 0.16) \ \text{SNU} \tag{11.16}$$

The measured event rate is significantly smaller than the rate predicted by the SSM (under the assumption that there are no neutrino oscillations):

$$(R_{\text{Cl}})_{SSM} = 8.6 \pm 1.2 \ \text{SNU} \tag{11.17}$$

11.2.3 Radiochemical GALLEX-GNO and SAGE Experiments

Neutrinos from all solar neutrino reactions, including low-energy neutrinos from the pp reaction, were detected in the radiochemical gallium GALLEX-GNO and SAGE

---

[8]The solar neutrino unit (SNU) is determined as follows:1 $SNU = 10^{-36}$events atom^{-1} s^{-1}.

experiments. In these experiments neutrinos were detected by the radiochemical method through the observation of the reaction

$$\nu_e + {}^{71}\text{Ga} \rightarrow e^- + {}^{71}\text{Ge}, \tag{11.18}$$

in which radioactive ^{71}Ge was produced. The threshold of this reaction is equal to 0.233 MeV. The half-life of ^{71}Ge is equal to 11.43 days.

The detector in the **GALLEX-GNO** experiment was a tank containing 100 tons of a water solution of gallium chloride (30.3 tons of ^{71}Ga). The experiment was done in the underground Gran Sasso Laboratory (Italy). During 1991–2003 there were 123 GALLEX and GNO exposure runs. The duration of one run was about 4 weeks. About 10 atoms of ^{71}Ge were produced during one run. Radioactive ^{71}Ge atoms were extracted from the detector by a chemical procedure and introduced into a small proportional counter in which Auger electrons, produced in the capture $e^- + {}^{71}\text{Ge} \rightarrow {}^{71}\text{Ga} + \nu_e$, were detected.

The measured event rate averaged over 123 runs is equal to

$$R_{\text{Ga}} = (67.5 \pm 5.1) \text{ SNU}. \tag{11.19}$$

The SSM event rate

$$(R_{\text{Ga}})_{\text{SSM}} = (128 \, {}^{+9}_{-7}) \text{ SNU}. \tag{11.20}$$

is about two times larger than the measured rate.

The major contribution to the SSM predicted event rate comes from the pp neutrinos (69.7 SNU). Contributions of ^{7}Be and ^{8}B neutrinos to the SSM event rate are equal to 34.2 SNU and 12.1 SNU, respectively.

In **SAGE** gallium experiment about 50 tons of ^{71}Ga in the liquid metal form were used. The experiment was done in the Baksan Neutrino Observatory (Caucasus mountains, Russia) in a hall with an overburden of 4700 m of water equivalent. Neutrinos were detected through the observation of the reaction (11.18). An exposure time in this experiment was about 4 weeks. The ^{71}Ge atoms, produced by the solar neutrinos, are chemically extracted from the target and are converted to GeH$_4$. Auger electrons, produced in decay of germanium, were detected in a small proportional counter.

The germanium production rate, measured in the SAGE experiment, averaged over 92 runs (1990–2001) was equal to

$$R_{\text{Ga}} = (70.8^{+5.3}_{-5.2}(\text{stat})^{+3.7}_{-3.2}(\text{syst})) \text{ SNU}. \tag{11.21}$$

As it is seen from (11.19) and (11.21), the rates measured in the SAGE and in the GALLEX-GNO experiments are in a good agreement.

11.2.4 Kamiokande and Super-Kamiokande Solar Neutrino Experiments

In radiochemical experiments the neutrino direction can not be determined. The first experiment in which the neutrino direction was measured was Kamiokande. It was proved that the detected neutrinos were coming from the sun.

In the **Kamiokande** experiment a 3000 ton water-Cherenkov detector was used. The experiment was done in the Kamioka mine (Japan) at a depth of about 1000 m (2700 m water equivalent).

In the Kamiokande experiment the solar neutrinos were detected through the observation of recoil electrons in the elastic neutrino-electron scattering

$$\nu_x + e \to \nu_x + e. \quad (x = e, \mu, \tau) \tag{11.22}$$

All types of flavor neutrinos could be detected via observation of the process (11.22). However, the cross section of $\nu_{\mu,\tau} - e$ scattering is significantly smaller than the cross section of $\nu_e - e$ scattering ($\sigma(\nu_{\mu,\tau}e \to \nu_{\mu,\tau}e) \simeq 0.16\,\sigma(\nu_e e \to \nu_e e)$). Thus, mainly the flux of solar ν_e was measured in the Kamiokande experiment.

Solar neutrinos were detected via the observation of the Cherenkov radiation of electrons in water. About 1000 large (50 cm in diameter) photomultipliers, which covered about 20% of the surface of the detector, were utilized in the experiment. Because of the contamination of Rn in the water it was necessary to apply a 7.5 MeV energy threshold for the recoil electrons.

At high energies recoil electrons are emitted in a narrow (about 15°) cone around the initial neutrino direction. In the experiment a strong correlation between the direction of recoil electrons and the direction to the sun was observed. This correlation was an important signature which allowed to suppress background and to prove that the observed events were due to solar neutrinos.

Because of the high threshold mainly ^{8}B neutrinos were detected in the Kamiokande experiment. The total flux of high energy ^{8}B neutrinos obtained from the data of the Kamiokande experiment was equal to

$$\Phi_\nu^K = (2.80 \pm 0.19 \pm 0.33) \cdot 10^6 \, \text{cm}^{-2} \, \text{s}^{-1}. \tag{11.23}$$

The ratio of the measured solar neutrino flux to the flux predicted by the SSM (under the assumption that there are no neutrino oscillations) was equal to $R^K = 0.51 \pm 0.04 \pm 0.06$.

The Kamiokande result was an important confirmation of the existence of the solar neutrino problem, discovered in Davis et al. in the Homestake experiment.[9]

---

[9]In 1987 the Kamiokande Collaboration (and also the IMB and Baksan Collaborations) observed neutrinos from the explosion of the supernova SN1987A. This was the first observation of supernova neutrinos. The experiment confirmed the general theory of the gravitational collapse.

The Kamiokande experiment was running during 9 years from 1987 till 1995. In 1996 the **Super-Kamiokande**, experiment of the next generation, started. In this experiment a huge 50 kton water-Cherenkov detector (fiducial volume 22.5 kton) was used.

There were four phases of the Super-Kamiokande experiment. The SK-I phase started in 1996 and finished in 2001. In this phase 11,146 photomultipliers (PMT) were used. In 2001 an accident happened in which about 60% of PMTs were destroyed. After about a year, the data-taking started with 5182 photomultipliers (SK-II). This phase finished in 2005. In 2006 SK-III started with 11,129 PMTs. The fourth phase of the experiment started in 2006 and finished in 2014.

During the SK-I phase the threshold for the kinetic energy of the recoil electrons was 6 MeV (first 280 days) and 4.5 MeV for the remaining days. In the SK-IV phase the recoil electron threshold was 3.49 MeV. Due to high threshold only ^8B and *hep* neutrinos where detected in the Super-Kamiokande experiment.

Improvement in the electronics, in the water circulation system, in calibration and in methods of analysis allowed to reach in the SK-IV much smaller systematic uncertainty than during other phases of the experiment. The measured in the SK-IV flux of the solar neutrinos is equal to

$$\Phi_\nu^{SK-IV} = (2.308 \pm 0.020 \text{ (stat.)} \pm 0.039 \text{ (syst.)} \cdot 10^6 \text{ cm}^{-2} \text{ s}^{-1}. \qquad (11.24)$$

Combining the results of all Super-Kamiokande phases it was found

$$\Phi_\nu^{SK} = (2.345 \pm 0.014 \text{ (stat.)} \pm 0.036 \text{ (syst.)} \cdot 10^6 \text{ cm}^{-2} \text{ s}^{-1}. \qquad (11.25)$$

Due to the earth matter effect the fluxes of solar neutrinos during day and during night must be different (in the night flux it must be more ν_e than in the day flux). The high statistics of the events allowed the Super-Kamiokande Collaboration to measure the day-night asymmetry. In the SK-IV it was found

$$A_{D-N} = (-3.6 \pm 1.6 \pm 0.6)\% \qquad (11.26)$$

No distortion of the spectrum of recoil electrons with respect to the expected spectrum was observed.[10] From analysis of the all SK data for the solar neutrino mass-squared difference the following value was obtained

$$\Delta m_S^2 = (4.8^{+1.5}_{-0.8}) \cdot 10^{-5} \text{ eV}^2. \qquad (11.27)$$

---

[10]The initial ^8B solar neutrino spectrum is determined by the weak decay $^8B \rightarrow e^+ + \nu_e + 2\alpha$. This spectrum can be obtained from the laboratory measurement of the α-spectrum. The fact that the electron spectrum, measured in the Super-Kamiokande experiment, is in an agreement with the expected spectrum means that in the high-energy ^8B region the probability of the solar ν_e to survive is a constant.

From the SK-IV data for the parameter $\sin^2 \theta_{12}$ it was found

$$\sin^2 \theta_{12} = 0.327^{+0.026}_{-0.031}. \tag{11.28}$$

From the analysis of all SK data it was obtained

$$\sin^2 \theta_{12} = 0.334^{+0.027}_{-0.023}. \tag{11.29}$$

Finally, from the fit of all solar and KamLAND data (assuming that the parameter $\sin^2 \theta_{13}$ is given by the reactor value (see later)) it was found

$$\Delta m_S^2 = (7.49^{+0.19}_{-0.18}) \cdot 10^{-5} \text{ eV}^2, \quad \sin^2 \theta_{12} = 0.307^{+0.013}_{-0.012}. \tag{11.30}$$

11.2.5 SNO Solar Neutrino Experiment

The fluxes of solar neutrinos, measured in the Homestake, GALLEX-GNO, SAGE, Kamiokande and Super-Kamiokande experiments, were significantly smaller than the fluxes, *predicted by the Standard Solar model.* From the analysis of the data of these experiments, strong indications in favor of neutrino transitions in matter, driven by neutrino masses and mixing, were obtained.

The first model-independent evidence for transitions of solar ν_e into ν_μ and ν_τ was obtained in the SNO solar neutrino experiment. The SNO detector was located in the Creighton mine (Sudbury, Canada) at a depth of 2092 m (5890 ± 94 m water equivalent). The detector consisted of the transparent acrylic vessel (a sphere, 12 m in diameter) containing 1 kton of pure heavy water D_2O. About 7 kton of H_2O shielded the vessel from external radioactive background. An array of 9456 PMTs detected Cherenkov radiation produced in the D_2O and H_2O.

A crucial feature of the SNO experiment was the detection of the solar neutrinos *via three different processes.*

1. The CC process

$$\nu_e + d \rightarrow e^- + p + p. \tag{11.31}$$

2. The NC process

$$\nu_x + d \rightarrow \nu_x + p + n \quad (x = e, \mu, \tau) \tag{11.32}$$

3. Elastic neutrino-electron scattering (ES)

$$\nu_x + e \rightarrow \nu_x + e. \tag{11.33}$$

The CC and ES processes were observed through the detection of the Cherenkov light produced by electrons in the heavy water. The NC process was observed via the detection of neutrons. There were three phases of the SNO experiment in which different methods of the detection of neutrons were used.

- The NC neutrons were captured in D_2O and produced 6.25 MeV γ-quanta in the reaction $n + d \rightarrow {}^3H + \gamma$. During Phase I the Cherenkov light of secondary Compton electrons and $e^+ - e^-$ pairs was detected.
- In Phase II of the SNO experiment about two tons of NaCl were dissolved in the heavy water. Neutrons were detected through the observation of γ-quanta from the capture of neutrons by ${}^{35}Cl$ nuclei. For thermal neutrons the cross section of this process is equal to 44 barn while the cross section of the process $nd \rightarrow {}^3H\gamma$ is equal to 0.5 mb. Thus, the addition of the salt significantly enhanced the NC signal.
- In Phase III an array of proportional counters was deployed in the heavy water. Neutrons were detected through the observation of the reaction $n + {}^3He \rightarrow p + {}^3H$ in which proton and ${}^3H$ had a total kinetic energy 0.76 MeV. Charged particles in the proportional counters produced ionization electrons and the induced by them voltage was recorded as a function of time. This technic allowed to reduce background significantly.

The SNO Collaboration started to collect data in 1999. The last phase was finished in 2006. The SNO threshold for the detection of the electrons from the CC and the ES processes was equal to $T_{thr} = 5.5$ MeV. The neutrino energy threshold for NC process is 2.2 MeV (the deuterium bounding energy). Thus, in the SNO experiments mostly high energy solar ${}^8B$ neutrinos were detected.

The initial spectrum of ν_e from the ${}^8B$ decay is known. It was obtained from the measurement of α-spectrum from the ${}^8B$ decay. The SNO and other solar neutrino data are compatible with the assumption that in the high-energy ${}^8B$ region $\nu_e \rightarrow \nu_e$ survival probability is a constant.

From the observation of the CC events for the flux ν_e the following value was obtained

$$\Phi_{\nu_e}^{CC} = (1.68 \pm 0.06 \text{ (stat.)}_{-0.09}^{+0.08} \text{ (syst.))} \cdot 10^6 \text{ cm}^{-2} \text{ s}^{-1}. \tag{11.34}$$

Because of the $\nu_e - \nu_\mu - \nu_\tau$ universality of the NC neutrino-hadron interaction the *observation of NC events allows to determine the total flux of all flavor neutrinos.* In the SNO experiment was found that the total flux of all flavor neutrinos is equal to

$$\Phi_{\nu_{e,\mu,\tau}}^{NC} = (5.25 \pm 0.16 \text{ (stat.)}_{-0.13}^{+0.11} \text{ (syst.))} \cdot 10^6 \text{ cm}^{-2} \text{ s}^{-1}. \tag{11.35}$$

The value (11.35) of the total flux of all flavor neutrinos is in agreement with the flux of the ${}^8B$ neutrinos predicted by the Standard Solar Model (see Table 11.1).

The SNO experiment solved the solar neutrino problem. If we compare the flux of ν_e with the total flux of ν_e, ν_μ and ν_τ, we come to the model independent conclusion that solar ν_e on the way from the sun to the earth are transformed into ν_μ and ν_τ.

From the three-neutrino analysis of the SNO and other solar neutrino data and also data of the reactor KamLAND experiment for the neutrino oscillation parameters the following values were obtained

$$\Delta m_S^2 = (7.41^{+0.21}_{-0.19}) \cdot 10^{-5} \text{ eV}^2, \quad \tan^2 \theta_{12} = 0.446^{+0.030}_{-0.029}, \quad \sin^2 \theta_{13} = 2.5^{+1.8}_{-1.5} \cdot 10^{-2}.$$
$$\tag{11.36}$$

11.2.6 Borexino Solar Neutrino Experiment

Due to very low background in the **Borexino** solar neutrino experiment low energy pp, ^7Be and pep neutrinos and also high energy ^8B neutrinos were detected.

The Borexino detector is located in the Gran Sasso underground laboratory (Italy). It consists of concentric shells of increasing radio-purity. The Inner Detector is a nylon vessel which contains 280 tons of liquid scintillator. It is surrounded by layers of buffer liquid and highly purified water which allows to suppress background of cosmic muons.

The Borexino collaboration are taking data since 2007. In the Phase I of the experiment (2207–2010) rates of solar pp, ^7Be and pep neutrinos were measured and a bound on the flux of CNO neutrinos was obtained. In the Phase II, started in 2011, the rate of pp neutrinos was determined.

The solar neutrinos are observed in the Borexino experiment through the detection of recoil electrons from the elastic neutrino-electron scattering

$$\nu_x + e \rightarrow \nu_x + e, \quad x = e, \mu, \tau. \tag{11.37}$$

The scintillation light is detected by 2212 PMTs uniformly distributed on the inner surface of the detector. The measurement of the scintillation light allows to determine the energy of the electrons. There is no information about the direction of the electrons. Because the energy threshold in the Borexino experiment must be low, the major requirement is an extremely low radioactive contamination of the scintillator (9–10 order of magnitude lower than the natural radioactivity).

In the reaction $^7Be + e^- \rightarrow^7 Li + \nu_e$ neutrino with the energy 0.86 MeV is produced. The signature of 0.86 MeV ν_e is a characteristic Compton-like spectrum with a shoulder at about 660 keV. The ^7Be rate was obtained from the fit of the data in which contributions of decays of ^{85}Kr, ^{210}Bi, ^{11}C and ^{210}Po were taken into account. For the interaction rate of the ^{7}Be neutrinos it was found the value

$$R_{7\text{Be}} = (46 \pm 1.5 \text{ (stat.)} \pm 1.5(\text{syst.})) \text{ cpd}/(100 \text{ tons}), \tag{11.38}$$

where cpd $\equiv$ counts/(day). Assuming that the probability of ^7Be electron neutrino to survive is given by the standard MSW value for the flux of ^7Be neutrinos from (11.38) we find

$$\Phi_{\nu_e}^{^7\text{Be}} = (4.43 \pm 0.22) \cdot 10^9 \text{ cm}^{-2} \text{s}^{-1}. \tag{11.39}$$

The *pep* neutrinos have an energy 1.44 MeV. Taking into account the major background from the ^{11}C decay in the Borexino experiment for the *pep* rate was obtained

$$R_{\text{pep}} = (3.1 \pm 0.6 \text{ (stat.)} \pm 0.3(\text{syst.})) \text{ cpd}/(100 \text{ tons}), \tag{11.40}$$

For the flux of *pep* neutrinos it was found

$$\Phi_{\nu_e}^{\text{pep}} = (1.63 \pm 0.35) \cdot 10^8 \text{ cm}^{-2} \text{ s}^{-1}. \tag{11.41}$$

From the fit of the data (in which the pep rate was fixed at the value (11.40)) for the rate of the CNO neutrinos the following upper bound was obtained

$$R_{\text{CNO}} < 7.4 \text{ cpd}/(100 \text{ tons}). \tag{11.42}$$

This bound implies the following bound for the CNO neutrinos

$$\Phi_{\nu_e}^{\text{CNO}} < 7.7 \cdot 10^8 \text{ cm}^{-2} \text{ s}^{-1}. \tag{11.43}$$

The flux of the low energy *pp* neutrinos (the maximal neutrino energy is 420 keV) is the major flux of the solar neutrinos. In the Phase II of the Borexino experiment the rate of the *pp* neutrinos was determined. The main problem was a background from the ^{14}C β decay. Taking into account in the fit of the data also decays of other background nuclei (^{210}Po, ^{210}Bi and others) the following rate of the *pp* neutrinos was obtained

$$R_{\text{pp}} = (144 \pm 13 \text{ (stat.)} \pm 10(\text{syst.})) \text{ cpd}/(100 \text{ tons}). \tag{11.44}$$

From (11.44) the following flux of the *pp* neutrinos was inferred

$$\Phi_{\nu_e}^{\text{pep}} = (6.6 \pm 0.7) \cdot 10^{10} \text{ cm}^{-2} \text{ s}^{-1}. \tag{11.45}$$

This flux is in an agreement with the flux predicted by SSM.

If we use the SSM flux, from (11.44) for the probability of the low energy *pp* neutrinos to survive we obtain the value

$$P(\nu_e \rightarrow \nu_e) = 0.64 \pm 0.12. \tag{11.46}$$

In the Borexino experiment the rate of the high-energy ^8B neutrinos was also determined. The threshold for recoil electron energy in this experiment was equal to 3 MeV. For the rate ^8B neutrinos it was found

$$R_{8\text{B}} = (0.22 \pm 0.04 \text{ (stat.)} \pm 0.01(\text{syst.})) \text{ cpd}/(100 \text{ tons}). \tag{11.47}$$

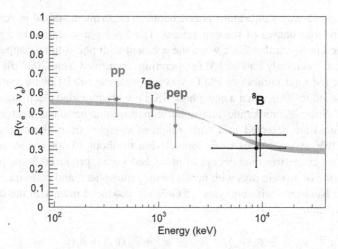

Fig. 11.2 Borexino experiment: ν_e survival probability of pp, ^7Be, pep and ^8B solar neutrinos. Curve is MSW prediction (arXiv:1707.9279)

The flux of ^8B neutrinos, determined from (11.47)

$$\Phi_{\nu_e}^{^8B} = (5.2 \pm 0.3) \cdot 10^6 \ \text{cm}^{-2}\,\text{s}^{-1}. \tag{11.48}$$

is in agreement with Super-Kamiokande and SNO data.

The results obtained in the Borexino experiment allow to obtain the ν_e survival probability of pp, ^7Be, pep and ^8B solar neutrinos (see Fig. 11.2). As it is seen from Fig. 11.2 the Borexino data are in agreement with MSW prediction. Up to now no indications in favor of a non-standard physics were obtained.

11.3 Super-Kamiokande Atmospheric Neutrino Experiment

The Super-Kamiokande is a multi-purpose detector. In the previous section we considered the Super-Kamiokande solar neutrino experiment. In this section we will consider the Super-Kamiokande atmospheric neutrino experiment. In the Super-Kamiokande atmospheric neutrino experiment the first model independent evidence in favor of neutrino oscillations was obtained (1998). This discovery opened a new era in the study of the problem of neutrino masses, mixing and oscillations.

The Super-Kamiokande detector is a 50 kton water Cerenkov detector which is optically separated into inner detector ID (32 kton, fiducial volume 22.5 kton) viewed by 11,146 inward-facing 50 cm PMTs and outer detector OD with 1885 20 cm PMTs which is used as a veto for events induced by the cosmic rays. Cherenkov radiation produced by charged particles, traveling through detector, is collected by PMTs.

The Super-Kamiokande atmospheric neutrino experiment started in April 1966. There were four phases of the experiment. The SK-I phase started in April 1996 and finished in November 2001 when the accident with photo-tubes happened. The SK-II phase, with only half of PMTs operating, continued from 2002 till 2005. In 2006 after the total number of PMTs was restored, the SK-III phase started. This phase finished in 2008 when a new phase SK-IV with upgraded electronics began.

In the Super-Kamiokande atmospheric neutrino experiment neutrinos (and antineutrinos) are detected in a wide range of energies from about 100 MeV to about 10 TeV and distances from about 10 km to about 13,000 km. Atmospheric neutrinos originate from the decays of pions and kaons, produced in the processes of interaction of cosmic rays with nuclei of the atmosphere, and consequent decays of muons. Neutrinos with energies $\lesssim 5$ GeV are produced mainly in the decays of pions and muons

$$\pi^{\pm} \to \mu^{\pm} + \nu_{\mu}(\bar{\nu}_{\mu}), \quad \mu^{\pm} \to e^{\pm} + \bar{\nu}_{\mu}(\nu_{\mu}) + \nu_e(\bar{\nu}_e) \qquad (11.49)$$

At higher energies the contribution of kaons becomes also important.

Neutrinos and antineutrinos are detected through the observation of electrons and muons produced in the CC processes

$$\nu_l(\bar{\nu}_l) + N \to l^-(l^+) + X \quad (l = e, \mu). \qquad (11.50)$$

Atmospheric neutrino events are divided into three categories

- Fully contained events (FC). Events are called FC if initial vertexes are in the ID fiducial volume and all energies are deposited in the inner detector. Such events are separated into two samples: sub-GeV ($E \leq 1.33$ GeV) and multi-GeV ($E > 1.33$ GeV).
- Partially contained events (PC). If a high energy muon escapes the inner detector and deposits part of its energy in the outer veto detector such an event is called a PC event.
- Upward going muons (Upμ). Upward going muon events are due to interaction of muon neutrinos in the rock outside of the Super-Kamiokande detector which produce muons entering into the detector from below. There are two categories of such events. Upward stopping muons are those muons which come to rest in the detector. Upward through-going muons are those muons which pass the whole detector.

FC events are produced by neutrinos with energies of a few GeV. PC events are produced by neutrinos with energies about an order of magnitude higher. The energies of neutrinos which produce upward stopping muons is about 10 GeV. Upward through-going muons are produced by neutrinos with an average energy of about 100 GeV.

During four phases of the Super-Kamiokande experiment it was observed 10,386 (10,493) sub-GeV μ (e)-like FC events, 4370 (4076) multi-GeV μ (e)-like FC events and 3003 PC events.

  *A model-independent evidence in favor of neutrino oscillations was obtained by
the Super-Kamiokande Collaboration through the investigation of the zenith-angle
dependence of the atmospheric electron and muon events.* The zenith angle θ is
determined in such a way that neutrinos going vertically downward have $\theta = 0$
and neutrinos coming vertically upward through the earth have $\theta = \pi$. Because of
the geomagnetic cutoff at small energies (0.3–0.5 GeV) the flux of downward going
neutrinos is lower than the flux of upward going neutrinos. At neutrino energies
$E \geq 0.9$ GeV the fluxes of muon and electron neutrinos are symmetric under the
change $\theta \to \pi - \theta$. Thus, if there are no neutrino oscillations at high energies the
numbers of electron and muon events must satisfy the relation

$$N_l(\cos\theta) = N_l(-\cos\theta) \quad l = e, \mu. \tag{11.51}$$

A significant violation of this relation was found in the Super-Kamikande
experiment.

  For the study of flavor neutrino oscillations it is crucial to distinguish electrons
and muons produced in the processes (11.50). In the Super-Kamiokande experiment
leptons are observed through the detection of the Cherenkov radiation. The shapes
of the Cherenkov rings of electrons and muons are completely different. In the case
of electrons the Cherenkov rings exhibit a more diffuse light than in the muon case.
The probability of a misidentification of electrons and muons is below 2%.

  First indication in favor of neutrino oscillations came from the measurement of
the ratio r of the $(\nu_\mu + \bar{\nu}_\mu)$ and $(\nu_e + \bar{\nu}_e)$ fluxes. This ratio can be predicted with
an accuracy of about 3%. In the SK-I phase, for the double ratio $R = \frac{r_{\text{meas}}}{r_{\text{MC}}}$ (r_{meas}
is the measured and r_{MC} is the predicted ratios) following value was obtained in the
sub-GeV region

$$R_{\text{sub-GeV}} = 0.658 \pm 0.016 \pm 0.035. \tag{11.52}$$

In the multi-GeV region was found

$$R_{\text{multi-GeV}} = 0.702 \pm 0.032 \pm 0.101. \tag{11.53}$$

If there are no neutrino oscillations the double ratio R must be equal to one.

  The most important Super-Kamiokande result was obtained from the measurement of the zenith-angle distribution of the electron and muon events. The latest
results of the measurement of these distributions are presented in Fig. 11.3. As is
seen from Fig. 11.3 the distributions of sub-GeV and multi-GeV electron events are
in agreement with the expected distributions. In the distributions of the sub-GeV
and multi-GeV muon events and upward stopping muon events a significant deficit
of upward-going muons is observed.

  This result can be explained by the disappearance of muon neutrinos due to
neutrino oscillations. As we have seen before, in the case of neutrino oscillations
the probability of ν_μ to survive depends on the distance between neutrino source
and neutrino detector. Downward going neutrinos ($\theta \simeq 0$) pass a distance of about

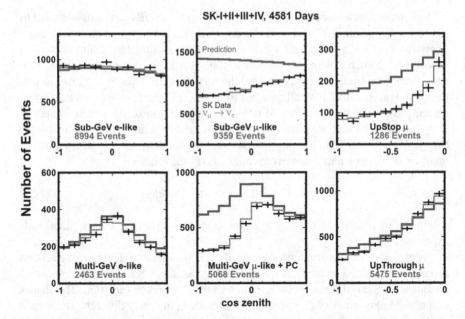

Fig. 11.3 Super-Kamiokande atmospheric neutrino experiment: zenith angle dependence of the numbers of electron and muon events. The MC prediction assuming that there are no oscillations and distribution of events obtained with best-fit values of the oscillation parameters are shown (arXiv:1412.5234v1 [hep-ex])

10–20 km. On the other side upward going neutrinos ($\theta \simeq \pi$) pass a distance of about 13,000 km (earth diameter). The measurement of the dependence of the numbers of the electron and muon events on the zenith angle θ allows to span distances from about 10 km to about 13,000 km. The energies of the atmospheric neutrinos are in the range 100 MeV–100 GeV Such wide ranges of energies and distances allow the Super-Kamikande Collaboration to study neutrino oscillations in details.

The Super-Kamiokande data can be explained by the ν_μ (and $\bar{\nu}_\mu$) disappearance due to dominant $\nu_\mu \leftrightarrows \nu_\tau$ oscillations. Taking into account that $\Delta m_S^2 \ll \Delta m_S^2$ and neglecting small contribution of $\sin^2 \theta_{13}$ we have

$$P(\nu_\mu \to \nu_\mu) = P(\bar{\nu}_\mu \to \bar{\nu}_\mu) = 1 - \sin^2 2\theta_{23} \sin^2 \frac{\Delta m_A L}{4E}. \qquad (11.54)$$

From analysis of the Super-Kamiokande data, obtained during SK-I phase, the following 90% CL ranges of the neutrino oscillation parameters were obtained

$$1.5 \cdot 10^{-3} < \Delta m_A^2 < 3.4 \cdot 10^{-3} \text{ eV}^2, \quad \sin^2 2\theta_{23} > 0.92. \qquad (11.55)$$

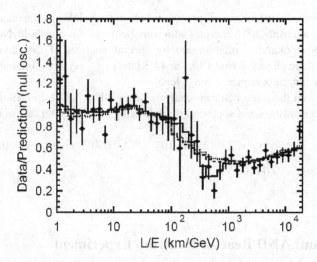

Fig. 11.4 Values of the probability $P(\nu_\mu \rightarrow \nu_\mu)$ as a function of the parameter $\frac{L}{E}$, determined from the data of the Super-Kamiokande atmospheric neutrino experiment. The best-fit two-neutrino oscillation curve is also plotted (arXiv:hep-ex/0404034)

In the standard Super-Kamiokande analysis of the data the dependence of the probability on $\frac{L}{E}$ is practically washed out because of the poor resolution. In order to reveal the oscillatory behavior of the probability, the Super-Kamiokande Collaboration performed a special analysis. A subset of events with high resolution in the variables L and E was chosen for the analysis. This allowed to determine the ν_μ survival probability as a function of $\frac{L}{E}$ and to reveal the first minimum of the survival probability (see Fig. 11.4). It is seen from Fig. 11.4 that the minimum of the survival probability is reached at

$$\left(\frac{L}{E}\right)_{\min} = \frac{2\pi\hbar c}{\Delta m_A^2 c^4} \simeq 500 \, \frac{\mathrm{km}}{\mathrm{GeV}} \qquad (11.56)$$

From this number we can estimate the neutrino mass-squared difference:

$$\Delta m_A^2 \simeq 2.5 \cdot 10^{-3} \, \mathrm{eV}^2. \qquad (11.57)$$

This value is in agreement with (11.55).

The detection in the Super-Kamiokande detector of ν_τ, produced in $\nu_\mu \leftrightarrows \nu_\tau$ oscillations, is a difficult problem.[11] This is connected with the fact the threshold

---

[11] ν_τ produced in $\nu_\mu \leftrightarrows \nu_\tau$ oscillations were observed in the long baseline experiment OPERA. The distance between the source of ν_μ (CERN) and the detector (Gran Sasso Laboratory) in this experiment was about 730 km. The production of τ in ν_τ-nucleon CC processes was detected in an emulsion. Five ν_τ events were observed.

for production of τ in CC $\nu_\tau - N$ processes is about 3.5 GeV and the majority of the atmospheric neutrinos have energies which are below of this threshold. Nevertheless the Super-Kamiokande Collaboration by special analysis of the data, obtained during first three phases, found 180.1 ± 44.3 (stat)$^{+17.8}_{-15.2}$ (syst) ν_τ-interactions. This result confirm ν_τ appearance at 3.8 σ level.

Finally from the three-neutrino analysis of the SK-I, SK-II and SK-III data for the NO (IO) neutrino mass spectrum the following 90% CL intervals were found

$$(1.9\,(1.7) < \Delta m_A^2 < 2.6\,(2.7)) \cdot 10^{-3}\ \text{eV}^2, \quad 0.407 < \sin^2\theta_{23} < 0.583,$$

$$\sin^2 2\theta_{13} < 0.04\,(0.09). \tag{11.58}$$

11.4 KamLAND Reactor Neutrino Experiment

In the **KamLAND** reactor experiment oscillations of reactor $\bar{\nu}_e$, driven by the solar mass-squared difference $\Delta m_{12}^2 \equiv \Delta m_S^2$ were observed. The KamLAND detector is located in the Kamioka mine (Japan) at a depth of about 1 km (2700 m water equivalent). It contains a 13 m-diameter transparent spherical nylon balloon filled with a 1 kton liquid scintillator. The balloon is suspended in 1800 m^3 non-scintillating purified mineral buffer oil. The internal detector (balloon and buffer oil) is contained in a 18 m-diameter stainless steel spherical vessel. On the inner surface of the vessel there are 1879 50-cm diameter PMTs (the PMT coverage is 34%). The internal detector is surrounded by 3.2 kton water with 225 PMTs (outer detector). This Cherenkov detector serves as a veto which provides shielding from cosmic-ray muons and external radioactivity.

In the KamLAND experiment $\bar{\nu}_e$ from about 50 power reactors were detected. A flux-averaged distance between reactors and the Kamioka mine was $\sim$180 km. About 80% of antineutrinos came from 26 reactors at distances 138–214 km.

Reactor $\bar{\nu}_e$s are produced in decays of nuclei, which are products of fission of ^{235}U (57%), ^{238}U (7.8%), ^{239}Pu (29.5%) and ^{241}Pu (5.7%). Each fission, in which about 200 MeV is produced, is accompanied by the emission of 6 $\bar{\nu}_e$. A reactor with power about 3 GW$_{th}$ emits about $6 \cdot 10^{20}$ $\bar{\nu}_e$/s.

Reactor antineutrinos are detected through the observation of the inverse β-decay

$$\bar{\nu}_e + p \to e^+ + n. \tag{11.59}$$

The threshold of this process is 1.8 MeV. Two γ-quanta from the annihilation of e^+ (prompt signal) and γ-quanta produced in a neutron capture by a proton or a ^{12}C nuclei (delayed signal with a mean delay time $(207.5 \pm 2.8) \cdot \mu$s) are detected in the experiment. The signature of the event in the KamLAND experiment (and in other reactor neutrino experiments) is a coincidence between the prompt and delayed signals. It provides a strong suppression of a radioactive background.

The prompt energy E_p is connected with the neutrino energy E by the relation

$$E \simeq E_p + \bar{E}_n + 0.8 \text{ MeV}. \tag{11.60}$$

where $\bar{E}_n$ is average neutron recoil energy ($\simeq 10 \text{ keV}$). The prompt energy is the sum of the positron kinetic energy and the annihilation energy ($2\, m_e$).

In the KamLAND experiment not only $\bar{\nu}_e$ from reactors but also $\bar{\nu}_e$, which are produced in decay chains of ^{238}U and ^{232}Th in the earth (geo-neutrinos), are detected. The prompt energy released in the interaction of geo-neutrinos with protons is less than 2.6 MeV. In order to avoid geo-neutrino background in the study of neutrino oscillations the KamLAND Collaboration imposed 2.6 MeV E_p-threshold.

As we have seen before, the neutrino oscillation length is given by the expression

$$L_{12} \simeq 2.5 \, \frac{E}{\Delta m_{12}^2} \text{ m}, \tag{11.61}$$

where E is the neutrino energy in MeV and Δm_{12}^2 is the neutrino mass-squared difference in eV^2. The average energy of the reactor antineutrinos is 3.6 MeV. For the solar neutrino mass-squared difference $\Delta m_{12}^2 \simeq 8 \cdot 10^{-5}$ eV^2 at $E = 3.6$ MeV we have $L_{12} \simeq 120$ km. From this estimate we conclude that distances between the KamLAND detector and Japanese reactors are appropriate to study neutrino oscillations driven by the solar neutrino mass-squared difference.

The KamLAND experiment started in 2002 and continued up to 2012. During this period 2611 $\bar{\nu}_e$ events were detected with an estimated background (excluding geo-neutrinos) 364.1 ± 30.5 events. Expected number of the events (assuming that there are no neutrino oscillations) is 3564 ± 145.

From the latest 3-neutrino analysis of all KamLAND data for the neutrino oscillation parameters Δm_S^2, $\tan^2 \theta_{12}$ and $\sin^2 \theta_{13}$ the following values were obtained

$$\Delta m_S^2 = (7.54^{+0.19}_{-0.18}) \cdot 10^{-5} \text{ eV}^2, \quad \tan^2 \theta_{12} = 0.481^{+0.092}_{-0.080}, \quad \sin^2 \theta_{13} = 0.010^{+0.033}_{-0.034}. \tag{11.62}$$

From a joint analysis of the KamLAND data and the data of all solar neutrino experiments a better accuracy for the parameter $\tan^2 \theta_{12}$ can be inferred[12]:

$$\Delta m_S^2 = (7.53^{+0.19}_{-0.18}) \cdot 10^{-5} \text{ eV}^2, \quad \tan^2 \theta_{12} = 0.437^{+0.029}_{-0.026}, \quad \sin^2 \theta_{13} = 0.023^{+0.015}_{-0.015}. \tag{11.63}$$

In Fig. 11.5 the ratio of the numbers of the observed and expected events (the $\bar{\nu}_e$ survival probability) as a function of L_0/E is plotted ($L_0 = 180$ km is

---

[12]This analysis is based on the assumption of the *CPT* invariance.

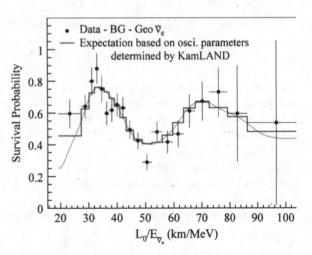

the flux-averaged distance between reactors and the KamLAND detector, E is
the neutrino energy). The curve is the calculated survival probability with best-
fit parameters obtained from the three-neutrino analysis of the KamLAND data.
Figure 11.5 illustrates oscillatory behavior of the $\bar{\nu}_e$ survival probability determined
from the data of the KamLAND experiment.

11.5   Measurement of the Angle θ_{13} in Reactor Experiments

11.5.1   Introduction: CHOOZ Reactor Experiment

From the data of first neutrino oscillation experiments an information about two
mass-squared differences (atmospheric Δm_A^2 and solar Δm_S^2) and two mixing
angles (θ_{23} and θ_{12}) was obtained. During many years only an upper bound for
the parameter $\sin^2 \theta_{13}$ was known. This bound was obtained from the data of the
reactor **CHOOZ** experiment.

The $\bar{\nu}_e$-survival probability, driven by the atmospheric mass-squared difference,
in the major two-neutrino approximation is given by the expression

$$P(\bar{\nu}_e \rightarrow \bar{\nu}_e) = 1 - \sin^2 2\theta_{13} \; \sin^2(\frac{\Delta m_A^2 L}{4E}) \qquad (11.64)$$

Thus the study of the disappearance of the reactor $\bar{\nu}_e$ allows to determine the
parameter $\sin^2 2\theta_{13}$. For the average energy of the reactor antineutrinos (3.6 MeV)
the corresponding oscillation length is equal to

$$L_{\text{osc}} \simeq 2.5 \frac{E}{\Delta m_A^2} \; m \simeq 3.6 \, \text{km}. \qquad (11.65)$$

In the CHOOZ experiment one antineutrino detector was exposed to two reactors
of the CHOOZ power station ($8.5\,\mathrm{GW}_{th}$). The distance between the detector
and reactors was about 1 km. The CHOOZ detector comprised 5 tons of Gd-loaded
liquid scintillator contained in an acrylic vessel. The antineutrinos were detected
through the observation of the classical reaction

$$\bar{\nu}_e + p \rightarrow e^+ + n. \tag{11.66}$$

From April 1997 till July 1998, in the CHOOZ experiment 3600 antineutrino events
were recorded. For the ratio R of the total number of detected and the expected
events it was found

$$R = 1.01 \pm 2.8\% \text{ (stat.)} \pm \pm 2.7\% \text{ (syst.).} \tag{11.67}$$

From the data of the experiment the following upper bound

$$\sin^2 2\theta_{13} \leq 0.16. \tag{11.68}$$

was obtained.

In 2012 three new reactor experiments Daya Bay, RENO and Double Chooz
started. The aim of these experiments was to measure the angle θ_{13} (or to
improve the CHOOZ bound). The parameter $\sin^2 2\theta_{13}$ was successfully measured
(it occurred that its value is close to the CHOOZ upper bound (11.68)). This finding
is extremely important for the future investigation of the problem of the neutrino
mixing. As we have seen before, the CP phase δ enter into the PMNS mixing matrix
the form $\sin \theta_{13}\, e^{-i\delta}$. Thus such fundamental effect of the three-neutrino mixing as
CP violation in the lepton sector can be studied only if the mixing angle θ_{13} is not
equal to zero (and relatively large). Another problem, the solution of which requires
nonzero θ_{13}, is the problem of the character of the neutrino mass spectrum (normal
or inverted ordering?). In the next subsections we will briefly discuss Daya Bay,
RENO and Double Chooz experiments.

11.5.2   Daya Bay Experiment

In the Daya Bay experiment antineutrinos from six commercial nuclear reactors,
located at Daya Bay and Ling Ao nuclear power stations (China), are detected.
Each reactor has a thermal power $2.9\,\mathrm{GW}_{th}$. Antineutrino detectors are disposed
in two near underground halls and one far underground hall (correspondingly, at the
distances 350–600 m and 1500–1950 m from the reactors). In each near hall there
are two antineutrino detectors. In the first phase of the experiment (December 2011–
July 2012) there were two antineutrino detectors in the far hall. In October 2012 two
additional antineutrino detectors were disposed in the far hall.

All eight antineutrino detectors have identical three-zone structure: 20 tons of Gd-loaded liquid scintillator in the inner zone ($\bar{\nu}_e$ detector), 22 tons of liquid scintillator in the middle zone and 37 tons of mineral oil in external zone. Scintillation light is detected by 192 8-in. PMTs.

Reactor $\bar{\nu}_e$'s are detected via the observation of the inverse β-decay

$$\bar{\nu}_e + p \rightarrow e^+ + n. \tag{11.69}$$

Detection of photons produced in $e^+ - e^-$ annihilation (prompt signal) allows to determine the positron energy ($E_{\text{pr}} = T + 2m_e$, T being positron kinetic energy). The neutron, produced in (11.69), thermalizes and is captured by a Gd nucleus, producing γ-rays with total energy 8 MeV, or a proton, producing γ-quantum with the energy 2.2 MeV (delayed signal).

During 217 days of the data-taking with six antineutrino detectors and 1013 days of the data-taking with eight antineutrino detectors in the Daya Bay experiment the total number of $2.5 \cdot 10^5$ $\bar{\nu}_e$-events were observed. Such large statistics allowed the Daya Bay Collaboration to obtain a very precise value of the parameter $\sin^2 \theta_{13}$. For the ratio of the $\bar{\nu}_e$ total rates in the far and near detectors it was found

$$R = 0.949 \pm 0.002 \text{ (stat.)} \pm 0.002 \text{ (syst.)} \tag{11.70}$$

From the three-neutrino analysis of observed rate and energy spectrum the following values of the parameters $\sin^2 2\theta_{13}$ and $|\Delta m_{ee}^2|$ were obtained

$$\sin^2 2\theta_{13} = 0.0841 \pm 0.0027 \text{ (stat.)} \pm 0.0019 \text{ (syst.)} \tag{11.71}$$

and

$$|\Delta m_{ee}^2| = (2.50 \pm 0.06 \text{ (stat.)} \pm 0.06 \text{ (syst.))} \cdot 10^{-3} \text{ eV}^2, \tag{11.72}$$

where

$$|\Delta m_{ee}^2| = \cos^2 \theta_{12} |\Delta m_{13}^2| + \sin^2 \theta_{12} |\Delta m_{23}^2|. \tag{11.73}$$

is the effective ("average") reactor mass-squared difference.

In the case of the normal ordering of the neutrino masses it was found

$$\text{NO} \quad \Delta m_A^2 = (2.45 \pm 0.06 \text{ (stat.)} \pm 0.06 \text{ (syst.))} \cdot 10^{-3} \text{ eV}^2. \tag{11.74}$$

For the inverted ordering it was obtained

$$\text{IO} \quad \Delta m_A^2 = (2.56 \pm 0.06 \text{ (stat.)} \pm 0.06 \text{ (syst.))} \cdot 10^{-3} \text{ eV}^2. \tag{11.75}$$

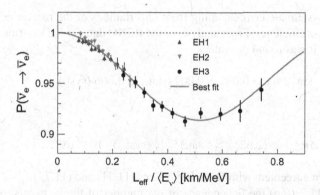

Fig. 11.6 Daya Bay experiment: the reactor $\bar{\nu}_e$ survival probability as function of the parameter L_{eff}/E. The points are the ratios of the observed and expected events. The solid line was calculated using the best-fit values of the parameters $\sin^2 2\theta_{13}$ and $|\Delta m_{ee}^2|$ (arXiv:1610.04802)

These values for the atmospheric mass-squared difference are in a good agreement with the values which were found from the data of the accelerator neutrino experiments which we will discuss later.

In Fig. 11.6 the reactor $\bar{\nu}_e$ survival probability as function of the parameter L_{eff}/E is presented. The points are the ratios of the observed and expected events. The curve was calculated using the best-fit values of the parameters $\sin^2 2\theta_{13}$ and $|\Delta m_{ee}^2|$. The curve demonstrates oscillation behavior of the event rates observed in the Daya Bay experiment.

11.5.2.1 RENO Experiment

The reactor neutrino experiment **RENO** started in August 2011. Antineutrinos from six reactors at Hanbit nuclear power plant (Korea) are detected by two underground detectors, located at the distances 294 and 1383 m from the center of reactor array. The thermal power of each reactor is about $2.8\,\text{GW}_{\text{th}}$.

Near and far detectors in the RENO experiment are identical. The innermost part of the detector is an acrylic vessel filled with 16 tons of Gd-doped liquid scintillator (target). It is contained in another acrylic vessel filled with liquid scintillator (γ-catcher in which γ-quanta, escaping from the target region, are detected). Outside the γ-catcher there is a 70-cm thick layer of mineral buffer oil which provides shielding from external radioactivity. Produced light is detected by 354 10-in. PMTs which are mounted on the inner wall of a stainless steel container. The container is surrounded by a veto water-Cherenkov detector which provides shielding against γ-quanta and neutrons from surrounding rocks.

During 500 days of the data-taking it was detected 290,775 (31,514) $\bar{\nu}_e$ events in the near (far) detector. Two identical detectors allow to perform a relative measurement of antineutrino rates and spectra. This relative measurement allowed to reduce systematic errors coming from uncertainties of the reactor neutrino flux and detection efficiency. From analysis of the RENO data for the neutrino oscillation parameters it was found the values

$$\sin^2 2\theta_{13} = 0.082 \pm 0.009 \text{ (stat.)} \pm 0.006 \text{ (syst.)} \tag{11.76}$$

and

$$|\Delta m_{ee}^2| = (2.62_{-0.23}^{+0.21} \text{ (stat.)}_{-0.13}^{+0.12} \text{ (syst.)}) \cdot 10^{-3} \text{ eV}^2, \tag{11.77}$$

which are in agreement with Daya Bay values (11.71) and (11.72).

In Fig. 11.7 (top) the dependence of the number of the $\bar{\nu}_e$ events, observed in the far RENO detector, on the prompt energy is plotted. The shaded histogram was found from near detector data. On the bottom the ratio of the number of $\bar{\nu}_e$ far detector events to the number of predicted events (assuming that there is no neutrino oscillations) is depicted (points). The shaded band is the ratio of the number of the far detector events to the number of MC predicted best-fit events.

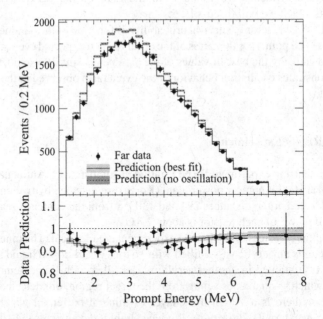

Fig. 11.7 Top: the dependence of the number of the $\bar{\nu}_e$ events, observed in the far RENO detector, on the prompt energy. The shaded histogram was found from near detector data. Bottom: points are the ratio of the number of $\bar{\nu}_e$ far detector events to the number of predicted events (assuming that there is no neutrino oscillations). The shaded band is the ratio of the number of the far detector events to the number of MC predicted best-fit events (arXiv:1610.04326v3)

11.5.2.2 Double Chooz Experiment

In the Double Chooz experiment $\bar{\nu}_e$'s from two reactors (the thermal power of each reactor is $4.25\,\mathrm{GW_{th}}$) are detected. The experiment is performed at CHOOZ-B power plant, Chooz, France. The Double Chooz experiment started in 2011 with one detector at the average distance 1050 m from reactors. In 2015 the near detector at the average distance 400 m was constructed. Both detectors have identical structure. The inner target detector has $10.3\,\mathrm{m^3}$ of Gd-doped liquid scintillator. It is surrounded by $22.4\,\mathrm{m^3}$ liquid scintillator (γ-catcher) and $100\,\mathrm{m^3}$ non-scintillating mineral buffer oil. Photons, produced in the target and γ-catcher, are detected by 390 10-in. diameter PMTs. Optically separated from three inter volumes there is an external veto detector.

Reactor antineutrinos are detected via classical reaction

$$\bar{\nu}_e + p \to e^+ + n. \tag{11.78}$$

Coincidence of a prompt signal from annihilation of e^+ with e^- in the liquid scintillator (energies from 1 to 11 MeV) and a delayed signal from capture of neutron by a Gd nucleus (photons with total energy 8 MeV) is a good signature of the event.

Disappearance of the reactor $\bar{\nu}_e$ was observed in the experiment. In the case of the three-neutrino mixing the probability of $\bar{\nu}_e$ to survive can be presented in the form

$$P(\bar{\nu}_e \to \bar{\nu}_e) \simeq 1 - \sin^2 2\theta_{13} \sin^2\left(\frac{\Delta m_{ee}^2 L}{4E}\right) - \cos^4\theta_{13} \sin^2 2\theta_{12} \sin^2\left(\frac{\Delta m_S^2 L}{4E}\right),$$
$$\tag{11.79}$$

where the effective reactor mass-squared difference Δm_{ee}^2 is given by (11.73).

From analysis of the data of 673 days of far detector and 151 days of near detector, using the constraint

$$|\Delta m_{ee}^2| = (2.44 \pm 0.09) \cdot 10^{-3}\ \mathrm{eV^2}, \tag{11.80}$$

obtained from the data of the accelerator MINOS experiment, in the Double Chooz experiment it was found

$$\sin^2 2\theta_{13} = 0.111 \pm 0.018\ (\text{stat.} + \text{syst.}). \tag{11.81}$$

11.6 Long-Baseline Accelerator Neutrino Experiments

11.6.1 K2K Accelerator Neutrino Experiment

In long baseline accelerator neutrino experiments there is a possibility to use beams of neutrinos and antineutrinos, to work with narrow band neutrino beams (off-axis neutrino beams) etc. This allow to study CP violation in the lepton sector, to reveal the character of the neutrino mass spectrum and to perform a high precision measurement of neutrino oscillation parameters.

Oscillations in the long baseline accelerator neutrino experiments are driven predominantly by the atmospheric mass-squared difference Δm_A^2. For a neutrino energy $E \simeq 1\,\mathrm{GeV}$ and $\Delta m_A^2 \simeq 2.5 \cdot 10^{-3}\,\mathrm{eV}^2$ the oscillation length L_{osc} is given by

$$L_{\mathrm{osc}} \simeq 2.5\ \frac{E}{\Delta m_A^2}\ \mathrm{m} \simeq 10^3\,\mathrm{km}\ . \tag{11.82}$$

In the first long baseline **K2K** experiment the distance between the neutrino source (KEK accelerator, Japan) and the neutrino detector (Super-Kamiokande) was about 250 km.

Protons with an energy of 12 GeV from the KEK-PS accelerator bombard an aluminum target in which secondary particles were produced. Positively charged particles (mainly π^+) were focused in horns and decayed in a 200 m-long decay pipe. After a beam dump in which all hadrons and muons were absorbed a neutrino beam was produced (there were 97.3% of ν_μ, 1.3% of ν_e and 1.4% of $\bar\nu_\mu$ in the beam). The neutrinos had energies in the range (0.5–1.5) GeV.

In the K2K experiment the disappearance of muon neutrinos was searched for. The two-neutrino probability of ν_μ to survive has the form

$$P(\nu_\mu \to \nu_\mu) \simeq 1 - \sin^2 2\theta_{23}\ \sin^2(1.27\ \Delta m_A^2 \frac{L}{E}), \tag{11.83}$$

where E is the neutrino energy in GeV, L is the source-detector distance in km and Δm_A^2 is the atmospheric neutrino mass-squared difference in eV^2.

From 1999 till 2004 in the K2K experiment 112 neutrino events were detected. For the number of the expected events (in the case if there were no neutrino oscillations) was found the value $158^{+9.2}_{-8.6}$. In the low energy region the distortion of the neutrino spectrum was observed.

From the two-neutrino analysis of the K2K data under the assumption that $\sin^2 2\theta_{23} = 1$ it was found the following 90% CL range for the parameter Δm_A^2:

$$1.9 \cdot 10^{-3}\,\mathrm{eV}^2 < \Delta m_A^2 < 3.5 \cdot 10^{-3}\,\mathrm{eV}^2. \tag{11.84}$$

The K2K experiment was the first experiment with artificially produced neutrinos which confirmed the existence of neutrino oscillations discovered in the atmospheric Super-Kamiokande neutrino experiment.

11.6.2 *MINOS Accelerator Neutrino Experiment*

In the long baseline **MINOS** experiment muon neutrinos produced at the Fermilab Main Injector facility were detected in the Sudan mine (Minnesota, USA) at a distance of 735 km. Data-taking started in the MINOS experiment in 2005 and finished in 2012. In 2013 the experiment MINOS+, successor of the MINOS, started.

Protons with an energy of 120 GeV, extracted from the Main Injector proton accelerator, bombarded a graphite target and produced (predominantly) pions and kaons. Positively (or negatively) charged particles were focused by two magnetic horns and directed into a 675 m long decay pipe. After the pipe there was an absorber for hadrons and 240 m of rock in which muons were stopped.

Muon neutrinos were produced in the decays $\pi^+(K^+) \to \mu^+ + \nu_\mu$. Electron neutrinos were produced in the decays $\mu^+ \to e^+ + \nu_e + \bar{\nu}_\mu$ and $K^+(K_L^0) \to e^+ + \nu_e + \pi^0(\pi^-)$. Antineutrinos were created in charge conjugated processes. The ν_μ-dominated beam consisted of ν_μ (91.7%), $\bar{\nu}_\mu$ (7%), ν_e and $\bar{\nu}_e$ (1.3%). The $\bar{\nu}_\mu$-enhanced beam consisted of ν_μ (58.1%), $\bar{\nu}_\mu$ (39.9%), ν_e and $\bar{\nu}_e$ (2.0%). For the MINOS experiment the Main Injector supplied $10.71 \cdot 10^{20}$ protons-on-target (POT) in order to produce the ν_μ-dominated beam and $3.36 \cdot 10^{20}$ POT in order to produce the $\bar{\nu}_\mu$-enhanced beam.

The majority of the MINOS data was obtained with the low-energy neutrino beam ($1 \leq E \leq 5$ GeV) which has a peak at 3 GeV. There were two identical neutrino detectors in the MINOS experiment. The near detector (ND) with a mass of about 1 kton was at a distance of about 1.04 km from the target and about 100 m underground. The far detector (FD) with a mass of 5.4 kton was at a distance of 735 km from the target and 705 m underground (2070 m water equivalent). The detectors were steel (2.54 cm thick)-scintillator (1 cm thick) calorimeters magnetized to 1.3 T (ND) and 1.4 T (FD). The measurement of the curvature of the muon tracks allows to distinguish ν_μ from $\bar{\nu}_\mu$ and to measure energy of muons which leave the detector. The energies of the muons which are stopped in the detector are determined by their ranges.

Muon neutrinos and antineutrinos were detected in the MINOS experiment via the observation of the CC process

$$\nu_\mu(\bar{\nu}_\mu) + \text{Fe} \to \mu^-(\mu^+) + X. \qquad (11.85)$$

Such events are characterized by tracks caused by muons and a hadronic showers. The neutrino energy is given by the sum of the muon energy and the energy of the hadronic shower. Electron neutrinos were detected via the observation of the CC process

$$\nu_e + \text{Fe} \to e^- + X. \qquad (11.86)$$

The signature of such events is an electromagnetic shower. Because electrons do not have track-like topology, ν_e and $\bar{\nu}_e$ events can not be separated.

In the near detector the initial neutrino spectrum was measured. This measurement allowed to predict the expected spectrum of the muon neutrinos in the far detector in the case if there were no neutrino oscillations and to determine the $\nu_\mu(\bar{\nu}_\mu)$ survival probability as a function of the neutrino energy. The search for $\nu_\mu \rightarrow \nu_e$ and $\bar{\nu}_\mu \rightarrow \bar{\nu}_e$ appearance was also performed.

In the MINOS experiment not only Main Injector beam generated neutrino events but also atmospheric neutrino events were observed (starting from 2003). In the final three-neutrino oscillation analysis all detected events were taken into account. In this analysis the average reactor neutrino value $\sin^2 \theta_{13} = 0.0242 \pm 0.0025$ was used. The solar-KamLAND values $\Delta m_S^2 = 7.54 \cdot 10^{-5}$ eV^2 and $\sin^2 \theta_{12} = 0.307$ were also kept fixed in the fit. In the case of the normal mass ordering from analysis of the MINOS data it was found

$$\Delta m_A^2 = (2.28 - 2.46) \cdot 10^{-3} \text{ eV}^2 \ (68\%), \quad \sin^2 \theta_{23} = (0.35 - 0.65) \ (90\%)$$
$$(11.87)$$

In the case of the inverted mass ordering it was obtained

$$\Delta m_A^2 = (2.32 - 2.53) \cdot 10^{-3} \text{ eV}^2 \ (68\%), \quad \sin^2 \theta_{23} = (0.34 - 0.67) \ (90\%)$$
$$(11.88)$$

The MINOS+ experiment is running in the high-energy region 3–10 GeV. During the first year of the operation, starting from September 2013, it was collected $2.99 \cdot 10^{20}$ POT. Oscillation parameters found from analysis of the MINOS+ data are in agreement with the oscillation parameters (11.87) and (11.88) obtained from the analysis of the MINOS data.

In Fig. 11.8 neutrino energy spectrum, obtained from the results of the MINOS and MINOS+ experiments, is presented. The curve is the prediction calculated with best-fit MINOS oscillation parameters. The histogram is the expected spectrum in the case if there is no neutrino oscillations.

11.6.3 T2K Experiment

The **T2K** experiment is performed at J-PARC accelerator facility in Tokai (Japan). In order to produce a neutrino beam a 30 GeV protons hit a graphite target and produce charged pions and kaons which are focused by three magnetic horns. Either positive or negative mesons are focused resulting in a beam of predominantly ν_μ or $\bar{\nu}_\mu$ produced in a 96 m long decay tube. The decay volume is followed by the beam dump and muon monitors. Neutrinos are detected by an on axis near detector and of axis, at 2.5° relative to the beam direction, near and far detectors. The off axis narrow band neutrino energy spectrum has a peak at 0.6 GeV.

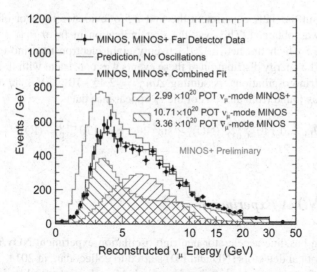

Fig. 11.8 Neutrino energy spectrum, obtained from the results of the MINOS and MINOS+ experiments. The curve is the prediction calculated with best-fit MINOS oscillation parameters. The histogram is the expected spectrum in the case of no neutrino oscillations (arXiv:1601.05233v3)

Two T2K near detectors INGRID (on axes) and ND280 (off axis) are located at the distance 280 m from the target. They measure the beam direction, composition, neutrino spectrum and the event rate. The 50 kton water-Cherenkov Super-Kamiokande detector is used as a far detector in the T2K experiment. It is located off axes at the distance 295 km from the target.[13]

From January 2010 till May 2013 in the T2K experiment only neutrino data were collected. From May 2014 till May 2016 predominantly antineutrino events were observed. This corresponds to a neutrino and antineutrino beam exposures on the far detector, correspondingly, $7.48 \cdot 10^{20}$ POT and $7.47 \cdot 10^{20}$ POT.

From the three-neutrino analysis of ν_μ and $\bar\nu_\mu$ disappearance data it was found

$$\sin^2 \theta_{23} = 0.514^{+0.055}_{-0.056}, \quad \Delta m_A^2 = (2.51 \pm 0.10) \cdot 10^{-3} \text{ eV}^2 \tag{11.89}$$

in the case of the normal mass ordering and

$$\sin^2 \theta_{23} = 0.511 \pm 0.055, \quad \Delta m_A^2 = (2.48 \pm 0.10) \cdot 10^{-3} \text{ eV}^2 \tag{11.90}$$

for the inverted mass ordering.

---

[13]For $E \simeq 0.6$ GeV and L=295 km we have $1.27 \frac{\Delta m_A^2 L}{E} \simeq \frac{\pi}{2}$. Thus, neutrino energy and the source-detector distance in the T2K experiment correspond to the first maximum of oscillations driven by the atmospheric mass-squared difference.

In 2011 in the T2K experiment the first 2.5 σ evidence in favor of $\nu_\mu \rightarrow \nu_e$ transition was obtained. This was the first indication in favor $\theta_{13} \neq 0$. During the exposure which finished in 2014 twenty eight electron neutrino events were observed. The energy distribution of these events was consistent with ν_e appearance due to neutrino oscillations. Assuming $\Delta m_A^2 = 2.45 \cdot 10^{-3}$ eV^2, $\sin^2 \theta_{23} = 0.5$, $\delta = 0$, it was found from analysis of the ν_e appearance data

$$\sin^2 2\theta_{13} = 0.140^{+0.038}_{-0.032} \text{ (NO)}, \quad \sin^2 2\theta_{13} = 0.170^{+0.045}_{-0.037} \text{ (IO)}. \qquad (11.91)$$

11.6.4 NOvA Experiment

A new long baseline accelerator neutrino oscillation experiment **NOvA** with near and far identical detectors (ND and FD) started to collect data in 2014. The energy spectrum of neutrinos, produced at FermiLab Main Injector facility, is measured by the 290-ton ND, located 1 km away from the Main Injector target, 100 m underground. The 14-kton FD is located on the surface, 14.6 mrad off axis, at the distance 810 km from the FermiLab (Ash River, Minnesota, USA). A narrow band neutrino beam with peak energy about 2 GeV (the first oscillation maximum) is utilized in the NOvA experiment. The flavor composition of the neutrino beam at the FD is estimated to be 97.8% ν_μ, 1.6% $\bar{\nu}_\mu$ and 0.6% $(\nu_e + \bar{\nu}_e)$ assuming that there are no oscillations.

Both ND and FD are segmented tracking calorimeters. Reflective cells of length 15.5 m (3.9 m) in the FD (ND) with a 3.9×6.6 cm^2 cross section are filled with liquid scintillator. Light, produced by charged particles, is collected in each cell by optical fiber and measured with an avalanche photodiode.

The study of ν_μ disappearance requires identification in FD of the reaction $\nu_\mu + N \rightarrow \mu^- + X$ and the measurement of the neutrino energy. The neutrino energy is given by a sum of the reconstructed muon energy and recoil hadronic energy. The investigation of ν_e appearance requires identification in FD of the process $\nu_e + N \rightarrow e^- + X$ and understanding background processes. The signature of this CC process in the NOvA detectors is an electromagnetic shower and associated hadronic recoil energy.

From the analysis of the NOvA ν_μ disappearance data, obtained from $6.05 \cdot 10^{20}$ POT exposure, in the case of the normal mass ordering it was found

$$\Delta m_A^2 = (2.67 \pm 0.11) \cdot 10^{-3} \text{ eV}^2 \qquad (11.92)$$

For the parameter $\sin^2 \theta_{23}$ two statistically-degenerate values were obtained

$$\sin^2 \theta_{23} = 0.404^{+0.030}_{-0.022}, \quad \sin^2 \theta_{23} = 0.624^{+0.022}_{-0.030} \qquad (11.93)$$

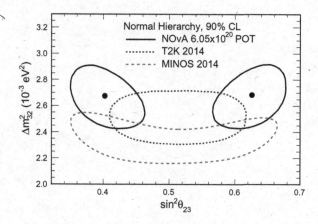

Fig. 11.9 NOvA experiment: 90% CL allowed region in the plane of the parameters $\sin^2 \theta_{23}$ and Δm_A^2. T2K and MINOS allowed regions are also shown (arXiv:1701.05891v1)

In the case of the inverted mass ordering it was found

$$\Delta m_A^2 = (2.72 \pm 0.11) \cdot 10^{-3} \text{ eV}^2 \qquad (11.94)$$

and

$$\sin^2 \theta_{23} = 0.398^{+0.030}_{-0.022}, \text{ or } \sin^2 \theta_{23} = 0.618^{+0.022}_{-0.030} \qquad (11.95)$$

In Fig. 11.9 NOvA 90% CL allowed region in the plane of the parameters $(\sin^2 \theta_{23}, \Delta m_A^2)$ are shown. T2K and MINOS allowed regions are also presented.

With the same exposure $6.05 \cdot 10^{20}$ POT 33 ν_e candidate events with a background 8.2 ± 0.8 events were observed in the NOvA experiment. Combing these data with NOvA ν_μ disappearance data and with the reactor value of the parameter $\sin^2 \theta_{13}$ the NOvA Collaboration concluded that inverted neutrino mass spectrum with $\theta_{23} < \frac{\pi}{4}$ is disfavored at 93% CL for all values of the CP phase δ.

Chapter 12
Neutrino in Cosmology

12.1 Basics of Cosmology

12.1.1 Introduction

All existing cosmological data are described by the standard Big-Bang cosmological model which we will briefly discuss in this section. The standard cosmology is based on

1. Cosmological Principle.
2. Friedman equations which are a consequence of the Einstein equations of the general relativity.

12.1.2 Cosmological Principle

According to **the cosmological principle** the Universe observed from any spacial position and at any time is *isotropic and homogeneous* at a large scale. This principle was formulated by Einstein as a theoretical suggestion. Present day cosmological observations (cosmic microwave background radiation, the large-scale structure of the Universe and others) are in agreement with the cosmological principle at a scale ~ 100 Mpc.[1]

---

[1] Mpc $= 3.26 \ 10^6$ light-year $= 3.09 \cdot 10^{22}$ m.

© Springer International Publishing AG, part of Springer Nature 2018
S. Bilenky, *Introduction to the Physics of Massive and Mixed Neutrinos*,
Lecture Notes in Physics 947, https://doi.org/10.1007/978-3-319-74802-3_12

12.1.3 Friedman-Robertson-Walker Metric: Hubble Law

Let $x^\alpha = (x^0, \mathbf{x})$ be a time-space coordinate of a point in some coordinate system. The square of the element of length (interval) has the following general form

$$ds^2 = g_{\alpha\beta}(x)\, dx^\alpha dx^\beta. \tag{12.1}$$

where $g_{\alpha\beta}(x) = g_{\beta\alpha}(x)$ is the metric tensor (or metric). The metric determines the geometry of a space. It plays a fundamental role in the General Theory of Relativity and Cosmology.

The metric depends on the coordinate system. The comoving coordinate system, the system in which matter is at rest, is the most natural reference system in cosmology. Because comoving observers see the same sequence of events they have the same time which is proper time. The Universe is isotopic and homogeneous only in the comoving system.

We will consider the metric of isotropic and homogeneous space in the comoving system. Because in such a space all directions are equivalent, we have $g_{0i} = 0$ ($i = 1, 2, 3$). Thus in an isotropic and homogeneous space interval can be presented in the form

$$ds^2 = dt^2 - g_{ik}\, dx^i dx^k. \tag{12.2}$$

Let us consider $t = $ const case. In the Euclidean space for the element of the length we have

$$dl^2 = \sum_{i=1}^{3} (dx^i)^2, \tag{12.3}$$

where x^1, x^2, x^3 are Cartesian coordinates. They are connected with the spherical coordinates ρ, θ and ϕ by the relations

$$x^1 = \rho \sin\theta \cos\phi, \quad x^2 = \rho \sin\theta \sin\phi, \quad x^3 = \rho \cos\theta. \tag{12.4}$$

In spherical coordinates we have

$$dl^2 = d\rho^2 + \rho^2(d\theta^2 + \sin^2\theta d\phi^2). \tag{12.5}$$

In the general case of the isotropic space we have

$$dl^2 = d\rho^2 + f^2(\rho)\, (d\theta^2 + \sin^2\theta d\phi^2), \tag{12.6}$$

Fig. 12.1 Robertson-Walker
geometry of a homogeneous
and isotropic space

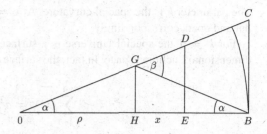

where $f(\rho)$ is a function of ρ. The condition of isotropy allows to determine possible functions $f(\rho)$. In fact, let us consider Fig. 12.1. Assuming that angles α and β are infinitesimally small, we have

$$CB = f(2\rho)\alpha = f(\rho)\beta, \quad DE = f(\rho - x)\alpha + f(x)\beta = f(\rho + x)\alpha. \quad (12.7)$$

From these relations we obtain the following relation

$$f(\rho - x) + f(x)\frac{f(2\rho)}{f(\rho)} = f(\rho + x) \quad (12.8)$$

Now if we take the derivative over x and put $x = 0$ from (12.8) we find

$$\frac{d\, f(\rho)}{d\rho} = \frac{f(2\rho)}{2f(\rho)}. \quad (12.9)$$

Notice that in deriving (12.9) we took into account that $f(\rho) \simeq \rho$ at $\rho \to 0$.
   It is obvious that

$$f_1(\rho) = \sin \rho, \quad f_0(\rho) = \rho, \quad f_{-1}(\rho) = \sinh \rho \quad (12.10)$$

are solutions of Eq. (12.9). It is possible to show that there are no other solutions of this equation. For arbitrary t we have

$$ds^2 = dt^2 - a^2(t)(d\rho^2 + f_k^2(\rho)\,(d\theta^2 + \sin^2\theta\, d\phi^2)), \quad (12.11)$$

where the functions $f_k(\rho)$ are given by (12.10) and $a(t)$ is function of t (scale factor). This function can not be determined from the requirements of isotropy. The metric (12.11) is called the Friedmann-Robertson-Walker metric.
   The Friedmann-Robertson-Watson metric can be presented in another form. Let us introduce the variable $r = f(\rho)$. We have

$$ds^2 = dt^2 - a^2(t)(\frac{dr^2}{1 - kr^2} + r^2\,(d\theta^2 + \sin^2\theta d\phi^2)), \quad k = 1, 0, -1. \quad (12.12)$$

The parameter k is the spacial curvature. At $k = +1, 0, -1$ the Universe is closed, flat and open, correspondingly.

For $k = 1$ the spacial Universe is a surface of a sphere of radius a in the 4-dimensional Euclidian space. In fact, the surface of a sphere is given by the equation

$$\sum_{i=1}^{4}(x^i)^2 = a^2. \tag{12.13}$$

For the metric on the sphere we have

$$dl^2 = \sum_{i=1}^{4}(dx^i)^2. \tag{12.14}$$

Equation (12.13) allows to exclude the x^4 coordinate. We find

$$dx^4 = -\frac{\sum_{i=1}^{3} x^i dx^i}{\sqrt{a^2 - \sum_{i=1}^{3}(x^i)^2}}. \tag{12.15}$$

Let us introduce the spherical coordinates

$$x^1 = r' \sin\theta \cos\phi, \quad x^2 = r' \sin\theta \sin\phi, \quad x^3 = r' \cos\theta. \tag{12.16}$$

For the metric we have

$$dl^2 = dr'^2 + r'^2(d\theta^2 + \sin^2\theta d\phi^2) + \frac{r'^2 dr'^2}{a^2 - r'^2} = \frac{a^2}{a^2 - r'^2}dr'^2 + r'^2(d\theta^2 + \sin^2\theta d\phi^2). \tag{12.17}$$

Finally, introducing the dimensionless variable $r = \frac{r'}{a}$, we find the following expression

$$dl^2 = a^2(t)(\frac{dr^2}{1 - r^2} + r^2(d\theta^2 + \sin^2\theta d\phi^2)). \tag{12.18}$$

The case $k = 0$ in (12.12) corresponds to the flat space. The case $k = -1$ can be obtained from the expression (12.17) if we change $a \rightarrow ia$. It corresponds to the space with negative curvature (hyperboloid in the 4-dimensional space).

The proper distance to an object is given by the relation

$$d(t) = a(t)r, \tag{12.19}$$

where r is the comoving distance which does not depend on time. From (12.19) we have

$$\dot{d}(t) = v(t) = \dot{a}(t)\, r. \tag{12.20}$$

Further, from (12.19) and (12.20) we find

$$v(t) = H(t)\, d(t), \tag{12.21}$$

where

$$H(t) = \frac{\dot{a}(t)}{a(t)}. \tag{12.22}$$

Thus, the relative velocity of two Galaxies is proportional to the distance between them. This relation was discovered by Hubble in 1929. It is called the Hubble law. The Hubble law is in a good agreement with experiment. The coefficient H is called the Hubble parameter. It is an important cosmological parameter.

Let us determine the red shift

$$z = \frac{\lambda_o - \lambda_e}{\lambda_e}, \tag{12.23}$$

where λ_o is the wavelength of the light observed at the time t_0 and λ_e is the wavelength of the light emitted at the time t_e ($t_e < t_0$). The cosmological change in time of all lengths is determined by the scale factor a(t). We have

$$\frac{\lambda_o}{\lambda_e} = \frac{a(t_0)}{a(t_e)} = z + 1. \tag{12.24}$$

All light spectra observed from different Galaxies are red-shifted: $\lambda_o > \lambda_e$. Thus, we have $a(t_0) > a(t_e)$. The observation of red shifts of the light emitted by Galaxies is the direct evidence in favor of the expansion of the Universe.

12.1.4 Friedman Equations

The evolution of the scale factor $a(t)$ is determined by the Einstein equation of the General Theory of Relativity

$$G_{\mu\nu} - \Lambda g_{\mu\nu} = 8\pi\, G T_{\mu\nu}. \tag{12.25}$$

Here

$$G_{\mu\nu} = R_{\mu\nu} - \frac{1}{2} R \tag{12.26}$$

is the conserved Einstein tensor, $T^{\mu\nu}$ is the conserved energy-momentum tensor, G is the gravitational constant and Λ is the cosmological constant. In the units $\hbar = c = 1$ the constant G has the dimension M^{-2} and the constant Λ has the dimension M^2. We have $\frac{1}{\sqrt{G}} = M_P$, where $M_P \simeq 1.2 \cdot 10^{19}$ GeV is the Planck mass. In (12.26) $R_{\mu\nu}$ is the Ricci curvature tensor which is determined by the metric tensor and its first and second derivatives and $R = R_{\mu\nu}g^{\mu\nu}$ is the scalar curvature.

The standard cosmology is based on the assumption that the Universe can be considered as a perfect fluid. In this case the energy-momentum tensor $T^{\mu\nu}$ is given by the expression

$$T_{\mu\nu} = (\rho + p)u_\mu u_\nu - p\, g_{\mu\nu}. \tag{12.27}$$

Here ρ, p and u_μ are energy density, pressure and velocity. In the comoving system $u = (1, 0, 0, 0)$ and ρ, p can depend only on t.

From the Einstein equations (12.25) and (12.27) for the isotropic and homogeneous Universe the following equations can be inferred

$$\left(\frac{\dot{a}}{a}\right)^2 = H^2 = \frac{8\pi G}{3}\rho - \frac{k}{a^2} + \frac{\Lambda}{3} \tag{12.28}$$

and

$$\frac{\ddot{a}}{a} = -\frac{4\pi G}{3}(\rho + 3p) + \frac{\Lambda}{3}. \tag{12.29}$$

These equations are called the **Friedman equations**.

The Friedman equation (12.28) can be interpreted in the following way. Let us consider a Galaxy with the mass m on a surface of a sphere of the radius r. The energy of the Galaxy is the sum of kinetic and gravitational potential energy:

$$E = \frac{mv^2}{2} - G\frac{m\,M}{r}, \tag{12.30}$$

where M is the total mass (energy) of the non relativistic matter and radiation inside of the sphere. Assuming that the mass density $\rho = \sum_i \rho_i$ is a constant we have $M = \frac{4\pi}{3}\rho\, r^3$. Further, using the Hubble law $v(t) = H(t)\, r(t)$, $H = \frac{\dot{a}(t)}{a(t)}$ and the relation $r(t) = a(t)\, r_c$, where r_c is the comoving distance between the center of the sphere and the Galaxy, from (12.30) we find

$$(\frac{\dot{a}}{a})^2 = \frac{8\pi G}{3}\rho - \frac{k}{a^2}, \tag{12.31}$$

where $k = -\frac{2E}{mr_c^2}$.

In order to include the contribution of the cosmological constant Λ we assume that not only the gravitational field acts on the Galaxy, we are considering, but also some additional field with the potential energy

$$U_\Lambda = -\frac{1}{6} m \Lambda r^2, \tag{12.32}$$

which is called **the dark energy**. This assumption means that *in addition to the standard attractive gravitational force the repulsive force* $\mathbf{F}_\Lambda = -\frac{\partial U_\Lambda}{\partial \mathbf{r}} = \frac{1}{3} m \Lambda \mathbf{r}$ *acts on the Galaxy.* The energy of the Galaxy is given in this case by the relation

$$E = \frac{1}{2} H^2 mr^2 - \frac{4\pi G}{3} \rho mr^2 - \frac{1}{6} \Lambda mr^2. \tag{12.33}$$

From (12.33) we have the equation

$$(\frac{\dot{a}}{a})^2 = \frac{8\pi}{3} G \rho - \frac{k}{a^2} + \frac{1}{3}\Lambda, \tag{12.34}$$

which coincides with the Friedman equation (12.28) if we identify the constant $k = -\frac{2E}{mr_c^2}$ with the curvature and assume that k takes the values $+1, 0, -1$.

   In Eq. (12.34) ρ is the density of non relativistic matter and radiation. Let us introduce the energy density determined by the cosmological constant Λ (vacuum density):

$$\rho_\Lambda = \frac{\Lambda}{8\pi G}. \tag{12.35}$$

The Friedman equation takes the form

$$(\frac{\dot{a}}{a})^2 = \frac{8\pi}{3} G \rho_{tot} - \frac{k}{a^2}, \tag{12.36}$$

where

$$\rho_{tot} = \rho + \rho_\Lambda \tag{12.37}$$

is the total density.

   In order to introduce pressure p_Λ, which is determined by the cosmological constant Λ, we will consider the Einstein equation. From (12.25) and (12.27) we have

$$G_{\mu\nu} = 8\pi G\, T_{\mu\nu} + \Lambda g_{\mu\nu} = 8\pi G\, [(\rho + p)u_\mu u_\nu - (p - \frac{\Lambda}{8\pi G})\, g_{\mu\nu}]. \tag{12.38}$$

We will present the Einstein equation in the form

$$G_{\mu\nu} = 8\pi G\, [(\rho_{tot} + p_{tot})u_\mu u_\nu - p_t\, g_{\mu\nu}], \tag{12.39}$$

where

$$\rho_{\text{tot}} = \rho + \rho_\Lambda, \quad p_{\text{tot}} = p + p_\Lambda \tag{12.40}$$

are the total density and total pressure.

From comparison of (12.38) and (12.39) we conclude that

$$\rho_\Lambda + p_\Lambda = 0, \quad p_\Lambda = -\frac{\Lambda}{8\pi G}. \tag{12.41}$$

Thus the pressure p_Λ is a negative quantity. From (12.41) for ρ_Λ we obtain the expression

$$\rho_\Lambda = \frac{\Lambda}{8\pi G} \tag{12.42}$$

which coincides with (12.35).

We will show now that the second Friedman equation (12.29) is a consequence of the first Eq. (12.28) and the first law of the thermodynamics

$$dU = dQ - p_{\text{tot}}dV. \tag{12.43}$$

where dU is the change of the internal energy of a system, dQ is the supplied heat, p_{tot} is the total pressure and dV is the change of the volume of the system. The expansion of the Universe is an adiabatic process. Thus $dQ = 0$ and

$$dU = -p_{\text{tot}}dV. \tag{12.44}$$

Here $U = \rho_{\text{tot}}V$, where ρ_{tot} is the total energy density. From (12.44) we find

$$\dot{\rho}_{\text{tot}} = -(\rho_{\text{tot}} + p_{\text{tot}})\frac{\dot{V}}{V}. \tag{12.45}$$

Finally, taking into account that $V(t) \sim a^3(t)$, we find

$$\dot{\rho}_{\text{tot}} = -3(\rho_{\text{tot}} + p_{\text{tot}})\frac{\dot{a}}{a}. \tag{12.46}$$

Let us consider now the first Friedman equation (12.28). Calculating derivative over t we find

$$\frac{\dot{a}}{a}\left(\frac{\ddot{a}}{a} - \left(\frac{\dot{a}}{a}\right)^2\right) = \frac{4\pi}{3}G\dot{\rho}_{\text{tot}} - \frac{k}{a^2}\frac{\dot{a}}{a}. \tag{12.47}$$

From (12.28) and (12.47) we easily find the equation

$$\frac{\ddot{a}}{a} = -\frac{4\pi}{3}G\left(\rho_{\text{tot}} + 3p_{\text{tot}}\right). \tag{12.48}$$

Taking into account that $\rho_{\text{tot}} + 3p_{\text{tot}} = \rho + 3p - \frac{\Lambda}{4\pi G}$, we come to the second Friedman equation

$$\frac{\ddot{a}}{a} = -\frac{4\pi}{3}G\left(\rho + 3p\right) + \frac{1}{3}\Lambda. \tag{12.49}$$

In the next subsection we will consider different solutions of the Friedman equations. Here we will show that using experimental data it is possible to draw important conclusions about the expanding Universe directly from the Friedman equations.

1. From Eq. (12.36) we have

$$1 = \frac{8\pi G}{3H^2}\rho_{\text{tot}} - \frac{k}{H^2 a^2} \tag{12.50}$$

This equation we can rewrite in the form

$$\frac{k}{H^2 a^2} = (\Omega_{\text{tot}} - 1). \tag{12.51}$$

Here

$$\Omega_{\text{tot}} = \frac{\rho_{\text{tot}}}{\rho_c}, \tag{12.52}$$

where

$$\rho_c = \frac{3H^2}{8\pi G}. \tag{12.53}$$

From (12.51) follows that in the case $\rho_{\text{tot}} = \rho_c$ ($\Omega_{\text{tot}} = 1$) the Universe is flat ($k = 0$). If $\rho_{\text{tot}} > \rho_c$ ($\Omega_{\text{tot}} > 1$) the Universe is closed. In the case $\rho_{\text{tot}} < \rho_c$ ($\Omega_{\text{tot}} < 1$) the Universe is open. From existing data it was found that the total density parameter Ω_{tot} at present time is equal to

$$\Omega_{\text{tot}} = 1.02 \pm 0.02 \tag{12.54}$$

Thus from existing data follows that *the Universe is flat.*

The density ρ_c is called the critical density. We have

$$\rho_c = 1.88 \cdot 10^{-29} \, h^2 \, \text{g cm}^{-3} = 1.05 \cdot 10^{-5} \, h^2 \, \text{GeV cm}^{-3}. \tag{12.55}$$

Here

$$h = \frac{H}{100 \text{ km s}^{-1} \text{ Mps}^{-1}} \qquad (12.56)$$

From analysis of the existing data follows that

$$h = 0.73 \pm 0.03. \qquad (12.57)$$

2. The second Friedman equation determine acceleration of the expansion rate of the Universe. Let us determine the cosmic deceleration parameter

$$q = -\frac{1}{H^2}\frac{\ddot{a}}{a} = -\frac{a\ddot{a}}{\dot{a}^2}. \qquad (12.58)$$

From (12.49) we find

$$q = \frac{4\pi}{3}\frac{G}{H^2}(\rho + 3p) - \frac{\Lambda}{3H^2} \qquad (12.59)$$

Terms in the right-hand side of (12.59) have *different signs*. If the first gravitational term dominates, the parameter q is positive and the expansion of the Universe is slowing down (due to gravitational attraction). If the second Λ-term dominates, q is negative and the expansion of the Universe is accelerating (due to repulsion caused by the cosmological constant).

From (12.42) and (12.53) we find

$$\Omega_\Lambda = \frac{\rho_\Lambda}{\rho_c} = \frac{\Lambda}{3H^2}. \qquad (12.60)$$

Further, we have

$$\rho = \rho_{mat} + \rho_{rad}, \quad p = p_{mat} + p_{rad}, \qquad (12.61)$$

where $\rho_{mat}(p_{mat})$ and $\rho_{rad}(p_{rad})$ is the matter and radiation density (pressure). Taking into account that $p_{mat} = 0$ and $p_{rad} = \frac{1}{3}\rho_{rad}$, from (12.59) we find

$$q = \frac{1}{2}\Omega_{mat} + \Omega_{rad} - \Omega_\Lambda. \qquad (12.62)$$

Here

$$\Omega_{mat} = \frac{\rho_{mat}}{\rho_c}, \quad \Omega_{rad} = \frac{\rho_{rad}}{\rho_c}. \qquad (12.63)$$

From analysis of the existing data follows that

$$\Omega_{mat} = 0.241 \pm 0.034, \quad \Omega_\Lambda = 0.759 \pm 0.034, \quad \Omega_{rad}h^2 = 2.47 \cdot 10^{-5}. \quad (12.64)$$

Thus, at present time the cosmological term in Eq. (12.59) dominates and *the expansion rate of the Universe is accelerating* ($q_0 < 0$).

12.1.5 Solutions of the Friedman Equation

After the Big Bang during large part of the history of the Universe radiation or non relativistic matter (dust) or dark energy (cosmological constant) dominated. It is instructive to obtain solutions of the Friedman equation under the assumption that only one component of the energy density presents.

For the perfect fluid the uniform pressure is a function of the density. For the non relativistic matter $p = 0$. In the case of the radiation $p = \frac{1}{3}\rho$. For the cosmological constant $p = -\rho$. Thus, for substances we are interested in, the equation of state has the form

$$\rho = w\, p, \quad (12.65)$$

where $w = 0$ for non relativistic matter, $w = \frac{1}{3}$ for radiation and $w = -1$ for dark energy (cosmological constant).

In the case of one component from (12.65) we have

$$\frac{\dot{\rho}}{\rho} = -3(1 + w)\frac{\dot{a}}{a}. \quad (12.66)$$

From this equation we find

$$\frac{d\ln\rho}{dt} = \frac{d\ln a^{-3(1+w)}}{dt}. \quad (12.67)$$

It is obvious that solution of this equation has the form

$$\rho(t) = C\, a^{-3(1+w)}(t), \quad (12.68)$$

where C is a constant.

From (12.68) we conclude that in the case of matter (usual and dark) we have

$$\rho_{mat} \propto a^{-3}. \quad (12.69)$$

For radiation (γ-quanta, neutrinos, ultra relativistic particles) from (12.68) we find

$$\rho_{\rm rad} \propto a^{-4}. \tag{12.70}$$

Finally for dark energy we obtain

$$\rho_\Lambda = {\rm const}. \tag{12.71}$$

Notice that (12.69) follows from the fact that $\rho_{\rm mat} \propto \frac{1}{V}$, $V \propto a^3$. The behavior (12.70) is due to the fact that the energy density of radiation is proportional to $\frac{\nu}{V} \propto \frac{1}{aV}$.

From (12.69)–(12.71) we can draw important conclusions about the history of the Universe. Namely, we can conclude from these relations that in the early Universe when the scale factor $a(t)$ is small radiation dominates. Because density of matter falls slower than the density of radiation at some time densities of radiation and matter become equal. After that time the matter starts to dominate. At later time the dark energy (constant density) dominates.

Let us obtain solutions of the Friedman equation for the flat Universe in the case of non relativistic matter or radiation. From (12.47) and (12.68) we have

$$\dot{a} = Ca^{-\frac{3(1+w)}{2}+1}. \tag{12.72}$$

From this equation we obviously find

$$\frac{d}{dt}a^{\frac{3(1+w)}{2}} = C_1, \tag{12.73}$$

where $C_1 = \frac{3(1+w)}{2}C$. After integration over t we obtain the following relation

$$a^{\frac{3(1+w)}{2}}(t) - a^{\frac{3(1+w)}{2}}(0) = C_1 t. \tag{12.74}$$

Further, taking into account that in the Big Bang Universe $a(0) = 0$, we find

$$a(t) = C_2 \, t^{\frac{2}{3(1+w)}}. \tag{12.75}$$

From this relation for the Hubble parameter we obtain the following expression

$$H(t) = \frac{\dot{a}(t)}{a(t)} = \frac{2}{3(1+w)} \frac{1}{t}. \tag{12.76}$$

In the case of matter $w = 0$ and we have

$$a(t) \propto t^{\frac{2}{3}}, \quad H(t) = \frac{2}{3t}. \tag{12.77}$$

In the case of radiation $w = \frac{1}{3}$ and from (12.75) and (12.76) we find

$$a(t) \propto t^{\frac{1}{2}}, \quad H(t) = \frac{1}{2t}.$$ (12.78)

If we take into account only the cosmological constant, the Friedman equation takes very simple form

$$\left(\frac{\dot{a}}{a}\right)^2 = H^2 = \frac{\Lambda}{3}$$ (12.79)

Thus if dark energy dominates, the Universe expands exponentially

$$a(t) \propto e^{\sqrt{\frac{\Lambda}{3}}t}$$ (12.80)

and the Hubble parameter is a constant

$$H = \sqrt{\frac{\Lambda}{3}}.$$ (12.81)

12.2 Early Universe; Neutrino Decoupling

In this section we consider radiation-dominated early Universe. During much of this period relativistic particles were in thermal equilibrium.[2] The equilibrium number density of a fermion (boson) of a type i is given by the expression[3]

$$n_i = \frac{g_i}{(2\pi)^3} \int \frac{d^3p}{e^{\frac{E_i(p)-\mu_i}{kT_i}} \pm 1} = \frac{g_i}{2\pi^2} \int_0^\infty \frac{p^2 dp}{e^{\frac{E_i(p)-\mu_i}{kT_i}} \pm 1}.$$ (12.82)

Here g_i is the number of the internal degrees of freedom, $E_i(p) = \sqrt{p^2 + m_i^2}$, m_i is the mass of the particle i, μ_i is the chemical potential, T_i is the temperature and $k \simeq 1.38 \cdot 10^{-16}$ erg/grad is the Boltzmann constant. For the equilibrium energy density we have

$$\rho_i = \frac{g_i}{2\pi^2} \int_0^\infty \frac{E_i(p) p^2 dp}{e^{\frac{E_i(p)-\mu_i}{kT_i}} \pm 1}.$$ (12.83)

---

[2]A condition for the thermal equilibrium we will discuss later.

[3]The factor $\frac{g_i}{(2\pi)^3}$ is the density of states $((2\pi)^3$ is due to the relation $h = 2\pi\hbar$).

In the early Universe the chemical potentials of all relativistic particles are small and can be neglected. In the ultra-relativistic case $kT_i \gg m_i$ for the number density and the energy density we find the following expressions

$$n_i = \frac{g_i}{2\pi^2}(kT_i)^3 \int_0^\infty \frac{x^2 dx}{e^x \pm 1}, \quad \rho_i = \frac{g_i}{2\pi^2}(kT_i)^4 \int_0^\infty \frac{x^3 dx}{e^x \pm 1}, \tag{12.84}$$

where $x = \frac{p}{kT_i}$. Integrals in (12.84) are connected with the Riemann zeta function $\zeta(n)$ determined by the relation

$$\zeta(n) = \frac{1}{\Gamma(n)} \int_0^\infty \frac{x^{n-1} dx}{e^x - 1}, \tag{12.85}$$

where $\Gamma(n)$ is the Gamma-function. For integer n we have $\Gamma(n) = (n-1)!$. For Bose particles from (12.84) and (12.85) we obtain the following expressions

$$n_i = \frac{g_i}{2\pi^2}(kT_i)^3 2\zeta(3), \quad \rho_i = \frac{g_i}{2\pi^2}(kT_i)^4 6\zeta(4). \tag{12.86}$$

Taking into account that

$$\zeta(3) = 1.202, \quad \zeta(4) = \frac{\pi^4}{90} \tag{12.87}$$

we find

$$n_i = \frac{\zeta(3)}{\pi^2} g_i \, (kT_i)^3, \quad \rho_i = \frac{\pi^2}{30} g_i \, (kT_i)^4 \quad \text{(bosons)}. \tag{12.88}$$

In order to calculate the number density and the energy density for the Fermi particles we will use the relation

$$\frac{1}{e^x + 1} = \frac{1}{e^x - 1} - \frac{2}{e^{2x} + 1}. \tag{12.89}$$

We obviously have

$$\int_0^\infty \frac{x^n}{e^x + 1} = (1 - \frac{1}{2^n}) \int_0^\infty \frac{x^n}{e^x - 1}. \tag{12.90}$$

Using this relation we easily find

$$n_i = \frac{3}{4} \frac{\zeta(3)}{\pi^2} g_i \, (kT_i)^3, \quad \rho_i = \frac{7}{8} \frac{\pi^2}{30} g_i \, (kT_i)^4 \quad \text{(fermions)}. \tag{12.91}$$

The total energy density can be presented in the form

$$\rho = \sum_i \rho_i = \frac{\pi^2}{30} g_* (kT)^4.$$

(12.92)

Here T is the photon temperature and

$$g_* = \sum_{\text{bosons}} g_i \left(\frac{T_i}{T}\right)^4 + \frac{7}{8} \sum_{\text{fermions}} g_i \left(\frac{T_i}{T}\right)^4.$$

(12.93)

is the effective number of degrees of freedom of ultra-relativistic particles. Let us consider particles with mass m. If $kT \ll m$ particles are non-relativistic, $E(p) \simeq m + \frac{p^2}{2m}$ and from (12.82) for the number density in the Bose and Fermi cases we have

$$n = \frac{g}{2\pi^2} e^{-\frac{(m-\mu)}{kT}} \int_0^\infty e^{-\frac{p^2}{2mkT}} p^2 dp.$$

(12.94)

After calculation of the integral in (12.94) we find

$$n = g \left(\frac{mkT}{2\pi}\right)^{3/2} e^{-\frac{(m-\mu)}{kT}}.$$

(12.95)

Thus the number density of non-relativistic particles in the Universe is Boltzmann-suppressed. For the energy density in the non-relativistic case we have

$$\rho = mn.$$

(12.96)

Let us consider now the entropy of the Universe. From the second law of thermodynamics

$$T dS = d(\rho V) + p \, dV$$

(12.97)

we have

$$dS = \frac{1}{T} d((\rho + p)V) - \frac{V}{T} dp.$$

(12.98)

Further, we find that in the case of equilibrium

$$dp = \frac{(\rho + p)}{T} dT.$$

(12.99)

Notice that this relation follows from the condition $\frac{\partial^2 S}{\partial V \partial T} = \frac{\partial^2 S}{\partial T \partial V}$. From (12.98) and (12.99) we have

$$dS = \frac{1}{T} d((\rho + p)V) - \frac{\rho + p}{T^2} V dT = d\left(\frac{\rho + p}{T} V\right). \tag{12.100}$$

In thermal equilibrium the total entropy is conserved. In fact, from (12.100) we have

$$\frac{dS}{dt} = \frac{V}{T}\left(\frac{d\rho}{dt} + (\rho + p)\frac{1}{V}\frac{dV}{dt}\right) + \frac{V}{T}\left(\frac{dp}{dt} - \frac{(\rho + p)}{T}\frac{dT}{dt}\right) = 0. \tag{12.101}$$

The first term in (12.101) vanishes due to the relation (12.45) and the second term vanishes due to the relation (12.99).

If there are several particles, for the total entropy density $s = \frac{S}{V}$ we have

$$s = \sum_i \frac{\rho_i + p_i}{kT_i}. \tag{12.102}$$

For relativistic particles pressure and density are connected by the relation $p_i = \frac{1}{3}\rho_i$. From (12.88) and (12.102) we find

$$s = \frac{2\pi^2}{45} g_{*s} (kT)^3, \tag{12.103}$$

where the effective number of degrees of freedom g_{*s} is given by the expression

$$g_{*s} = \sum_{\text{bosons}} g_i \left(\frac{T_i}{T}\right)^3 + (\frac{7}{8}) \sum_{\text{fermions}} g_i \left(\frac{T_i}{T}\right)^3. \tag{12.104}$$

The total entropy of the expanding Universe $S = sV$ is a conserved quantity. Taking into account that $V(t) \propto a^3(t)$ from (12.103) we conclude that the temperature of the expanding Universe drops as $a^{-1}(t)$

$$kT \propto g_{*s}^{-1/3} a^{-1}(t). \tag{12.105}$$

As we discussed before, in the early Universe ultra-relativistic particles dominate. At this stage the contribution of the curvature and the cosmological constant terms in the Friedman equation can be neglected. For the Hubble parameter we have in this case

$$H = \sqrt{\frac{8\pi G}{3}\rho} \simeq \sqrt{\frac{4\pi^3}{45}} g_*^{1/2} \frac{(kT)^2}{M_P}. \tag{12.106}$$

Here $M_P = \frac{1}{\sqrt{G}} \simeq 1.22 \cdot 10^{19}$ GeV is the Planck mass. From (12.106) we obtain the following expression for the Hubble parameter

$$H = 0.21 \, g_*^{1/2} \left(\frac{kT}{\text{MeV}} \right)^2 \text{s}^{-1}. \tag{12.107}$$

The time of expansion t is connected with the Hubble parameter by the relation $t = \frac{1}{2H}$ (see (12.80)). Using the relation (12.107), we find the following relation which connect the time of expansion with the temperature

$$t = \frac{1}{2H} = 2.38 \, g_*^{-1/2} \left(\frac{\text{MeV}}{kT} \right)^2 \text{s}. \tag{12.108}$$

Let us notice that the effective number of the degrees of freedom depends on the temperature. For example, at $kT \geq 100$ GeV all Standard Model particles are relativistic and internal degrees of freedom of photons (2), $W^\pm$, Z^0 bosons (3×3), gluons (8×2), the Higgs boson (1), quarks and antiquarks ($6 \times (4 \times 3)$), charged leptons and antileptons (3×4), neutrinos and antineutrinos (3×2) contribute to g_*. We have in this case $g_* = g_b + \frac{7}{8} g_f = 106.75$. As the temperature decreases different particles became non relativistic and annihilate. At 1 MeV $\leq kT \leq 100$ MeV electrons and positrons, photons and neutrinos are relativistic particles. Taking into account that in this range of temperatures $T_\nu = T_e = T$ we have $g_* = 2 + \frac{7}{8} \cdot 10 = 10.75$. At $1 \text{ eV} \ll T \ll 1$ MeV the only relativistic particles are the photon, and neutrinos. Taking into account that $T_\nu = (\frac{4}{11})^{1/3} T_\gamma$ (see later) we have, at such temperatures, $g_* = 3.36$.

If interaction rate Γ of reactions, which are responsible for the thermal equilibrium, is much larger than the Hubble parameter H, which characterizes the expansion rate of the Universe, the thermal equilibrium is reached before the effect of expansion becomes important. When the Universe expands the temperature drops and at some temperature the interaction rate Γ for some particles become comparable with the expansion rate H. At such temperatures the equilibrium will not be maintained and the particles decouple with a freeze-out abundance. Different particles have different interaction rates and decouple at different times.

In order to determine the interaction rate let us consider the reaction $a + b \rightarrow c + d$. The cross section of the reaction is given by the relation

$$\sigma_{fi} = \frac{w_{fi}}{j_i}, \tag{12.109}$$

where w_{fi} is the number of transitions in unit volume during unit time and $j_i = n_a n_b v$ (n_a(n_b) being the number density of the particles a (b) and v is the relative velocity). The interaction rate of the particle a (b) is determined by the relation

$$\Gamma_a = n_b \sigma v, \quad \Gamma_b = n_a \sigma v. \tag{12.110}$$

At high energies we have $n_a \simeq n_b = n$ and $\Gamma_a \simeq \Gamma_b = \Gamma$. From (12.110) it follows that Γ has the dimension $L^{-3}L^2\frac{L}{T} = T^{-1}$ (or $[M]$ in the $\hbar = c = 1$ units).

We will consider neutrino decoupling. In the early Universe neutrino equilibrium is kept by the reactions

$$e^+ + e^- \rightleftarrows v_l + \bar{v}_l, \quad v_l(\bar{v}_l) + e^\pm \rightleftarrows v_l(\bar{v}_l) + e^\pm \quad (l = e, \mu, \tau) \qquad (12.111)$$

Averaged cross sections of the weak processes (12.111) are of the order

$$\sigma \simeq G_F^2 (kT)^2. \qquad (12.112)$$

Taking into account that for the ultra-relativistic particles $n \simeq (kT)^3$, for the neutrino interaction rate we have

$$\Gamma \simeq G_F^2 (kT)^5. \qquad (12.113)$$

The Hubble parameter is given by

$$H \simeq G^{1/2} (kT)^2 = \frac{(kT)^2}{M_P}. \qquad (12.114)$$

With the expansion of the Universe the interaction rate drops more rapidly than the Hubble parameter. The neutrino freeze-out temperature T_v^{dec} can be estimated from the relation

$$\frac{\Gamma}{H} \simeq 1. \qquad (12.115)$$

From this relation we have

$$kT_v^{\text{dec}} \simeq (\frac{1}{M_P G_F^2})^{1/3} \simeq 1 \text{ MeV}. \qquad (12.116)$$

Notice that from more accurate calculation it was found that $T_v^{\text{dec}} = 0.8$ MeV. After the decoupling neutrinos preserve Fermi-Dirac distribution, the Universe is transparent for neutrinos and the neutrino temperature evolves as a^{-1}.

After the neutrino decoupling γ's and $e^\pm$ are in thermal equilibrium. When the temperature drops below electron mass the electrons and positrons begin to annihilate. The released energy heats up only γ's because neutrinos are decoupled. Thus, after $e^+ - e^-$ annihilation the temperature of photons will be higher than the neutrino temperature.

In fact, the effective number of degrees of freedom of γ's and $e^{\pm}$ is equal to $g_{*s} = 2 + \frac{7}{8} \cdot 4 = \frac{11}{2}$. After the annihilation of electrons and positrons we have $g_{*s} = 2$. Because the total entropy of relativistic particles which are in equilibrium is conserved we have

$$\frac{11}{2}(Ta)_b^3 = 2(Ta)_a^3, \qquad (12.117)$$

where subscript b(a) means before (after) $e^+ - e^-$ annihilation.

Further, from the conservation of the total entropy of decoupled neutrinos we find

$$\frac{(T_\nu)_a}{(T_\nu)_b} = \frac{(a)_b}{(a)_a}. \qquad (12.118)$$

Finally, taking into account that $(T_\nu)_b = (T)_b$, from (12.117) and (12.118) we find the following relation between neutrino temperature and photon temperature

$$T_\nu = \left(\frac{4}{11}\right)^{1/3} T. \qquad (12.119)$$

This relation holds also at present. From the study of the cosmic microwave background radiation (CMB) it was found that $T = 2.725\,\mathrm{K}$. Thus, the neutrino temperature at present is equal to $T_\nu = 1.945\,\mathrm{K}$ and $kT_\nu = 1.676 \cdot 10^{-4}\,\mathrm{eV}$.

At least two massive neutrinos at present are non-relativistic. In fact, for the normal mass ordering we have

$$m_2 = \sqrt{m_1^2 + \Delta m_{12}^2} \gtrsim 8.6 \cdot 10^{-3}\,\mathrm{eV}, \; m_3 \simeq \sqrt{m_1^2 + \Delta m_A^2} \gtrsim 4.9 \cdot 10^{-2}\,\mathrm{eV}$$

$$(12.120)$$

Thus, we have $m_{2,3} \gg kT_\nu$. Analogously, in the case of the inverted mass ordering we have $m_{2,1} \gtrsim 4.9 \cdot 10^{-2}\,\mathrm{eV}$ and $m_{2,1} \gg kT_\nu$.

The entropy density of γ's and ν's at $kT \ll m_e$ is given by the expression

$$s = \frac{2\pi^2}{45} g_{*s}(kT)^3, \quad g_{*s} = 2 + \frac{7}{8} \cdot 2N_{\mathrm{eff}} \frac{4}{11} = 2 + \frac{7}{11} N_{\mathrm{eff}} = 3.94. \qquad (12.121)$$

If there are only flavor neutrinos $N_{\mathrm{eff}} = 3$. However, if we take into account that neutrino decoupling was not quite complete when $e^+ - e^-$ annihilation started, we have $N_{\mathrm{eff}} = 3.046$.

For the energy density of γ's and ν's we have

$$\rho = \frac{\pi^2}{30} g_*(kT)^4, \quad g_* = 2 + \frac{7}{8} \cdot 2N_{\mathrm{eff}} (\frac{4}{11})^{4/3} = 2 + \frac{7}{11}(\frac{4}{11})^{1/3} N_{\mathrm{eff}} = 3.36.$$

$$(12.122)$$

12.3 Neutrino Background

After $e^+ - e^-$ annihilation, the number density of γ's is given by the expression

$$n_\gamma = \frac{\zeta(3)}{\pi^2} g_\gamma (kT)^3, \quad g_\gamma = 2. \tag{12.123}$$

For the number density of neutrinos we have

$$n_\nu = \frac{3}{4} n_\gamma N_{\text{eff}} \frac{4}{11} \tag{12.124}$$

Taking into account that the photon temperature at present is equal to $T = 2.725\,\text{K}$ from (12.123) for the present photon number density we find

$$n_\gamma \simeq 410\,\text{cm}^{-3}. \tag{12.125}$$

From (12.124) follows that number density of all flavor neutrinos at present is equal to

$$n_\nu \simeq 336\,\text{cm}^{-3}. \tag{12.126}$$

These numbers can be compared with the number density of baryons in the Universe

$$n_B \simeq 2.5 \cdot 10^{-7}\,\text{cm}^{-3}. \tag{12.127}$$

Thus, *photons and neutrinos are the most abundant particles in the Universe.*

Because the neutrino number density is so large it is possible to obtain some information on neutrino masses directly from the measurement of the cosmological parameters. Let us assume that all neutrinos at present are non-relativistic and neutrino mass spectrum is quasi-degenerate ($m_1 \simeq m_2 \simeq m_3 \simeq \frac{1}{3} \sum_i m_i$). We have in this case

$$\Omega_\nu = \frac{\sum_i m_i n_i}{\rho_c} \simeq \frac{\sum_i m_i\, n_\nu}{3\,\rho_c}, \tag{12.128}$$

where $\rho_c = \frac{3H^2}{8\pi G}$ is the critical density. From (12.126) we find[4]

$$\Omega_\nu \simeq \frac{\sum_i m_i}{94\, h^2\,\text{eV}}. \tag{12.129}$$

---

[4] $\frac{\rho_c}{\frac{1}{3} n_\nu} = \frac{1.05 \cdot 10^4\,\text{eV}\,\text{cm}^{-3}\,h^2}{112\,\text{cm}^{-3}} \simeq 94\,h^2\,\text{eV}$, where h is determined by the relation $H = 100\,h\,\text{km}\,\text{s}^{-1}\,\text{Mpc}^{-1}$.

Neutrinos can be a part of the dark matter. It is obvious, however, that

$$\Omega_\nu \leq \Omega_{DM}. \tag{12.130}$$

From the analysis of the existing data follows that $\Omega_{DM} \simeq 0.26$ and $h \simeq 0.68$. Thus, from (12.129) and (12.130) we find the following bound (Gerstein and Zeldovich)

$$\sum_i m_i \lesssim 11 \, \text{eV}. \tag{12.131}$$

Comparable bound can be obtained from the tritium β-spectrum data. In fact, from the Troitsk and Meinz data we have $m_\beta = (\sum_i |U_{ei}|^2 m_i^2)^{1/2} \leq 2.3 \, \text{eV}$. In the case of the degenerate spectrum from this bound it follows that $\sum_i m_i \leq 6.9 \, \text{eV}$. We will see later that from CMB and other cosmological data an about one order of magnitude better bound can be derived.

Unlike the relic photons, the relic neutrinos have not been observed. Their observation is an extremely challenging problem. The cross section of the neutrino-nucleon scattering is so small ($\sim 10^{-62} \, \text{cm}^2$) that the direct observation of this process does not look possible.

12.4 Big Bang Nucleosynthesis

The measurement of the primordial abundances of the light elements D, ^{3}He, ^{4}He and ^{7}Li, produced in the end of the first 3 min of the evolution of the Universe, provides one of the most important confirmations of the Big Bang theory. It is very impressive that these abundances span nine orders of magnitude range. The detailed study of the primordial nucleosynthesis allows to obtain information about the number of neutrinos. We will discuss here briefly the primordial nucleosynthesis.

The synthesis of light elements is determined by conditions in the early Universe at $kT \simeq 1 \, \text{MeV}$ corresponding to $t \simeq 1 \, \text{s}$. At higher energies the thermodynamic equilibrium between protons and neutrons was due to the neutrino processes

$$\nu_e + n \rightleftarrows e^- + p, \quad \bar{\nu}_e + p \rightleftarrows e^+ + n. \tag{12.132}$$

Assuming that $\mu_p = \mu_n$ from (12.95) for the ratio of neutron and proton number densities we find the following expression

$$\frac{n_n}{n_p} = \left(\frac{m_n}{m_p}\right)^{3/2} e^{\frac{-\Delta m}{kT}}. \tag{12.133}$$

Here m_n and m_p are neutron and proton masses and $\Delta m = m_n - m_p = 1.293\,\text{MeV}$. In the first factor in Eq. (12.133) neutron-proton mass difference can be neglected. We have

$$\frac{n_n}{n_p} \simeq e^{\frac{-\Delta m}{kT}}. \tag{12.134}$$

Let us estimate the temperature at which neutrons and protons are decoupled from equilibrium due to the weak reactions (12.132). The $n \rightleftarrows p$ conversion rate is determined by G_F^2 and is given by

$$\Gamma_{n \rightleftarrows p} \simeq G_F^2 (k\,T)^5. \tag{12.135}$$

The freeze-out temperature can be estimated from the relation

$$G_F^2 (k\,T_{\text{freeze}})^5 \simeq H \simeq \frac{(kT_{\text{freeze}})^2}{M_P g_*^{-1/2}}, \tag{12.136}$$

where M_P is the Planck mass and g_* is the number of relativistic degrees of freedom which depends on the number of neutrinos. Thus, we have

$$kT_{\text{freeze}} \simeq \left(\frac{1}{G_F^2 M_P g_*^{-1/2}} \right)^{1/3} \simeq 1\,\text{MeV}. \tag{12.137}$$

At such temperatures from (12.133) for the ratio of the neutron and proton number densities we find

$$\frac{n_n}{n_p} \simeq \frac{1}{6}. \tag{12.138}$$

After the freeze-out neutrons decay $(n \to p + e^+ + \bar{\nu}_e)$ with half-life $\tau_n = 880.3 \pm 1.1\,\text{s}$.

The rates of reactions of the nucleosynthesis depend on the barion (nucleon) number density which is usually normalized to the photon density

$$\eta = \frac{n_B}{n_\gamma}, \tag{12.139}$$

where

$$n_B = n_p + n_n + A n_A \tag{12.140}$$

and n_A is the number density of nuclei with atomic number A.

The Big Bang nucleosynthesis starts with the production of deuterium in the reaction

$$n + p \rightarrow D + \gamma. \tag{12.141}$$

However, at $kT \simeq 1\,\text{MeV}$ because of the large number density of γ's and the small deuterium bounding energy ($\epsilon_D \simeq 2.23\,\text{MeV}$) deuterium nuclei, produced in the process (12.141), are dissociated in the inverse reaction

$$\gamma + D \rightarrow n + p. \tag{12.142}$$

The nucleosynthesis temperature can be estimated from the condition

$$\eta^{-1} e^{-\frac{\epsilon_D}{kT}} \leq 1, \tag{12.143}$$

i.e. from the requirement that the number density of photons with energy above the threshold of the reaction (12.142) is not larger than the barion number density. From (12.143) it follows that the nucleosynthesis starts at $kT \simeq 0.1\,\text{MeV}$. Due to the decay of neutrons by the time of the beginning of the nucleosynthesis the neutron-proton ratio drops to

$$\frac{n_n}{n_p} \simeq \frac{1}{7}. \tag{12.144}$$

During the nucleosynthesis practically all neutrons will be bound in ^4He, a light nucleus with the largest binding energy ($\epsilon_{^4\text{He}} \simeq 28.3\,\text{MeV}$), through the chain of two-body reactions

$$n + p \rightarrow D + \gamma, \quad p + D \rightarrow {}^3\text{He} + \gamma, \quad D + {}^3\text{He} \rightarrow {}^4\text{He} + p. \tag{12.145}$$

In order to estimate the mass fraction of ^4He we take into account that $n_{^4\text{He}} \simeq \frac{n_n}{2}$. We have

$$Y_p = \frac{4 n_{^4\text{He}}}{n_B} \simeq \frac{2\frac{n_n}{n_p}}{1 + \frac{n_n}{n_p}} \simeq 0.25. \tag{12.146}$$

This estimate is compatible with observed primordial abundance of ^4He.

When rates of the reactions (12.145) are smaller than the expansion rate (at the small number density of deuterium) the abundances of D and ^3He are frozen out. The predicted abundances of D and ^3He are decreased with the increase of η and are in the range (10^{-5}–10^{-4}).

In the nucleosynthesis a small amount of ^7Li is produced in the reactions

$$^3\text{H} + {}^4\text{He} \rightarrow {}^7\text{Li} + \gamma, \quad {}^3\text{He} + {}^4\text{He} \rightarrow {}^7\text{Be} + \gamma, \quad {}^7\text{Be} + e^- \rightarrow {}^7\text{Li} + \nu_e. \tag{12.147}$$

The predicted abundance of ^{7}Li depends on η and lies in the range $(10^{-10}-10^{-9})$. Notice that stable nuclei with atomic numbers 5 or 8 do not exist. This prevent production in the Big Bang nucleosynthesis of nuclei heavier than ^{7}Li in $p + {}^4$He, $n + {}^4$He and ^{4}He $+ {}^4$He reactions.

Light elements were produced in the Big Bang nucleosynthesis during the first 3 min after the Big Bang. Due to effects of the subsequent stellar nucleosynthesis, the estimation of the systematic errors in the present-day measurements of primordial abundances of light elements is a complicated problem. For the abundances of D, the following value was found

$$\frac{D}{H} = (2.53 \pm 0.26)\, 10^{-5}. \tag{12.148}$$

For the abundances of ^{4}He from the latest measurements it was obtained

$$Y_p = 0.245 \pm 0.004. \tag{12.149}$$

For abundance of ^{7}Li from different data it was found

$$\frac{Li}{H} = (1.7 \pm 0.3)\, 10^{-10}, \quad \frac{Li}{H} = (2.19 \pm 0.28)\, 10^{-10}, \quad \frac{Li}{H} = (1.86 \pm 0.23)\, 10^{-10}. \tag{12.150}$$

No reliable data for the primordial abundance of ^{3}He exists at the moment. The primordial abundances of the light elements as functions of the parameter η were calculated in the framework of the Standard Model with three types of neutrinos. Thus from the measured primordial abundances of the light elements we can determine the value(s) of the parameter η. Using the most precise $\frac{D}{H}$ data it was found

$$5.8 \cdot 10^{-10} \le \eta \le 6.6 \cdot 10^{-10} \ (95\% \ CL) \tag{12.151}$$

From these inequalities we obtain the following range for the parameter Ω_b

$$0.021 \le \Omega_b h^2 \le 0.024 \ (95\% \ CL) \tag{12.152}$$

The parameter $\Omega_b h^2$ can be determined from CMB temperature fluctuations (see later). From the latest Planck data it was found the value

$$\Omega_b h^2 = 0.0223 \pm 0.0002, \tag{12.153}$$

which corresponds $\eta = (6.09 \pm 0.06) \cdot 10^{-10}$. Impressive agreement between (12.152) and (12.153) is a strong argument in favor of the Big Bang Cosmology and the Standard Model (with three neutrinos).

The agreement of the theory of the Big Bang nucleosynthesis with the measurements allows to limit the number of possible additional light neutrino types. At temperatures $kT' \simeq 1\,\text{MeV}$ the energy density of the Universe is determined by photons, $e^{\pm}$, neutrinos and antineutrinos. The effective number of the relativistic degrees of freedom can be written as

$$g_* = 2 + 4\frac{7}{8} + 2\frac{7}{8}\, N_\nu, \tag{12.154}$$

where N_ν is the number of neutrino types. If $N_\nu > 3$ the expansion rate

$$H = \sqrt{\frac{8\pi}{3} G g_*}\, (k\,T)^2 \tag{12.155}$$

will be larger than in the standard case and, as a result, the decoupling temperature will also be larger. If the decoupling temperature will be larger the ratio $\frac{n_n}{n_p}$ will be larger and the primordial abundance of ^{4}He will be larger than in the case of $N_\nu = 3$. From the primordial abundance of ^{4}He together with the CMB value of the parameter η the value

$$N_\nu = 3.14^{+0.70}_{-0.65} \tag{12.156}$$

was found for the number of neutrino types.

12.5 Large Scale Structure of the Universe and Neutrino Masses

The most direct information about the sum of neutrino masses $\sum_i m_i$ can be inferred from the study of the suppression of the matter fluctuations caused by massive neutrinos. Such information can be obtained from the investigation the Large Scale Structure (LSS) of the Universe. We will first define the density fluctuation function

$$\delta(\mathbf{x}) = \frac{\rho(\mathbf{x}) - <\rho>}{<\rho>}, \tag{12.157}$$

where $\rho(\mathbf{x})$ is the matter density and $<\rho>$ is the volume average density. The Fourier component of the function $\delta(\mathbf{x})$ is given by

$$\delta(\mathbf{k}) = \frac{1}{(2\pi)^3} \int e^{-i\mathbf{k}\mathbf{x}} \delta(\mathbf{x})\, d^3x. \tag{12.158}$$

Let us define the matter power spectrum $P(k)$ as follows

$$P(\mathbf{k}) = <|\delta(\mathbf{k})|^2>. \tag{12.159}$$

Because all directions of fluctuations are equivalent we have

$$P(\mathbf{k}) = P(k).\tag{12.160}$$

The power spectrum at present is given by the relation

$$P(k, t_0) = T^2(k)\, P(k),\tag{12.161}$$

where $P(k)$ is the primordial spectrum and $T(k)$ is the transfer function which is determined by the evolution of initial perturbations. The primordial spectrum $P(k)$ is determined by the initial conditions in the Universe. It is usually assumed that it has a power-law form

$$P(k) = A\, k^{n_s},\tag{12.162}$$

where A is a constant. If $n_s = 1$ the primordial power spectrum is the scale invariant Harrison-Zeldovich spectrum. From the fit of the latest Planck data it was found that $n_s = 0.968 \pm 0.006$. In the calculation of the function $T(k)$ complicated effects, connected with the growth of the original density perturbations, must be taken into account. The measurement of the LSS of the Universe allows us to determine the power spectrum of visible matter $P_g(k, t_0)$. The power spectrum of all matter (visible and dark) $P_m(k, t_0)$ can be different from the power spectrum of visible galaxies. Thus, for the comparison of the measurements and theory we need to know the bias parameter

$$b^2(k) = \frac{P_g(k, t_0)}{P_m(k, t_0)}.\tag{12.163}$$

Notice that this parameter can be determined from higher order correlations.

The contribution of neutrinos to the matter density of the Universe is small ($\frac{\Omega_\nu}{\Omega_m} < 7\%$). Nevertheless from analysis of the modern high precision cosmological data a rather stringent limit on the sum of neutrino masses $\sum_i m_i$ can be obtained. We will present now some qualitative arguments in favor of the high sensitivity of the LSS data to the sum of neutrino masses.

The growth of density fluctuations induced by the gravitational attraction has the form

$$\delta\rho \backsim a^p.\tag{12.164}$$

If all matter is able to cluster, $p = 1$. In general we have

$$p = \Omega_*^{3/5},\tag{12.165}$$

where Ω_* is fraction of matter which can cluster. On the scale where neutrinos are not clustering we have

$$\Omega_* = \frac{\Omega_m - \Omega_\nu}{\Omega_m} \simeq 1 - f_\nu, \qquad (12.166)$$

where Ω_m is the density of all matter, Ω_ν is the neutrino density and

$$f_\nu = \frac{\Omega_\nu}{\Omega_m}. \qquad (12.167)$$

Density fluctuations start growing at the beginning of the mater dominated era (scale factor a_M) and they stop growing at the time when the dark energy dominated era starts (scale factor a_Λ). The growth of the fluctuations during this time is given by the factor

$$\left(\frac{a_\Lambda}{a_M} \right)^{\Omega_*^{3/5}} \simeq \left(\frac{a_\Lambda}{a_M} \right) e^{-\frac{3}{5} f_\nu \ln \frac{a_\Lambda}{a_M}}, \qquad (12.168)$$

where the exponent gives the suppression of the growth of the fluctuations due to neutrino non-clustering. The suppression of the power spectrum at a scale where neutrinos do not cluster is given by

$$\frac{P(k, \sum m_i)}{P(k, 0)} \simeq e^{-\frac{6}{5} f_\nu \ln \frac{a_\Lambda}{a_M}}. \qquad (12.169)$$

For the fraction of neutrinos f_ν we have (see (12.129))

$$f_\nu = \frac{\sum m_i}{94 \text{ eV } \Omega_m h^2}. \qquad (12.170)$$

From analysis of Planck CMB data it was found that $\Omega_m h^2 \simeq 0.14$. Thus, we have

$$f_\nu \simeq \frac{\sum m_i}{13 \text{ eV}}. \qquad (12.171)$$

High sensitivity of LSS data to the parameter $\sum m_i$ is connected with the fact that the ratio $\left(\frac{a_\Lambda}{a_M} \right)$ is large. It was found that

$$\frac{P(k, \sum m_i)}{P(k, 0)} \simeq e^{-10 f_\nu}. \qquad (12.172)$$

After decoupling neutrinos form a collisionless fluid in which separate particles free-stream with a thermal velocity v_{th}. The neutrino free-streaming wave length λ_{FS} is determined by the distance which neutrinos pass during the Hubble time $\frac{1}{H}$

$$\lambda_{FS} = 2\pi \sqrt{\frac{2}{3} \frac{v_{th}}{H}}. \tag{12.173}$$

Taking into account that $H = \sqrt{\frac{8\pi G}{3} \rho_m}$ for the neutrino free-streaming wave number we have

$$k_{FS} = 2\pi \frac{a}{\lambda_{FS}} = \sqrt{\frac{4\pi G \rho_m a^2}{v_{th}^2}}. \tag{12.174}$$

For non-relativistic neutrinos with mass m we have

$$v_{th} \simeq \frac{\bar{p}}{m} \simeq \frac{3.15 \, kT_\nu}{m} = \frac{3.15 \, kT_\nu^0}{m} \frac{a_0}{a} = 3.15 \, (1+z)(\frac{4}{11})^{1/3} \frac{kT_\gamma^0}{m}, \tag{12.175}$$

where $T_\gamma^0 = 2.725 \, \text{K}$ is the CMB radiation temperature at present.

The minimal wave number, which corresponds to the time when neutrinos became non relativistic, is given by the relation

$$k_{nr} \simeq 0.018 \, \Omega_m^{1/2} \left(\frac{\sum m_i}{1\text{eV}} \right)^{1/2} h \, \text{Mpc}^{-1}. \tag{12.176}$$

At $k > k_{nr}$ the power spectrum is suppressed by a factor given by Eq. (12.172). This suppression is due to the fact that neutrinos do not cluster on the scales $k > k_{nr}$. In the region $k < k_{nr}$, corresponding to scales larger than the horizon, there is no suppression of the power spectrum.

From analysis of the SDSS and BOSS Galaxy distribution data the following upper bound was obtained

$$\sum m_i \leq 0.6 \, \text{eV} \tag{12.177}$$

The bound on the sum of neutrino masses depends on the values of other cosmological parameters. For example, a change of the spectral index n_s in the primordial power spectrum (12.162) can partially mimic the effect of neutrino masses. The joint analysis of the SDSS data and the Cosmic Microwave Background (CMB) radiation data, which strongly constrain the values of the cosmological parameters, allows to obtain more stringent and reliable bound on the parameter $\sum m_i$. We will briefly discuss these data in the next subsection.

12.6 Cosmic Microwave Background Radiation

The measurements of anisotropies of the Cosmic Microwave Background (CMB) radiation provide a profound confirmation of the Big Bang cosmology. These measurements allow one to obtain the most precise values of cosmological parameters.

The spectrum of the CMB radiation is an ideal Planck spectrum which is characterized by the temperature. The mean measured temperature is $\bar{T} = (2.72548 \pm 0.00057)\,\mathrm{K}$. The CMB radiation is almost isotropic in the sky with small anisotropies of the order $\frac{\Delta T}{T} \sim 10^{-5}$. Starting from the pioneering COBE satellite, the CMB anisotropies were measured in different experiments. After the COBE satellite CMB anisotropies were studied by the Wilkinson Microwave Anisotropy Probe (WMAP). The Planck is the satellite of third generation dedicated to the observation of the sky in the microwaves. It operated from 2009 till 2013.

The CMB radiation was generated at the time when the Universe was cool enough to allow the recombination of hydrogen atoms (recombination era, about 380,000 years after the Big Bang). The temperature at that time dropped to 3000 K. On the way to us due to elastic Compton scattering on hot electron gas CMB photons became linearly polarized. This polarization was measured in the WMAP, Planck and other experiments. Anisotropy of the CMB radiation is due to perturbations of the gravitational potential (Sachs-Wolfe effect) and other effects.

The CMB anisotropies can be expanded in spherical harmonics

$$\delta T(\mathbf{n}) = \sum_{l=1}^{\infty} \sum_{m=-l}^{m=l} a_{lm}\, Y_{ml}(\mathbf{n}). \tag{12.178}$$

Here $\delta T(\mathbf{n}) = T(\mathbf{n}) - \bar{T}$, where $\bar{T}$ is the average temperature, and $\mathbf{n}$ is the unit vector. The angular temperature angular power spectrum is defined as follows

$$C_l^{TT} = \frac{1}{2l+1} \sum_m <a_{lm}^* a_{lm}>. \tag{12.179}$$

Analogously are determined angular power spectra C_l^{EE}, C_l^{TE} and C_l^{BB} for polarization anisotropies (E is the electric mode and B is the magnetic mode).

CMB photons during their traveling to the Earth are experienced gravitational effects of the matter distribution (CMB lensing). The CMB lensing has distinctive effect on the angular power spectra. It is strongly affected by neutrino masses: massive neutrinos suppress the clustering on scales smaller than the horizon at the nonrelativistic transition and affect the lensing potential. From the latest analysis of the Planck data it was found (95% CL)[5]

$$\sum_i m_i < 0.72\,\mathrm{eV} \quad \text{Planck}\ TT + lowP. \tag{12.180}$$

---

[5] low P is TE and EE polarization data at $l < 30$.

If the Planck TT + $lowPB$ data are analyzed together with Barion Acoustic Oscillation (BAO) data,[6] the acoustic scale degeneracy will be broken and the bound on the parameter $\sum_i m_i$ is much stronger

$$\sum_i m_i < 0.21\ \text{eV} \quad \text{Planck}\ TT + lowP + \text{BAO}. \tag{12.181}$$

Additional polarization data make a relatively small improvement to the bounds:

$$\sum_i m_i < 0.49\ \text{eV} \quad \text{Planck}\ TT, TE, EE + lowP;$$

$$\sum_i m_i < 0.17\ \text{eV} \quad \text{Planck}\ TT, TE, EE + lowP + \text{BAO}. \tag{12.182}$$

Analysis of the Planck CMB data allows to obtain an information about the effective number of neutrinos N_{eff}.[7] No evidence in favor of additional neutrino degree of freedom was found (68% CL):

$$N_{\text{eff}} = 3.13 \pm 0.32 \quad \text{Planck}\ TT + lowP;$$

$$N_{\text{eff}} = 3.15 \pm 0.23 \quad \text{Planck}\ TT + lowP + \text{BAO};$$

$$N_{\text{eff}} = 2.99 \pm 0.20 \quad \text{Planck}\ TT, TE, EE + lowP;$$

$$N_{\text{eff}} = 3.04 \pm 0.18 \quad \text{Planck}\ TT, TE, EE + lowP + \text{BAO}. \tag{12.183}$$

The Planck CMB data are perfectly described by the Standard Cosmological Model ΛCDM with six parameters: the barion density $\Omega_b\,h^2$, the cold dark matter density $\Omega_c\,h^2$, the spectral index n_s plus additional three parameters which describe dynamics of the evolution of the Universe. From analysis of the Planck data it was found

$$\Omega_b\,h^2 = 0.02226 \pm 0.00023, \quad \Omega_c\,h^2 = 0.1186 \pm 0.0020, \quad n_s = 0.968 \pm 0.006. \tag{12.184}$$

In this analysis the normal neutrino mass hierarchy with $\sum_i m_i = 0.06\,\text{eV}$ was assumed. The bounds (12.180)–(12.182) were obtained under the assumption that

---

[6]In the early Universe baryons and photons can be treated as a fluid. The combination of effects of gravity and pressure of radiation creates longitudinal acoustic oscillations in the photon-baryon fluid. The oscillations of the photons induce peaks in the power spectra at different angular momenta.

[7]The effective number of neutrinos, larger than the standard one ($N_{\text{eff}} = 3.046$), would lead to a faster expansion of the Universe and earlier recombination era.

$\sum_i m_i$ is an additional variable parameter. Marginalizing over neutrino masses and using BAO measurements from SDSS, BOSS and 6dF surveys it was found

$$h = 0.678 \pm 0.009. \tag{12.185}$$

Within the ΛCDM model the following values of the cosmological parameters can be derived

$$\Omega_m = 0.308 \pm 0.012, \quad \Omega_\Lambda = 0.692 \pm 0.012, \quad \Omega_k = 0.000 \pm 0.005. \tag{12.186}$$

We considered mainly Planck CMB data. Other cosmological data are also described by the six parameters ΛCDM model. However, there is a tension between Planck and other analysis. For example, from HST data it was found

$$h = 0.732 \pm 0.017. \tag{12.187}$$

Summarizing, from analysis of the cosmological data more stringent bounds on neutrino masses than from laboratory experiments can be inferred. However, cosmological results are model dependent.

12.7 Supernova Neutrinos

Neutrinos play a crucial role in gravitational collapse of a core of a massive star (supernova explosion). 99% of energy released during a supernova explosion is emitted in the form of neutrinos of different types. In 1987 for the first time neutrinos from the supernova SN 1987A were detected by the Kamiokande II, IMB and Baksan detectors. The observation of supernova neutrinos and the first observation of solar neutrinos by R. Davis et al. opened a new field in astronomy, neutrino astronomy. In 2002 R. Davis and M. Koshiba were awarded with Nobel Prize for "for pioneering contributions to astrophysics, in particular for the detection of cosmic neutrinos".

We will briefly discuss here main stages of the gravitational collapse of a type II supernova and main neutrino emission phases. A massive star with a mass larger than eight solar masses evolves through a chain of fusion reactions. It starts with transition $4\, p \rightarrow {}^4\text{He}$ which take place at the temperature $2 \cdot 10^7$ K. When the temperature reaches $2 \cdot 10^8$ K the transition $3\, {}^4\text{He} \rightarrow {}^{12}\text{C}$ becomes possible. Then ${}^{16}\text{O}$ nuclei are produced in the transition ${}^4\text{He} + {}^{12}\text{C} \rightarrow {}^{16}\text{O}$ etc. At the temperature $3.5 \cdot 10^9$ K ${}^{56}\text{Fe}$ nuclei are produced in the transition $2\, {}^{28}\text{Si} \rightarrow {}^{56}\text{Fe}$. ${}^{56}\text{Fe}$ are the most tightly bound nuclei. With production of ${}^{56}\text{Fe}$ nuclei the chain of the thermonuclear transitions is terminated. As a result of the evolution a star has an onion-like structure with an iron core surrounded by shells of silicon, neon, oxygen, carbon, helium and hydrogen.

The iron core typically has a mass of about one solar mass, a radius of a few thousand km, a central density of about 10^{10}g cm^{-3} and a central temperature of about 1 MeV.

The gravitational contraction of the iron core with the mass smaller than the Chandrasekhar limit (1.44 $M_\odot$ where $M_\odot$ is the solar mass) is balanced by the electron degeneracy pressure. When the mass of the core becomes larger than the Chandrasekhar limit the electron degeneracy pressure can not prevent the gravitational attraction and the collapse of the core starts.

The released energy is given by the gravitational binding energy of the core

$$E_B \simeq \frac{3}{5} \frac{G M_c^2}{R_c}, \tag{12.188}$$

where M_c is the mass of the core and R_c is the radius of the core (after collapse). At $M_c \simeq M_\odot$ and $R_c \simeq 10$ km the energy $\simeq 3 \cdot 10^{53}$ erg is released during the gravitational collapse.

At an earlier stage of the collapse due to increase of the temperature the photo-dissociation of iron nuclei

$$\gamma + {}^{56}\text{Fe} \rightarrow 13\alpha + 4n \tag{12.189}$$

take place. At the same time because energies of electrons are increased the electron capture processes

$$e^- + p \rightarrow \nu_e + n, \quad e^- + (A, Z) \rightarrow \nu_e + (A, Z - 1) \tag{12.190}$$

start. At this earlier stage of the collapse the produced ν_e's freely escape the star.

The processes (12.190) reduce the electron pressure and accelerate the collapse. When the density of the core reaches about $\sim 10^{12}$ g cm^{-3} neutrinos become trapped and star can not loose lepton number due to neutrino emission. Neutrino trapping is mainly due to coherent NC scattering of neutrinos on heavy nuclei. At this stage, the inner part of the core ($\sim 0.8 M_\odot$) collapses with subsonic velocity proportional to the radius (homologous collapse). The outer part of the core collapses with supersonic free-fall velocities.

When the density of the core reaches nuclear density of about 10^{14} g cm^{-3} the pressure of the degenerate nucleons stops the collapse of the inner part of the core. The stop of the collapse of the inner core creates shock wave which propagates outward through the outer part of the core.

The shock wave propagating through infalling matter of the outer part of the core dissociates nuclei into protons and neutrons. The capture rate of electrons on protons is larger than on nuclei. As a result a huge number of ν_e's are produced behind moving shock front in the process $e^- + p \rightarrow \nu_e + n$. When the shock reaches the region with a density of about 10^{11} g cm^{-3} neutrinos can leave the star (*neutronization burst*). Emitted ν_e's carry away about 10^{51} erg during a few milliseconds.

Because of the loss of energy through photodisintegration of nuclei and neutrino emission the shock wave is weakened and stalled at 100–200 km radius, during tens of milliseconds after the bounce.

The most popular mechanism of the revival of the stalled shock wave is the energy transfer by neutrinos emitted by a proto-neutron star. Current numerical stimulations demonstrate the viability of this neutrino-heating mechanism. The energy input by neutrinos drives the shock way outward leading to the supernova explosion in about 0.5 s after the bounce.

Before the shock front re-accelerates outward, by massive accretion of infalling material (about 0.1 $M_\odot$ s^{-1}) a remnant begins to form. The remnant evolves to a neutron star or a black hole depending on the mass of the progenitor star. The hot accretion mantle around the core emits all types of neutrinos and antineutrinos (*accretion phase*). During this phase within 20 ms about $2 \cdot 10^{51}$ erg is emitted.

After about 1 s accretion ends and the remnant enters into Kelvin-Helmholtz cooling phase. During this phase remaining gravitational bounding energy is transferred into neutrinos and antineutrinos via NC reactions

$$e^- + e^+ \to \nu_l + \bar{\nu}_l, \quad e^\pm + N \to e^\pm + N + \nu_l + \bar{\nu}_l \ (l = e, \mu, \tau)$$

$$N + N \to N + N + \nu_l + \bar{\nu}_l, \quad \gamma + e^\pm \to e^\pm + \nu_l + \bar{\nu}_l, \ldots \quad (12.191)$$

Neutrinos are trapped in the inner high-density part of the proto-neutron star. Because ν_e and $\bar{\nu}_e$ have both CC and NC interactions and $\nu_{\mu,\tau}$ and $\bar{\nu}_{\mu,\tau}$ have only NC interaction there are three different neutrino-spheres with radii from about 50 to 100 km. It takes a fraction of a second for the trapped neutrinos to diffuse to that part of the neutrino-sphere where they can leave the star. Neutrinos are emitted from the neutrino-spheres with black-body spectrum and average energies in the range (10–20) MeV (*cooling phase*). The emission of thermal neutrinos of all flavors continues for a few seconds. These neutrinos carry out practically all energy produced in the supernova explosion. The luminosity of all types of neutrinos and antineutrinos during the cooling phase are practically the same.

On 23 February 1987 in the Large Magellanic Claud (a nearby galaxy) at a distance of about 51.4 kpc from the earth a supernova SN 1987A was observed. In three underground neutrino detectors Kamiokande II, IMB and Baksan at the same time (up to uncertainties in time calibrations) neutrino bursts with neutrino energies of about 15 MeV over a time interval about 12 s were observed. The neutrino events were detected about 3 h before the first optical observation of SN1987A was done.[8]

In all three detectors antineutrinos were observed via the reaction

$$\bar{\nu}_e + p \to e^+ + n. \quad (12.192)$$

---

[8] This corresponds to the general theory of the supernova explosion: neutrinos are produced during 10 s after the core collapse and visible light is produced later after the shock reaches the surface of the star.

The cross section of this reaction

$$\sigma \simeq 8.5 \cdot 10^{-44} \left(\frac{E}{\text{MeV}} \right)^2 \text{cm}^2 \qquad (12.193)$$

is much larger than the cross sections of other possible reactions. Eleven antineutrino events were observed in the Kamiokande II detector, eight events in the IMB detector and five candidate-events in the Baksan detector. The total measured released energy $((4.7 \pm 1.5) \cdot 10^{53}$ erg (Kamiokande II) and $2.9 \pm 1.0) \cdot 10^{53}$ erg (IMB)) is consistent with the standard estimate.

From one to three core-collapse supernovae per century is expected in our galaxy. When this happens, in modern and future neutrino detectors thousands of supernova neutrino events will be detected. The detailed investigation of supernova neutrinos will be very important for the study of the mechanism of supernova explosions and for obtaining information about neutrino properties. In the case of the supernova neutrinos not only the usual MSW effect due to coherent neutrino-electron scattering but also nonlinear matter effects due to neutrino-neutrino scattering become important. The consideration of all these important effects is out of the scope of this book.

12.8   Baryogenesis Through Leptogenesis

From existing cosmological data follows that our Universe predominantly consists of matter. The baryon-antibaryon asymmetry of the Universe η_B is determined as follows

$$\eta_B = \frac{n_B - n_{\bar{B}}}{n_\gamma}, \qquad (12.194)$$

where n_B and $n_{\bar{B}}$ is the baryon and antibaryon number densities and n_γ is the photon number density. Taking into account that $n_{\bar{B}} \ll n_B$ we have $\eta_B \simeq \eta$, where $\eta = \frac{n_B}{n_\gamma}$. The parameter η was determined from the measurement of the primordial abundances of deuterium and other light elements and from the measurement of the anisotropy of CMB radiation. From very precise recent Planck data it was found

$$\eta_B = (6.10 \pm 0.04) \cdot 10^{-10}. \qquad (12.195)$$

In the Big Bang Universe there was no initial baryon asymmetry. Baryon asymmetry can be generated only during the evolution of the Universe. There exist several approaches to the generation of the baryon asymmetry of the Universe. We will consider here briefly the generation of the baryon asymmetry through the lepton asymmetry, produced by CP violating decays of heavy Majorana particles. This approach was inspired by the seesaw mechanism of the neutrino mass generation which connect the smallness of neutrino masses with the existence of heavy Majorana leptons.

In order that the baryon asymmetry was created in the evolution of the Universe, the following three conditions, formulated by A. Saharov, must be satisfied:

- The baryon number has to be violated.
- C and CP must be violated.
- Departure from thermal equilibrium must take place.

In fact, if the baryon number is conserved, from the state with $B = 0$ we can not obtain the present state with $B \neq 0$. In order that particles and antiparticles behave differently, C and CP must be violated. From the CPT invariance it follows that the masses of a baryon and an antibaryon are equal and, consequently, in the thermal equilibrium the number densities of baryons and antibaryons are the same.

In principle, the Standard Model could ensure all three Sakharov's conditions.

1. The baryon and lepton numbers are not conserved in the SM in processes of transitions between different vacua which have different topological charges and are separated by a potential barrier. The heights of the barriers are given by the sphaleron energy (a saddle point of the energy of gauge and Higgs fields). At $T \ll 100\,\text{GeV}$ the rate of such tunnel transitions is determined by the instanton action and is negligibly small ($\Gamma \simeq e^{-\frac{4\pi}{\alpha}} \simeq 10^{-165}$). At temperatures higher than $\simeq 100\,\text{GeV}$, transitions over the barrier due to thermal fluctuations become important and the rate of $(B + L)$-violating processes can be significant. The interaction rate of such sphaleron processes are larger that the expansion rate of the Universe in the temperature range $10^2\,\text{GeV} \lesssim kT \lesssim 10^{12}\,\text{GeV}$. $B - L$ in the sphaleron processes is conserved.
2. The SM interactions violate C. The CP invariance is violated via the Kobayashi-Maskawa phase δ_{KM} which enter into the CKM quark mixing matrix.
3. The electroweak phase transition (transition from $SU_L(2) \times U_Y(1)$ symmetric phase of massless SM particles to the phase with broken symmetry and massive $W^{\pm}$'s, Z^0's, quarks and leptons) which occurred at $kT \simeq 100\,\text{GeV}$ in principle could be out-of-equilibrium transition.

However, *the SM can not explain the baryon asymmetry of the Universe*:

- the CP asymmetry in the Standard Model is different from zero only if all three families are involved. This means that masses of all quarks and all mixing angles must enter into the CP asymmetry. As a result, the CP asymmetry in the SM is suppressed by the smallness of masses of light quarks with respect to the scale of the electroweak breaking and by the smallness of the product $s_{12}s_{23}s_{13}$. The estimated Standard Model CP asymmetry ($\sim 10^{-18}$) is too small to explain the asymmetry (12.195),
- the departure from equilibrium can be satisfied if the mass of the Higgs boson is less than $\sim 70\,\text{GeV}$. From the data of the LHC experiments follows that $m_H \simeq 125\,\text{GeV}$,

- because in the SM $(B - L)$ is conserved, the sphaleron processes would wash out the baryon asymmetry.

Thus, *explanation of the baryon asymmetry of the Universe requires a new beyond the Standard Model Physics.* This new physics must insure: (1) a new source of CP violation; (2) $(B - L)$ violation; (3) out of equilibrium processes.

The leptogenesis is a scenario in which new physical processes generate a lepton asymmetry of the Universe which is partially converted into a baryon asymmetry through sphaleron processes.

Let us assume that there exist heavy Majorana leptons N_i $(i = 1, 2, \ldots)$ with masses M_i much larger than the electroweak vacuum expectation value $v \simeq 246\,\text{GeV}$ and that the fields $N_i(x) = N_i^c(x)$ are singlets of the electroweak $SU_L(2) \times U_Y(1)$ group. Further we assume that N_i interact with the SM leptons and Higgs. The $SU_L(2) \times U_Y(1)$ invariant Yukawa interaction and the mass term of the heavy Majorana leptons N_i have the form

$$\mathscr{L} = -\frac{1}{2} \sum_i M_i \bar{N}_i N_i - (\sum_{l,i} \bar{L}_{lL} \tilde{\phi} \, y_{li} N_{iR} + \text{h.c.}). \tag{12.196}$$

Here L_{lL} and $\tilde{\phi} = i\tau_2 \phi^*$ are lepton and conjugated Higgs doublets (see Chap. 3), y_{li} are dimensionless complex constants.

The Lagrangian (12.196)

1. is not invariant under the global phase transformation of the fermion fields and does not conserve the lepton number L,
2. conserves the baryon number B and, consequently, does not conserve $(B - L)$.
3. in the case of a complex matrix y it violates CP.

In the second order of the perturbation theory the Lagrangian (12.196) induces the Weinberg effective Lagrangian

$$\mathscr{L} = -\sum_{l',l,i} \bar{L}_{l'L} \tilde{\phi} \, y_{l'i} \frac{1}{M_i} y_{li} C \tilde{\phi}^T \bar{L}_{lL}^T + \text{h.c.} \tag{12.197}$$

which after the spontaneous $SU_L(2) \times U_Y(1)$ symmetry breaking generate the Majorana neutrino mass term

$$\mathscr{L}^{\text{M}} = -\frac{1}{2} \sum_{l',l} \bar{\nu}_{l'L} M_{l'l}^{\text{M}} C \bar{\nu}_{lL}^T + \text{h.c.} \tag{12.198}$$

where M^{M} is the seesaw mass matrix

$$M^{\text{M}} = y \frac{v^2}{M} y^T. \tag{12.199}$$

As we discuss in Chap. 5, the seesaw mechanism is the most plausible mechanism of the small neutrino mass generation. Thus, existence of heavy Majorana leptons N_i and interaction (12.196) allows not only to explain phenomena of light neutrino masses but also to ensure conditions 1–3 above for the leptogenesis.

Let us notice that if small neutrino masses imply existence of heavy Majorana leptons and of the interaction (12.196), then leptogenesis is a likely phenomenon. Whether it can explain observed baryon asymmetry of the Universe is a complicated quantitative problem with many unknown parameters involved. We will make a few illustrative remarks.

The leptogenesis starts in the Early Universe at $kT \simeq M_i$ when out of equilibrium decays of Majorana leptons into CP-conjugated final states LH and $\bar{L}\bar{H}$ occur. The CP asymmetry is defined as follows

$$\epsilon_i = \frac{\Gamma(N_i \to LH) - \Gamma(N_i \to \bar{L}\bar{H})}{\Gamma(N_i \to LH) + \Gamma(N_i \to \bar{L}\bar{H})}. \tag{12.200}$$

In the tree approximation the phases of the matrix y do not enter into the expression for the decay probability and $\Gamma_0(N_i \to LH) = \Gamma_0(N_i \to \bar{L}\bar{H})$. In order to reveal CP violation we must take into account loop diagrams. The CP phases enter into the interference between the tree and loop diagrams.

In the case of the hierarchical spectrum of N_i the most important contribution to the leptogenesis give decays of the lightest Majorana particles N_1 For the CP asymmetry we have

$$\epsilon_1 = \frac{3}{16\pi} \sum_i \frac{\text{Im}(yy^\dagger)^2_{1i}}{(yy^\dagger)_{11}} \frac{M_1}{M_i} \tag{12.201}$$

The decays of N_1's are out of equilibrium if the decay rate is smaller than the expansion rate at the time of the leptogenesis:

$$\Gamma_{N_1} \lesssim H(T \simeq M_1). \tag{12.202}$$

Taking into account that $\Gamma_{N_1} \simeq \frac{1}{8\pi}(YY^\dagger)_{11}M_1$ and that the Hubble constant is given by the relation $H(T) = 1.66\, g_*^{1/2} \frac{T^2}{M_P}$, from (12.202) we obtain the condition

$$\tilde{m}_1 = \frac{(YY^\dagger)_{11}v^2}{M_1} \lesssim 10^{-3} \text{ eV}. \tag{12.203}$$

The baryon asymmetry η_B is the product of the CP asymmetry ϵ_1, a wash out parameter η[9] and a factor which takes into account the sphaleron conversion of the

---

[9]The calculation of the parameter η requires the numerical solution of the Boltzmann equation for leptogenesis. Approximately we have $\eta \simeq \frac{10^{-3} \text{ eV}}{\tilde{m}_1}$.

lepton asymmetry into the baryon asymmetry. It was found that

$$\eta_B = -\frac{1}{103}\epsilon_1\eta.$$ (12.204)

From the requirement of an agreement with the observed asymmetry we have

$$M_1 \geq 10^9 \text{ GeV}.$$ (12.205)

Notice that if we consider a one-flavor lepton asymmetry for the lightest neutrino mass we obtain the bound

$$m_1 \leq 10^{-1} \text{ eV}.$$ (12.206)

However, if we include all three flavors, large flavor effects are possible and the estimate (12.206) will be not valid.

Let us also notice that in the case of the quasidegenerate spectrum of masses of heavy Majorana particles the picture of the leptogenesis becomes completely different. If $M_2 - M_1 \simeq \Gamma_N$ the lepton asymmetry is enhanced resonantly. The baryon asymmetry of the Universe could be explained for $M_i \simeq 1$ TeV.

In conclusion let us stress that the leptogenesis is an attractive possibility of the explanation of the baryon asymmetry of the Universe. It is a consequence of the seesaw mechanism of the neutrino mass generation which is apparently the most plausible possibility of the explanation of the smallness of the neutrino masses. However, masses of heavy Majorana leptons and Yukawa coupling constants are unknown parameters. Processes induced by the interaction (12.196), which drive leptogenesis, can not be observed in a laboratory. Thus, it is not possible to test leptogenesis in a model independent way. However, the observation of the neutrinoless double β-decay of heavy nuclei would prove that massive neutrinos are Majorana particles. This observation would be a strong argument in favor of the seesaw mechanism of the neutrino mass generation and leptogenesis.

Chapter 13
Conclusion and Prospects

Existence of neutrino was predicted by Pauli in 1930 from the requirement of the conservation of energy in the β-decay of nuclei. In 1934 Fermi built the first phenomenological four-fermion theory of the decay of $n \to p + e^- + \bar{\nu}_e$ which could describe wide range of nuclear β-decay (Fermi) transitions.

It took more than 20 years to confirm the Pauli's prediction: in the fifties the neutrino was discovered in the Reines-Cowan reactor neutrino experiment in which the process $\bar{\nu}_e + p \to e^+ + n$ was observed.

The first theory of the massless two-component neutrino was proposed by Landau, Lee and Young and Salam in 1957 soon after the parity violation in the weak interaction was discovered. The two-component neutrino theory was perfectly confirmed in 1958 in the spectacular Goldhaber et al. experiment in which neutrino helicity was measured. Soon after the confirmation of the two-component neutrino theory Feynman-Gell-Mann and Marshak-Sudarshan proposed the universal current × current $V - A$ theory of the weak interaction which could describe all existed at that time β-decay and other weak interaction data.

At this earlier stage of the development of the weak interaction theory B. Pontecorvo put forward courageous idea of small neutrino masses, neutrino mixing and neutrino oscillations (1957–1958). One flavor neutrino was known at that time. Pontecorvo considered the only possible in this case transitions of flavor neutrino and antineutrino into sterile states: $\nu_L \rightleftarrows \bar{\nu}_L$ and $\bar{\nu}_R \rightleftarrows \nu_R$.

In 1962 in the first experiment with neutrinos from accelerator the second flavor neutrino ν_μ was discovered (Brookhaven). At that time Maki, Nakagawa and Sakata proposed idea that flavor fields ν_e and ν_μ are "mixtures" of the fields of neutrinos ν_1 and ν_2 with masses m_1 and m_2.

In the nineties in the LEP experiments at CERN it was established that the number of the flavor neutrinos is equal to three. The third flavor neutrino ν_τ was observed in the DONUT experiment at Fermilab.

In 1967 the Glashow-Weinberg-Salam unified theory of the weak and electro-magnetic interactions (the Standard Model) was proposed. This theory predicted existence of charged and neutral vector bosons $W^{\pm}$ and Z^0 and a new type of the weak interaction-Neutral Currents. The discovery of NC processes $\nu_\mu(\bar{\nu}_\mu) + N \to \nu_\mu(\bar{\nu}_\mu) + X$ at CERN in 1973 was the beginning of triumphal confirmation of the Standard Model. Detailed study of different NC induced processes fully confirm predictions of the Standard Model. In 1983 at CERN vector $W^{\pm}$ and Z^0 bosons were discovered.

Before discussion of the problem of neutrino masses and mixing, which is the main subject of this book, we will make the following general remarks. Neutrinos are the only Standard Model elementary particles with equal to zero electric charges. There are two major consequences of $Q_\nu = 0$.

1. *Neutrinos have no direct electromagnetic interaction.* In the region of energies $E \ll m_W$ neutrinos have only well known Standard Model weak interaction determined by the Fermi constant $G_F \simeq 1.1666 \cdot 10^{-5}\,\mathrm{GeV}^{-2}$. As a result detection of neutrinos usually is a challenge. It requires large sophisticated (often underground) detectors, intensive neutrino sources, effective ways of background suppression etc. However, if the problem of neutrino detection is solved *neutrino become an unique tool*

   • for the investigation of the quark structure of nucleon (deep inelastic neutrino-nucleon scattering),
   • for the study of internal invisible region of the sun in which thermonuclear solar energy is produced (solar neutrino experiments)
   • for the investigation of the mechanism of the gravitational collapse (detection of supernova neutrinos) etc.

2. Quarks and leptons are Dirac particles. *Neutrinos are the only elementary particles which can be Dirac or truly neutral Majorana particles.* Modern understanding of neutrino masses is based on the assumption that *neutrinos are Majorana particles.*

The first indication in the favor of neutrino masses, mixing and oscillations was obtained in the Davis solar neutrino experiment in the seventies. The Davis finding (observed flux of the solar neutrinos is two to three times smaller than the predicted flux) was confirmed the Kamiokande, GALLEX and SAGE solar neutrino experiments in the eighties and nineties. Neutrino oscillations were discovered in the 1998 in the atmospheric Super-Kamiokande experiment, in 2002 in the SNO solar neutrino experiment and in the KamLAND reactor neutrino experiment. The discovery of neutrino oscillations was confirmed by the accelerator neutrino experiments K2K, MINOS, T2K and NOvA.

From all existing data it follows that CC and NC neutrino interactions are the Standard Model interactions. Are neutrinos in the SM massless or massive particles? From our point of view this fundamental question is still open. However, there exist strong arguments that the *SM neutrinos are massless two-component particles.*

1. Masses of all particles generated by the standard Higgs mechanism are proportional to the Higgs vacuum expectation value $v = (\sqrt{2}G_F)^{-1/2} \simeq 246\,\text{GeV}$. If neutrino masses m_i are generated by the Higgs mechanism in this case we have

$$m_i = y_i v \quad (i = 1, 2, 3), \tag{13.1}$$

where y_i are dimensionless Yukawa couplings which can be determined from the relation (13.1) if neutrino masses are known. At present absolute values of neutrino masses are unknown. From neutrino oscillation data and existing upper bounds on neutrino masses we can conclude, however, that the heaviest neutrino mass m_3 is in the range $5 \cdot 10^{-2} \leq m_3 \leq 1\,\text{eV}$. Thus for y_3 we have the range

$$2 \cdot 10^{-13} \leq y_3 \leq 4 \cdot 10^{-12}. \tag{13.2}$$

For other particles of the third generation we have

$$y_t \simeq 0.7, \quad y_b \simeq 1.7 \cdot 10^{-2}, \quad y_\tau \simeq 0.7 \cdot 10^{-2}. \tag{13.3}$$

From comparison of (13.2) and (13.3) we conclude that *it is very unlikely that neutrino masses are of the same SM origin as masses of leptons and quarks.*

2. In the framework of such general principles as local gauge invariance, unification of the weak and electromagnetic interactions and spontaneous symmetry breaking *in Standard Model the simplest possibilities are realized* (local $SU_L(2) \times U_Y(1)$ symmetry with lepton and quarks left-handed doublets and right-handed singlets etc.). *Massless, two-component, left-handed SM neutrinos is the simplest, most economical possibility.*

Thus, it is very plausible that neutrino masses are generated by a new beyond the Standard Model mechanism. The simplest and the most economical mechanism was prosed by Weinberg in 1979. The Weinberg mechanism of the neutrino mass generation is equivalent to the seesaw mechanism. It is based on the assumption that at a scale Λ, much larger than the electroweak scale v, the total lepton number L is violated. As a result neutrinos ν_i ($i = 1, 2, 3$) with masses m_i are Majorana particles. The Majorana masses m_i are given by the expression

$$m_i = \bar{y}_i \frac{v^2}{\Lambda}, \tag{13.4}$$

where $\bar{y}_i$ are dimensionless (unknown) constants.

Expressions (13.1) and (13.4) are significantly different: in (13.4) enters additional factor

$$\frac{v}{\Lambda} = \frac{\text{electroweak scale}}{\text{scale of a new Physics}} \tag{13.5}$$

which can naturally explain why neutrino masses are so much smaller than lepton and quark masses.

In order to confirm such a beyond the SM mechanism of the neutrino mass generation it is necessary to prove that L is not conserved and neutrino with definite masses are Majorana particles. The most sensitive experiments on the test of the lepton number violation are experiments on the search for neutrinoless double β-decay of even-even nuclei

$$(A, Z) \to (A, Z + 2) + e^- + e^-. \tag{13.6}$$

If L is violated $0\nu\beta\beta$-decay is allowed but its probability is extremely small. This is connected with the fact that

- the process (13.6) is the second order in the Fermi constant G_F process,
- the probability of the $0\nu\beta\beta$-decay is proportional to very small effective Majorana mass $|m_{\beta\beta}|^2 = |\sum_i U_{ei}^2 m_i|^2$.

Many experiments on the search for the neutrinoless double β-decay were performed. Up to now no indications in favor of the $0\nu\beta\beta$-decay were obtained. From existing data follows that $|m_{\beta\beta}| \leq (1.4 - 4.5) \, 10^{-1} \, \text{eV}$. New more precise experiments with about one ton detectors are in preparation. In future experiments the sensitivity $|m_{\beta\beta}| \simeq$ a few $10^{-2} \, \text{eV}$ will be reached.

Data of experiments on the investigation of neutrino oscillations are perfectly described in the framework of the three-neutrino mixing. Neutrino transition probabilities are characterized in this case by six parameters: atmospheric Δm_A^2 and solar Δm_S^2 mass-squared differences, three mixing angles $\theta_{12}, \theta_{23}, \theta_{13}$ and CP phase δ. From analysis of the existing data first five neutrino oscillations parameters have been determined with accuracies from $\sim 2\%$ ($\Delta m_{S,A}^2$) to $\sim 11\%$ ($\sin^2 \theta_{23}$). Important aim of future neutrino oscillation experiments (T2K, NOvA, JUNO, RENO-50, DUNE and others) is to measure these parameters with $\sim 1\%$ accuracy.

Measurement of the angle θ_{13} in the T2K, Daya Bay, RENO and Double Chooze experiments opened the way of the solution of the following fundamental problems of the three-neutrino mixing

1. What is three-neutrino mass spectrum, Normal or Inverted?
2. Does CP violated in the lepton sector and what is the precise value of the phase δ?

These challenging problems will be resolved in the accelerator T2K, NOvA, DUNE, reactor JUNO and RENO-50, atmospheric Hyper-Kamiokande, IceCube, PINGU, KM3NeT and other future experiments.

The next problem, which will be solved in the nearest years, is the problem of *the transitions of the flavor neutrinos into sterile states?* At the moment exist indications in favor of "anomalous" neutrino oscillations obtained in several short-baseline experiments. The first and the most detailed indications in favor of sterile neutrinos were obtained in the accelerator LSND experiment in which short-baseline transitions $\bar{\nu}_\mu \rightarrow \bar{\nu}_e$ were observed. Many years later indications in favor of $\bar{\nu}_\mu \rightarrow \bar{\nu}_e$ transitions were obtained in another accelerator MiniBooNE experiments. Recently data of old short-baseline reactor experiments were reanalyzed with a new reactor neutrino fluxes. From this analysis indications in favor of sterile neutrinos were also found (reactor neutrino anomaly). Finally indications in favor of transition of ν_e's to the sterile neutrino states were obtained in the GALLEX and SAGE source experiments.

In order to explain indications in favor of sterile neutrinos we need to assume that the number of neutrinos with definite masses is larger than the number of the flavor neutrinos (three) and that additional mass-squared difference (or differences) is much larger than Δm_A^2. For example, the results of the LSND experiment can be explained if $\Delta m_{14}^2 = 1.2 \ \mathrm{eV}^2$ and $sin^2 2\theta_{14} = 3 \cdot 10^{-3}$ (best-fit values).

Let us notice that

- From global analysis of data of all short baseline experiments a disagreement (tension) between appearance and disappearance results was found.
- In recent IceCube, Daya Bay, MINOS and NEOS experiments no indications in favor of transitions of flavor neutrinos into sterile states were obtained. However, these data can not exclude the whole region of parameters which is allowed by previous experiments.

The problem of the existence of sterile neutrinos is one of the most urgent problem of modern neutrino physics. Let us stress that if the minimal, beyond the Standard Model, lepton number violating Weinberg mechanism of neutrino mass generation is realized, it must be no light sterile neutrinos.

At the moment many new reactor, accelerator and source neutrino short-baseline experiments on the search for transitions of flavor neutrinos into sterile states are in preparation. There is no doubts that the problem of existence of light sterile neutrinos will be solved in nearest years.

At present only upper bounds of the absolute values of neutrino masses are known from Mainz and Troitsk tritium experiments and from cosmological data. With new tritium and other experiments (KATRIN, Project-8 and others) and with new more precise cosmological data a large progress in our information about neutrino masses is expected.

Summarizing, the discovery of neutrino oscillations, driven by small neutrino mass-squared differences and neutrino mixing, opened a new exciting field of the investigation of a beyond the Standard Model Physics. It will take many years of research and development of new technologies in order to answer fundamental questions of modern neutrino physics.

Appendix A
Diagonalization of a Complex Matrix

Let us consider a hermitian operator $\hat{A}$. Eigenstates and eigenvalues of the operator $\hat{A}$ are given by the equation

$$\hat{A} \, |i\rangle = a_i \, |i\rangle \, . \tag{A.1}$$

We will assume that the states $|i\rangle$ are normalized

$$\langle i|i\rangle = 1. \tag{A.2}$$

From the condition $\hat{A} = (\hat{A})^\dagger$ follows that $a_i = a_i^*$ and that states belonging to different eigenvalues are orthogonal

$$\langle i'|i\rangle = 0, \quad a_{i'} \neq a_i. \tag{A.3}$$

Further the states $|i\rangle$ form a full system. Thus we have

$$\sum_i |i\rangle\langle i| = 1. \tag{A.4}$$

From (A.1) and (A.4) we easily find

$$\hat{A} = \sum_i |i\rangle \, a_i \, \langle i|. \tag{A.5}$$

Let $|\alpha\rangle$ be another normalized and orthogonal full system of states:

$$\langle \alpha'|\alpha\rangle = \delta_{\alpha'\alpha}. \tag{A.6}$$

© Springer International Publishing AG, part of Springer Nature 2018
S. Bilenky, *Introduction to the Physics of Massive and Mixed Neutrinos*,
Lecture Notes in Physics 947, https://doi.org/10.1007/978-3-319-74802-3

From (A.5) we have

$$\langle \alpha' | \hat{A} | \alpha \rangle = \sum_i \langle \alpha' | i \rangle \, a_i \, \langle i | \alpha \rangle. \qquad (A.7)$$

Further, taking into account that

$$\langle i | \alpha \rangle = \langle \alpha | i \rangle^* \qquad (A.8)$$

we can rewrite the relation (A.7) in the following matrix form

$$A = U \, a \, U^\dagger, \qquad (A.9)$$

where A is the matrix with the matrix elements $\langle \alpha' | \hat{A} | \alpha \rangle$, $a_{i'i} = a_i \, \delta_{i'i}$ and

$$U_{\alpha i} = \langle \alpha | i \rangle. \qquad (A.10)$$

It is easy to see that U is an unitary matrix. In fact, we have

$$\sum_i \langle \alpha | i \rangle \langle i | \alpha' \rangle = \sum_i U_{\alpha i} U^\dagger_{i\alpha'} = (U \, U^\dagger)_{\alpha \alpha'} = \delta_{\alpha \alpha'}. \qquad (A.11)$$

From (A.1) we find the following equation for eigenvalues and eigenfunctions of the matrix A

$$\sum_{\alpha'} A_{\alpha \alpha'} \, u^i_{\alpha'} = a_i \, u^i_\alpha, \qquad (A.12)$$

where $\langle \alpha | i \rangle = u^i_\alpha$. Equation (A.13) has nonzero solution if the condition

$$\mathrm{Det}(A - a) = 0 \qquad (A.13)$$

is satisfied. This equation determines the eigenvalues of the matrix A.

The result (A.9) is very well known from quantum mechanics: hermitian matrix can be bring to the diagonal form with the help of a unitary transformation. In order to present in the standard form mass terms of leptons and quarks, generated by the SM Higgs mechanism, and also the Dirac neutrino mass term we need to bring to the diagonal form *arbitrary complex matrix*. We will present here a simple method of the diagonalization of a general, complex $n \times n$ matrix M. It is obvious that $M \, M^\dagger$ is a hermitian matrix. In fact, we have $(M \, M^\dagger)^\dagger = M \, M^\dagger$. Thus the matrix $M \, M^\dagger$ can be presented in the form

$$M \, M^\dagger = U \, m^2 \, U^\dagger. \qquad (A.14)$$

Here U is a unitary matrix and $m_{ik}^2 = m_i^2 \, \delta_{ik}$, where m_i^2 is the eigenvalue of the matrix $M \, M^\dagger$. The eigenvalues m_i^2 and the matrix U can be found from the solution of the equation

$$\sum_{\alpha'} (M \, M^\dagger)_{\alpha\alpha'} \, \chi_{\alpha'}^i = m_i^2 \, \chi_\alpha^i. \qquad (A.15)$$

We have $U_{\alpha i} = \chi_\alpha^i$. It is easy to see that $m_i^2 > 0$. In fact, we have

$$m_i^2 = (\chi^i)^\dagger M \, M^\dagger \chi^i = \sum_\alpha |\sum_{\alpha'} (\chi_{\alpha'}^i)^* M_{\alpha'\alpha}|^2 > 0. \qquad (A.16)$$

The matrix M can always be presented in the form

$$M = U \, m \, V^\dagger, \qquad (A.17)$$

where $m_i = +\sqrt{m_i^2}$ and[1]

$$V^\dagger = m^{-1} \, U^\dagger \, M. \qquad (A.18)$$

We will show now that the matrix V is an unitary matrix. From (A.18) we find

$$V = M^\dagger \, U \, m^{-1}. \qquad (A.19)$$

Further from (A.18), (A.19) and (A.14) we have

$$V^\dagger \, V = m^{-1} \, U^\dagger \, M \, M^\dagger \, U \, m^{-1} = m^{-1} \, U^\dagger \, U \, m^2 \, U^\dagger \, U \, m^{-1} = 1. \qquad (A.20)$$

Thus, we have shown that a complex $n \times n$ nonsingular matrix M can be diagonalized by the *bi-unitary transformation (A.17)* and presented the way how matrices U, m and V can be found.

---

[1] We assumed that all eigenvalues of the matrix $M \, M^\dagger$ are different from zero. Thus, the diagonal matrix m^{-1} does exists.

Appendix B
Diagonalization of a Complex Symmetrical Matrix

In the case of the Majorana and the Dirac and Majorana mass terms mixing matrices are symmetric. We will consider in this Appendix the diagonalization of a general, $n \times n$ complex, symmetric matrix

$$M = M^T . \tag{B.1}$$

We have shown in Appendix A that any complex matrix M can be presented in the form

$$M = V_1 \, m \, V_2^\dagger , \tag{B.2}$$

where $V_{1,2}$ are unitary matrices and $m_{ik} = m_i \, \delta_{ik}$, $m_i > 0$. From (B.2) it follows that

$$M^T = V_2^{\dagger\,T} \, m \, V_1^T . \tag{B.3}$$

From (B.2) and (B.3) we have

$$M \, M^\dagger = V_1 \, m^2 \, V_1^\dagger , \qquad M^T \, M^{T\,\dagger} = V_2^{\dagger\,T} \, m^2 \, V_2^T . \tag{B.4}$$

Taking into account that M is a symmetrical matrix, from (B.4) we find

$$V_1 \, m^2 \, V_1^\dagger = V_2^{\dagger\,T} \, m^2 \, V_2^T . \tag{B.5}$$

From this relation it follows that

$$V_2^T \, V_1 \, m^2 = m^2 \, V_2^T \, V_1 , \tag{B.6}$$

© Springer International Publishing AG, part of Springer Nature 2018
S. Bilenky, *Introduction to the Physics of Massive and Mixed Neutrinos*,
Lecture Notes in Physics 947, https://doi.org/10.1007/978-3-319-74802-3

i.e. that the commutator of the matrix $V_2^T V_1$ and the diagonal matrix m^2 is equal to zero. We assume that $m_i \neq m_k$ for all $i \neq k$. From (B.6) it follows in this case that $V_2^T V_1$ is a diagonal matrix. Further, taking into account that $V_2^T V_1$ is a unitary matrix we conclude that

$$V_2^T V_1 = S(\alpha) , \tag{B.7}$$

where

$$S_{ik}(\alpha) = e^{i \alpha_i} \delta_{ik} . \tag{B.8}$$

From (B.7) it follows that the matrices V_2 and V_1 are connected by the following relation

$$V_2^\dagger = S^*(\alpha) V_1^T . \tag{B.9}$$

Finally from (B.2) and (B.9) we find

$$M = U \, m \, U^T , \tag{B.10}$$

where $U = V_1 (S^*(\alpha))^{1/2} = V_1 S^*(\frac{\alpha}{2})$ is a unitary matrix. Thus, we have proved that a complex, symmetrical $n \times n$ matrix can be diagonalized with the help of *one unitary matrix*.

Appendix C
Diagonalization of a Real Symmetrical 2 × 2 Matrix

In the simplest case of two neutrinos in matter we need to diagonalize the effective Hamiltonian which (after subtracting the trace of H) has the form

$$\mathscr{H} = \begin{pmatrix} -a & b \\ b & a \end{pmatrix},$$
(C.1)

where a and b are real quantities. For the eigenfunctions and the eigenvalues of the matrix $\mathscr{H}$ we have the following equation

$$\mathscr{H}\, u_i = E_i\, u_i .$$
(C.2)

The eigenvalues E_i can be found from the equation

$$\mathrm{Det}(\mathscr{H} - E) = 0 .$$
(C.3)

Obviously we find

$$E_{1,2} = \mp\sqrt{a^2 + b^2} .$$
(C.4)

Further, we have

$$\mathscr{H} = O\, E\, O^T ,$$
(C.5)

where O is a real orthogonal 2×2 matrix which has the following general form

$$O = \begin{pmatrix} \cos\theta & \sin\theta \\ -\sin\theta & \cos\theta \end{pmatrix}.$$
(C.6)

© Springer International Publishing AG, part of Springer Nature 2018
S. Bilenky, *Introduction to the Physics of Massive and Mixed Neutrinos*,
Lecture Notes in Physics 947, https://doi.org/10.1007/978-3-319-74802-3

From (C.4)–(C.6) we find the following equations for the angle θ

$$a = \sqrt{a^2 + b^2} \, \cos 2\theta, \quad b = \sqrt{a^2 + b^2} \, \sin 2\theta . \tag{C.7}$$

From these relations we find

$$\tan 2\theta = \frac{b}{a} , \quad \cos 2\theta = \frac{a}{\sqrt{a^2 + b^2}} . \tag{C.8}$$

From these relations follow that if diagonal element of the Hamiltonian vanishes $(a = 0)$ in this case

$$\theta = \pi/4 \ (\text{maximal mixing}) \tag{C.9}$$

and $E_2 - E_1$ reaches the minimum. Notice that $a = 0$ is the condition for the MSW resonance in matter.

References

1. J. Chadwick, Intensitätavertailung im magnetischen Spektren der β-Strahlen von Radium B+C. Verh. Deutsch. Phys. Ges. **16**, 383 (1914)
2. C.D. Ellis, W.A. Wooster, Proc. R. Soc. A **117**, 109 (1927)

The Discovery of the Neutron

3. J. Chadwick, Possible existence of a neutron. Nature **192**, 312 (1932)

The First Theory of the β-Decay

4. E. Fermi, An attempt of a theory of beta radiation. Z. Phys. **88**, 161 (1934)
5. G. Gamow, E. Teller, Selection rules for the β-disintegration. Phys. Rev. **49**, 895 (1936)
6. F. Perrin, C. R. **197**, 1625 (1933)

Majorana Neutrino

7. E. Majorana, Teoria simmetrica dell'elettrone e del positrone. Nuovo Cim. **14**, 171 (1937)
8. G. Racah, On the symmetry of particle and antiparticle. Nuovo Cim. **14**, 322 (1937)

Neutrinoless Double β-Decay

9. W.H. Furry, On transition probabilities in double beta-disintegration. Phys. Rev. **56**, 1184 (1939)

Intermediate Vector Boson

10. O. Klein, in *Proceedings of Symposium on Les Nouvelles Theories de la Physique, Warsaw* (World Scientific, Singapore, 1991); Reprinted in O. Klein Memorial Lectures, vol. 1, ed. by G. Ekspong

Radiochemical Method of the Neutrino Detection

11. B. Pontecorvo, Inverse β-process. Report PD-205, Chalk River Laboratory (1946)

Neutrino Discovery

12. F. Reines, C.L. Cowan, Detection of the free neutrino. Phys. Rev. **92**, 830 (1953)
13. F. Reines, C.L. Cowan, The neutrino. Nature **178**, 446 (1956)
14. F. Reines, C.L. Cowan, Free anti-neutrino absorption cross-section. 1: Measurement of the free anti-neutrino absorption cross-section by protons. Phys. Rev. **113**, 273 (1959)

© Springer International Publishing AG, part of Springer Nature 2018

S. Bilenky, *Introduction to the Physics of Massive and Mixed Neutrinos*,
Lecture Notes in Physics 947, https://doi.org/10.1007/978-3-319-74802-3

Discovery of the Parity Violation in the Weak Interaction

15. T.D. Lee, C.N. Yang, Question of parity conservation in weak interactions. Phys. Rev. **104**, 254 (1956)
16. C.S. Wu et al., An experimental test of parity conservation in beta decay. Phys. Rev. **105**, 1413 (1957)
17. R.L. Garwin, L.M. Lederman, W. Weinrich, Observation of the failure of conservation of parity and charge conjugation in meson decays: the magnetic moment of the free muon. Phys. Rev. **105**, 1415 (1957)
18. V.L. Telegdi, A.M. Friedman, Nuclear emulsion evidence for parity nonconservation in the decay chain $\pi^+ - \mu^+ - e^+$. Phys. Rev. **105**, 1681 (1957)

Two-Component Neutrino Theory

19. L.D. Landau, On the conservation laws for weak interactions. Nucl. Phys. **3**, 127 (1957)
20. T.D. Lee, C.N. Yang, Parity nonconservation and a two component theory of the neutrino. Phys. Rev. **105**, 1671 (1957)
21. A. Salam, On parity conservation and neutrino mass. Nuovo Cim. **5**, 299 (1957)

Measurement of the Neutrino Helicity

22. M. Goldhaber, L. Grodzins, A.W. Sunyar, Helicity of neutrinos. Phys. Rev. **109**, 1015 (1958)

The Current$\times$Current, $V - A$ Theory of the Weak Interaction

23. R.P. Feynman, M. Gell-Mann, Theory of the Fermi interaction. Phys. Rev. **109**, 193 (1958)
24. E.C.G. Sudarshan, R.E. Marshak, Chirality invariance and the universal Fermi interaction. Phys. Rev. **109**, 1860 (1958)

Accelerator Neutrinos and Discovery of the Muon Neutrino

25. B. Pontecorvo, Electron and muon neutrinos. Sov. Phys. JETP **10**, 1236 (1960)
26. M. Schwartz, Feasibility of using high-energy neutrinos to study the weak interactions. Phys. Rev. Lett. **4**, 306 (1960)
27. M.A. Markov, *Neutrino* (Nauka, Moscow, 1964, in Russian); On High-Energy Neutrino Physics preprint JINR-D577 (1960, in Russian)
28. G. Feinberg, Decay of μ-meson in the intermediate-meson theory. Phys. Rev. **110**, 148 (1958)
29. G. Danby, J.-M. Gaillard, K. Goulianos, L.M. Lederman, N. Mistry, M. Schwartz, J. Steinberger, Observation of high-energy neutrino reactions and the existence of two kinds of neutrinos. Phys. Rev. Lett. **9**, 36 (1962)

The Standard Model

30. S.L. Glashow, Partial-symmetries of weak interactions. Nucl. Phys. **22**, 579 (1961)
31. S. Weinberg, A model of leptons. Phys. Rev. Lett. **19**, 1264 (1967)
32. A. Salam, Weak and electromagnetic interactions, in *Proceedings of the Eighth Nobel Symposium*, ed. by N. Svartholm (Wiley, New York, 1968)

The Brout-Englert-Higgs Mechanism of the Spontaneous Symmetry Breaking

33. P.W. Higgs, Broken symmetries and the masses of gauge bosons. Phys. Rev. Lett. **13**, 508 (1964)
34. P.W. Higgs, Broken symmetries, massless particles and gauge fields. Phys. Lett. **12**, 132 (1964)
35. P.W. Higgs, Spontaneous symmetry breakdown without massless bosons. Phys. Rev. **145**, 1156 (1966)
36. F. Englert, R. Brout, Broken symmetry and the mass of gauge vector mesons. Phys. Rev. Lett. **13**, 321 (1964)

37. G.S. Guralnik, C.R. Hagen, T.W.B. Kibble, Global conservation laws and massless particles. Phys Rev. Lett. **13**, 585 (1964)
38. T.W.B. Kibble, Symmetry breaking in nonAbelian gauge theories. Phys. Rev. **155**, 1554 (1967)

CKM Quark Mixing

39. N. Cabibbo, Unitary symmetry and leptonic decays. Phys. Rev. Lett. **10**, 531 (1963)
40. S.L. Glashow, J. Iliopoulos, L. Maiani, Weak interactions with lepton—hadron symmetry. Phys. Rev. D **2**, 1258 (1970)
41. M. Kobayashi, K. Maskawa, CP violation in the renormalizable theory of weak interaction. Prog. Theor. Phys. **49**, 652 (1973)

Discovery of the Neutral Current Processes

42. F.J. Hasert et al., Observation of neutrino-like interactions without muon or electron in the Gargamelle neutrino experiment. Phys. Lett. B **46**, 138 (1973)
43. A.C. Benvenuti et al., Observation of muonless neutrino induced inelastic interactions. Phys. Rev. Lett. **32**, 800 (1974)

Discovery of the $W^{\pm}$ and Z^0 Bosons

44. G. Arnison et al. [UA1 Collaboration], Further evidence for charged intermediate vector bosons at the SPS collider. Phys. Lett. B **129**, 273 (1983)
45. G. Arnison et al. [UA1 Collaboration], Experimental observation of isolated large transverse energy electrons with associated missing energy at s**(1/2) = 540-GeV. Phys. Lett. B **122**, 103 (1983)
46. G. Arnison et al. [UA1 Collaboration], Associated production of an isolated large transverse momentum lepton (electron or muon), and two jets at the CERN p anti-p collider. Phys. Lett. B **147**, 493 (1984)
47. P. Bagnaia et al. [UA2 Collaboration], Evidence for $Z^0 \rightarrow e^+ + e^-$ at the CERN anti-p p collider. Phys. Lett. B **129**, 130 (1983)

Discovery of the Higgs Boson

48. S. Chatrchyan et al. [CMS Collaboration], Observation of a new boson at a mass of 125 GeV with the CMS experiment at the LHC. Phys. Lett. B **716**, 30 (2012)
49. G. Aad et al. [ATLAS Collaboration], Observation of a new particle in the search for the Standard Model Higgs boson with the ATLAS detector at the LHC. Phys. Lett. B **716**, 1 (2012)

PMNS Neutrino Mixing

50. B. Pontecorvo, Mesonium and anti-mesonium. Sov. Phys. JETP **6**, 429 (1957); Zh. Eksp. Teor. Fiz. **33**, 549 (1957)
51. B. Pontecorvo, Inverse β-processes and non conservation of lepton charge. Sov. Phys. JETP **7**, 172 (1958); Zh. Eksp. Teor. Fiz. **34**, 247 (1958)
52. Z. Maki, M. Nakagawa, S. Sakata, Remarks on the unified model of elementary particles. Prog. Theor. Phys. **28**, 870 (1962)

Neutrino Oscillations in Vacuum

53. B. Pontecorvo, Neutrino experiments and the question of the lepton charge conservation. Sov. Phys. JETP **26**, 984 (1968); Zh. Eksp. Teor. Fiz. **53**, 1717 (1967)
54. V.N. Gribov, B. Pontecorvo, Neutrino astronomy and lepton charge. Phys. Lett. B **28**, 493 (1969)
55. S.M. Bilenky, B. Pontecorvo, Quark-lepton analogy and neutrino oscillations. Phys. Lett. B **61**, 248 (1976)

56. H. Fritzsch, P. Minkowski, Vector—like weak currents, massive neutrinos, and neutrino beam oscillations. Phys. Lett. B **62**, 72 (1976)
57. S. Eliezer, A.R. Swift, Experimental consequences of $v_e - v_\mu$ mixing for neutrino beams. Nucl. Phys. B **105**, 45 (1976)
58. S.M. Bilenky, B. Pontecorvo, Again on neutrino oscillations. Nuovo Cim. Lett. **17**, 569 (1976)

Neutrino Transitions in Matter. MSW Effect

59. L. Wolfenstein, Neutrino oscillations in matter. Phys. Rev. D **17**, 2369 (1978)
60. S.P. Mikheev, A.Yu. Smirnov, Resonance enhancement of oscillations in matter and solar neutrino spectroscopy. J. Nucl. Phys. **42**, 913 (1985)
61. S.P. Mikheev, A.Yu. Smirnov, Resonant amplification of neutrino oscillations in matter and solar neutrino spectroscopy. Nuovo Cim. **C9**, 17 (1986)

Seesaw Mechanism

62. P. Minkowski, $\mu \rightarrow e\gamma$ at a rate of one out of 1-billion muon decays?. Phys. Lett. B **67**, 421 (1977)
63. M. Gell-Mann, P. Ramond, R. Slansky, in *Complex Spinors and Unified Theories in Supergravity*, ed. by F. van Nieuwenhuizen, D. Freedman (North Holland, Amsterdam, 1979), p. 315
64. T. Yanagida, Horizontal symmetry and masses of neutrinos, in *Proceedings of the Workshop on Unified Theory and the Baryon Number of the Universe* (KEK, Japan, 1979)
65. S.L. Glashow, NATO Adv. Study Inst. Ser. B Phys. **59**, 687 (1979)
66. R.N. Mohapatra, G. Senjanović, Neutrino masses and mixings in gauge models with spontaneous parity violation. Phys. Rev. D **23**, 165 (1981)

Effective Lagrangian Mechanism

67. S. Weinberg, Baryon and lepton nonconserving processes. Phys. Rev. Lett. **43**, 1566 (1979)

Discovery of Neutrino Oscillations

68. Y. Fukuda et al. [Super-Kamiokande Collaboration], Evidence for oscillation of atmospheric neutrinos. Phys. Rev. Lett. **81**, 1562 (1998)
69. T. Kajita, Nobel lecture: discovery of atmospheric neutrino oscillations. Rev. Mod. Phys. **88**(3), 030501 (2016)
70. Q.R. Ahmad et al. [SNO Collaboration], Measurement of the rate of $v_e + d \rightarrow p + p + e^-$ interactions produced by ^8B solar neutrinos at the Sudbury Neutrino Observatory. Phys. Rev. Lett. **87**, 071301 (2001)
71. Q.R. Ahmad et al. [SNO Collaboration], Direct evidence for neutrino flavor transformation from neutral current interactions in the Sudbury Neutrino Observatory. Phys. Rev. Lett. **89**, 011301 (2002)
72. A.B. McDonald, Nobel lecture: the Sudbury Neutrino Observatory: observation of flavor change for solar neutrinos. Rev. Mod. Phys. **88**(3), 030502 (2016)
73. T. Araki et al. (KamLAND), Measurement of neutrino oscillation with KamLAND: evidence of spectral distortion. Phys. Rev. Lett. **94**, 081801 (2005). hep-ex/0406035
74. K. Eguchi et al. (KamLAND), First results from KamLAND: evidence for reactor anti-neutrino disappearance. Phys. Rev. Lett. **90**, 021802 (2003). hep-ex/0212021

Solar Neutrino Experiments and First Indications in Favor of Neutrino Masses and Mixing

75. R. Davis Jr., D.S. Harmer, K.C. Hoffman, Search for neutrinos from the sun. Phys. Rev. Lett. **20**, 1205 (1968)
76. K.S. Hirata et al. [Kamiokande-II Collaboration], Results from one thousand days of real time directional solar neutrino data. Phys. Rev. Lett. **65**, 1297 (1990)
77. A.I. Abazov et al., Search for neutrinos from sun using the reaction Ga-71 (electron-neutrino e-) Ge-71, Phys. Rev. Lett. **67**, 3332 (1991)

78. V.N. Gavrin et al., Latest results from the Soviet-American gallium experiment. AIP Conf. Proc. **272**, 1101 (2008)
79. P. Anselmann et al. [GALLEX Collaboration], Solar neutrinos observed by GALLEX at Gran Sasso. Phys. Lett. B **285**, 376 (1992)

Measurement of the Angle θ_{13}

80. K. Abe et al. [T2K Collaboration], Indication of electron neutrino appearance from an accelerator-produced off-axis muon neutrino beam. Phys. Rev. Lett. **107**, 041801 (2011)
81. F.P. An et al. [Daya Bay Collaboration], Improved measurement of electron antineutrino disappearance at Daya bay. Chin. Phys. C **37**, 011001 (2013)
82. J.K. Ahn et al. [RENO Collaboration], Observation of reactor electron antineutrino disappearance in the RENO experiment. Phys. Rev. Lett. **108**, 191802 (2012)
83. Y. Abe et al. [Double Chooz Collaboration], Indication of reactor $\bar{\nu}_e$ disappearance in the double chooz experiment. Phys. Rev. Lett. **108**, 131801 (2012)

Neutrino Experiments

Atmospheric Neutrino Experiments

84. K. Abe et al. (Super-Kamiokande), Atmospheric neutrino oscillation analysis with external constraints in Super-Kamiokande I-IV. arXiv:1710.09126 [hep-ex]
85. E. Richard et al. (Super-Kamiokande), Measurements of the atmospheric neutrino flux by Super-Kamiokande: energy spectra, geomagnetic effects, and solar modulation. Phys. Rev. D **94**, 052001 (2016). arXiv:1510.08127
86. K. Abe et al. (Super-Kamiokande), A measurement of the appearance of atmospheric tau neutrinos by Super-Kamiokande. Phys. Rev. Lett. **110**, 181802 (2013). arXiv:1206.0328
87. R. Wendell et al. (Super-Kamiokande), Atmospheric neutrino oscillation analysis with subleading effects in Super-Kamiokande I, II, and III. Phys. Rev. D **81**, 092004 (2010). arXiv:1002.3471
88. I.Y. Ashie et al. (Super-Kamiokande), Evidence for an oscillatory signature in atmospheric neutrinooscillation. Phys. Rev. Lett. **93**, 101801 (2004). arXiv:hep-ex/0404034
89. M.G. Aartsen et al. (IceCube), Measurement of atmospheric neutrino oscillations at 6-56 GeV with IceCube DeepCore. arXiv:1707.07081 [hep-ex]
90. M.G. Aartsen et al. (IceCube), Measurement of the ν_μ energy spectrum with IceCube-79. arXiv:1705.07780
91. M.G. Aartsen et al. (IceCube), Determining neutrino oscillation parameters from atmospheric muon neutrino disappearance with three years of IceCube DeepCore data. Phys. Rev. D **91**, 072004 (2015). arXiv:1410.7227
92. M.G. Aartsen et al. (IceCube), Measurement of atmospheric neutrino oscillations with IceCube. Phys. Rev. Lett. **111**, 081801 (2013). arXiv:1305.3909

Solar Neutrino Experiments

93. M. Agostini et al. (Borexino), Improved measurement of ^{8}B solar neutrinos with 1.5 kt y of Borexino exposure. arXiv:1709.00756 [hep-ex]
94. M. Agostini et al. (Borexino), First simultaneous precision spectroscopy of pp, ^{7}Be, and pep solar neutrinos with borexino phase-II. arXiv:1707.09279 [hep-ex]
95. O.Y. Smirnov et al. (Borexino), Measurement of neutrino flux from the primary proton-proton fusion process in the sun with borexino detector. Phys. Part. Nucl. **47**, 995 (2016). arXiv:1507.02432
96. G. Bellini et al. (Borexino), Neutrinos from the primary proton-proton fusion process in the sun. Nature **512**, 383 (2014)
97. G. Bellini et al. (Borexino), Final results of borexino phase-I on low energy solar neutrino spectroscopy. Phys. Rev. D **89**, 112007 (2014). arXiv:1308.0443
98. G. Bellini et al. (Borexino), First evidence of pep solar neutrinos by direct detection in borexino. Phys. Rev. Lett. **108**, 051302 (2012). arXiv:1110.3230

99. G. Bellini et al. (Borexino), Precision measurement of the ^{7}Be solar neutrino interaction rate in Borexino. Phys. Rev. Lett. **107**, 141302 (2011). arXiv:1104.1816

100. B. Aharmim et al. (SNO), Combined analysis of all three phases of solar neutrino data from the Sudbury Neutrino Observatory. Phys. Rev. C **88**, 025501 (2013). arXiv:1109.0763

101. B. Aharmim et al. (SNO), Measurement of the ν_e and total ^{8}B solar neutrino fluxes with the Sudbury Neutrino Observatory phase-III data set. Phys. Rev. C **87**, 015502 (2013). arXiv:1107.2901

102. B. Aharmim et al. (SNO), Low energy threshold analysis of the phase I and phase II data sets of the Sudbury Neutrino Observatory. Phys. Rev. C **81**, 055504 (2010). arXiv:0910.2984

103. B. Aharmim et al. (SNO), An independent measurement of the total active ^{8}B solar neutrino flux using an array of ^{3}He proportional counters at the Sudbury Neutrino Observatory. Phys. Rev. Lett. **101**, 111301 (2008). arXiv:0806.0989

104. A.B. McDonald, The Sudbury Neutrino Observatory: observation of flavor change for solar neutrinos. Ann. Phys. **528**, 469 (2016)

105. K. Abe et al. (Super-Kamiokande), Solar neutrino measurements in Super-Kamiokande-IV. Phys. Rev. D **94**, 052010 (2016). arXiv:1606.07538

106. A. Renshaw et al. (Super-Kamiokande), First indication of terrestrial matter effects on solar neutrino oscillation. Phys. Rev. Lett. **112**, 091805 (2014). arXiv:1312.5176

107. K. Abe et al. (Super-Kamiokande), Solar neutrino results in Super-Kamiokande-III. Phys. Rev. D **83**, 052010 (2011). arXiv:1010.0118

108. J.P. Cravens et al. (Super-Kamiokande), Solar neutrino measurements in Super-Kamiokande-II. Phys. Rev. D **78**, 032002 (2008). arXiv:0803.4312

109. J. Hosaka et al. (Super-Kamkiokande), Solar neutrino measurements in Super-Kamiokande-I. Phys. Rev. D **73**, 112001 (2006). hep-ex/0508053

110. M. Altmann et al. (GNO), Complete results for five years of GNO solar neutrino observations. Phys. Lett. B **616**, 174 (2005). arXiv:hep-ex/0504037

111. W. Hampel et al. (GALLEX), GALLEX solar neutrino observations: results for GALLEX IV. Phys. Lett. B **447**, 127 (1999)

112. J.N. Abdurashitov et al. (SAGE), Measurement of the solar neutrino capture rate by the Russian-American gallium solar neutrino experiment during one half of the 22-year cycle of solar activity. J. Exp. Theor. Phys. **95**, 181 (2002). arXiv:astro-ph/0204245

113. Y. Fukuda et al. (Kamiokande), Solar neutrino data covering solar cycle 22. Phys. Rev. Lett. **77**, 1683 (1996)

114. K. Lande et al. (Homestake), The homestake solar neutrino program. Nucl. Phys. Proc. Suppl. **77**, 13 (1999)

115. B.T. Cleveland et al. (Homestake), Measurement of the solar electron neutrino flux with the homestake chlorine detector. Astrophys. J. **496**, 505 (1998)

Accelerator Neutrino Experiments

116. P. Adamson et al. (MINOS), Combined analysis of ν_μ disappearance and $\nu_\mu \rightarrow \nu_e$ appearance in MINOS using accelerator and atmospheric neutrinos. Phys. Rev. Lett. **112**, 191801 (2014). arXiv:1403.0867

117. P. Adamson et al. (MINOS), Electron neutrino and antineutrino appearance in the full MINOS data sample. Phys. Rev. Lett. **110**, 171801 (2013). arXiv:1301.4581

118. P. Adamson et al. (MINOS), An improved measurement of muon antineutrino disappearance in MINOS. Phys. Rev. Lett. **108**, 191801 (2012). arXiv:1202.2772

119. K. Abe et al. (T2K), Measurement of neutrino and antineutrino oscillations by the T2K experiment including a new additional sample of ν_e interactions at the far detector. arXiv:1707.01048

120. K. Abe et al. (T2K), Updated T2K measurements of muon neutrino and antineutrino disappearance using 1.5e21 protons on target. Phys. Rev. D **96**, 011102 (2017). arXiv:1704.06409

121. K. Abe et al. (T2K), Sensitivity of the T2K accelerator-based neutrino experiment with an extended run to 20×10^{21} POT. arXiv:1607.08004

122. K. Abe et al. (T2K), Measurements of neutrino oscillation in appearance and disappearance channels by the T2K experiment with 6.6E20 protons on target. Phys. Rev. D **91**, 072010 (2015). arXiv:1502.01550

123. K. Abe et al. (T2K), Precise measurement of the neutrino mixing parameter θ_{23} from muon neutrino disappearance in an off-axis beam. Phys. Rev. Lett. **112**, 181801 (2014). arXiv:1403.1532

124. P. Adamson et al. (NOvA), Measurement of the neutrino mixing angle θ_{23} in NOvA. Phys. Rev. Lett. **118**, 151802 (2017). arXiv:1701.05891

125. P. Adamson et al. (NOvA), First measurement of muon-neutrino disappearance in NOvA. Phys. Rev. D **93**, 051104 (2016). arXiv:1601.05037

126. P. Adamson et al. (NOvA), First measurement of electron neutrino appearance in NOvA. Phys. Rev. Lett. **116**, 151806 (2016). arXiv:1601.05022

127. M.H. Ahn et al. (K2K), Measurement of neutrino oscillation by the K2K experiment. Phys. Rev. D **74**, 072003 (2006). hep-ex/0606032

KamLAND Reactor Neutrino Experiment

128. A. Gando et al. (KamLAND), Reactor on-off antineutrino measurement with KamLAND. Phys. Rev. D **88**, 033001 (2013). arXiv:1303.4667 [hep-ex]

129. S. Abe et al. (KamLAND), Precision measurement of neutrino oscillation parameters with KamLAND. Phys. Rev. Lett. **100**, 221803 (2008). arXiv:0801.4589 [hep-ex]

Reactor Neutrino Experiments

130. F.P. An et al. (Daya Bay), Measurement of electron antineutrino oscillation based on 1230 days of operation of the Daya Bay experiment. Phys. Rev. D **95**, 072006 (2017). arXiv:1610.04802

131. F.P. An et al. (Daya Bay), New measurement of θ_{13} via neutron capture on hydrogen at Daya Bay. Phys. Rev. D **93**, 072011 (2016). arXiv:1603.03549

132. F.P. An et al. (Daya Bay), A new measurement of antineutrino oscillation with the full detector configuration at Daya Bay. Phys. Rev. Lett. **115**, 111802 (2015). arXiv:1505.03456

133. F.P. An et al. (Daya Bay), Spectral measurement of electron antineutrino oscillation amplitude and frequency at Daya Bay. Phys. Rev. Lett. **112**, 061801 (2014). arXiv:1310.6732

134. S.H. Seo et al. (RENO), Spectral measurement of the electron antineutrino oscillation amplitude and frequency using 500 live days of RENO data. arXiv:1610.04326 [hep-ex]

135. J.H. Choi et al. (RENO), Observation of energy and baseline dependent reactor antineutrino disappearance in the RENO experiment. Phys. Rev. Lett. **116**, 211801 (2016). arXiv:1511.05849

136. S.-B. Kim et al. (RENO), Observation of reactor electron antineutrino disappearance in the RENO experiment. Phys. Rev. Lett. **108**, 191802 (2012). arXiv:1204.0626

137. Y. Abe et al. (Double Chooz), Measurement of θ_{13} in Double Chooz using neutron captures on hydrogen with novel background rejection techniques. J. High Energy Phys. **01**, 163 (2016). arXiv:1510.08937 [hep-ex]

138. Y. Abe et al. (Double Chooz), Improved measurements of the neutrino mixing angle θ_{13} with the Double Chooz detector. J. High Energy Phys. **10**, 086 (2014). arXiv:1406.7763 [hep-ex]; Erratum: JHEP 02, 074 (2015)

139. Y. Abe et al. (Double Chooz), Background-independent measurement of θ_{13} in Double Chooz. Phys. Lett. B **735**, 51 (2014). arXiv:1401.5981 [hep-ex]

Neutrinoless Double β-Decay

140. C. Alduino et al. (CUORE), First results from CUORE: a search for lepton number violation via $0\nu\beta\beta$ decay of ^{130}Te. arXiv:1710.07988 [nucl-ex]

141. K. Alfonso et al. (CUORE), Search for neutrinoless double-beta decay of ^{130}Te with CUORE-0. Phys. Rev. Lett. **115**, 102502 (2015). arXiv:1504.02454

142. J.B. Albert et al., Search for neutrinoless double-beta decay with the upgraded EXO-200 detector. arXiv:1707.08707 [hep-ex]

143. J.B. Albert et al. (EXO-200), Searches for double beta decay of ^{134}Xe with EXO-200. arXiv:1704.05042

144. J.B. Albert et al. (EXO-200), Search for majorana neutrinos with the first two years of EXO-200 data. Nature **510**, 229 (2014). arXiv:1402.6956

145. M. Agostini et al. (GERDA), Background free search for neutrinoless double beta decay with GERDA Phase II. arXiv:1703.00570

146. M. Agostini et al. (GERDA), Limit on the radiative neutrinoless double electron capture of ^{36}Ar from GERDA phase I. Eur. Phys. J. C **76**, 652 (2016). arXiv:1605.01756

147. A.M. Bakalyarov et al. (C03-06-23.1), Results of the experiment on investigation of germanium-76 double beta decay. Experimental data of Heidelberg-Moscow collaboration November 1995–August 2001. Phys. Part. Nucl. Lett. **2**, 77 (2005). hep-ex/0309016

148. C.E. Aalseth et al. (IGEX), The IGEX Ge-76 neutrinoless double-beta decay experiment: prospects for next generation experiments. Phys. Rev. D **65**, 092007 (2002), hep-ex/0202026

149. A. Gando et al. (KamLAND-Zen), Search for majorana neutrinos near the inverted mass hierarchy region with KamLAND-Zen. Phys. Rev. Lett. **117**, 082503 (2016). arXiv:1605.02889

150. A. Gando et al. (KamLAND-Zen), Limit on neutrinoless betabeta decay of Xe-136 from the first phase of KamLAND-Zen and comparison with the positive claim in Ge-76. Phys. Rev. Lett. **110**, 062502 (2013). arXiv:1211.3863

151. R. Arnold et al., Search for neutrinoless quadruple-β decay of ^{150}Nd with the NEMO-3 detector. Phys. Rev. Lett. **119**, 041801 (2017). arXiv:1705.08847

152. R. Arnold et al. (NEMO-3), Measurement of the $2\nu\beta\beta$ decay half-life and search for the $0\nu\beta\beta$ decay of ^{116}Cd with the NEMO-3 detector. arXiv:1610.03226

Neutrino Mass Measurement

153. Ch. Kraus et al., Final results from phase II of the Mainz neutrino mass search in tritium β decay. Eur. Phys. J. C **40**, 447 (2005). hep-ex/0412056

154. V.N. Aseev et al. (Troitsk), An upper limit on electron antineutrino mass from Troitsk experiment. Phys. Rev. D **84**, 112003 (2011). arXiv:1108.5034 [hep-ex]

155. V.N. Aseev et al. (Troitsk), An upper limit on electron antineutrino mass from Troitsk experiment. Phys. Rev. D **84**, 112003 (2011). arXiv:1108.5034 [hep-ex]

156. A. Osipowicz et al. (KATRIN), KATRIN: a next generation tritium beta decay experiment with sub-eV sensitivity for the electron neutrino mass. hep-ex/0109033

157. J. Angrik et al. (KATRIN), KATRIN design report 2004 (2005)

158. F. Fraenkle, Status of the neutrino mass experiments KATRIN and Project 8. PoS **EPS-HEP2015**, 084 (2015)

159. R.G. Hamish Robertson (KATRIN), KATRIN: an experiment to determine the neutrino mass from the beta decay of tritium, 2013 Snowmass. arXiv:1307.5486 [physics]

Global Analysis of Neutrino Data

160. M.C. Gonzalez-Garcia, M. Maltoni, T. Schwetz, Global analyses of neutrino oscillation experiments. Nucl. Phys. B **908**, 199 (2016). arXiv:1512.06856

161. F. Capozzi, E. Lisi, A. Marrone, D. Montanino, A. Palazzo, Status and prospects of global analyses of neutrino mass-mixing parameters. J. Phys. Conf. Ser. **888**(1), 012037 (2017)

162. F. Capozzi, E. Lisi, A. Marrone, D. Montanino, A. Palazzo, Neutrino masses and mixings: status of known and unknown 3ν parameters. Nucl. Phys. B **908**, 218 (2016). https://doi.org/10.1016/j.nuclphysb.2016.02.016. arXiv:1601.07777 [hep-ph]

163. F. Capozzi, E. Di Valentino, E. Lisi, A. Marrone, A. Melchiorri, A. Palazzo, Global constraints on absolute neutrino masses and their ordering. Phys. Rev. D **95**(9), 096014 (2017). https://doi.org/doi:10.1103/PhysRevD.95.096014. arXiv:1703.04471 [hep-ph]

164. P.F. de Salas, D.V. Forero, C.A. Ternes, M. Tortola, J.W.F. Valle, Status of neutrino oscillations (2017). arXiv:1708.01186 [hep-ph]

Future Experiments

165. B. Abi et al. (DUNE), The single-phase ProtoDUNE technical design report. arXiv:1706.07081 [physics]
166. J. Strait et al. (DUNE), Long-Baseline Neutrino Facility (LBNF) and Deep Underground Neutrino Experiment (DUNE) Conceptual Design Report Volume 3: Long-Baseline Neutrino Facility for DUNE June 24, 2015. arXiv:1601.05823 [physics]
167. R. Acciarri et al. (DUNE), Long-Baseline Neutrino Facility (LBNF) and Deep Underground Neutrino Experiment (DUNE) Conceptual Design Report Volume 2: The Physics Program for DUNE at LBNF. arXiv:1512.06148 [physics]
168. F. An et al. (JUNO), Neutrino physics with JUNO. J. Phys. G **43**, 030401 (2016). arXiv:1507.05613 [physics]
169. J. Migenda [Hyper-Kamiokande Proto- Collaboration], Astroparticle physics in Hyper-Kamiokande. arXiv:1710.08345 [physics.ins-det]
170. M. Yokoyama [Hyper-Kamiokande Proto Collaboration], The Hyper-Kamiokande experiment. arXiv:1705.00306 [hep-ex]

Reviews

Neutrino Oscillations

171. P. Huber, Prospects for neutrino oscillation parameters. PoS **NOW2016**, 025 (2017). arXiv:1612.04843
172. P. Hernandez, *Neutrino Physics*. arXiv:1708.01046 [hep-ph]; 8th CERN-Latin-American School of High-Energy Physics (CLASHEP2015): Ibarra, Ecuador, March 05-17, 2015
173. S. Bilenky, Neutrino oscillations: from an historical perspective to the present status. Nucl. Phys. B **908**, 2 (2016). arXiv:1602.00170
174. Y. Wang, Z.-Z. Xing, Neutrino masses and flavor oscillations. Adv. Ser. Direct. High Energy Phys. **26**, 371 (2016). arXiv:1504.06155
175. P. Vogel, L.J. Wen, C. Zhang, Neutrino oscillation studies with reactors. Nat. Commun. **6**, 6935 (2015). arXiv:1503.01059
176. S.J. Parke, Neutrinos: theory and phenomenology. Phys. Scr. T **158**, 014013 (2013). arXiv:1310.5992; *Nobel Symposium on LHC results, May 13–19, 2013* (Krusenberg, Uppsala)
177. A. de Gouvea et al. (Intensity Frontier Neutrino Working Group), *Neutrinos*. arXiv:1310.4340
178. M.C. Gonzalez-Garcia, M. Maltoni, Phenomenology with massive neutrinos. Phys. Rep. **460**, 1 (2008). arXiv:0704.1800
179. A. Strumia, F. Vissani, Neutrino masses and mixings and..., hep-ph/0606054
180. R.D. McKeown, P. Vogel, Neutrino masses and oscillations: triumphs and challenges. Phys. Rep. **394**, 315 (2004). arXiv:hep-ph/0402025
181. W.M. Alberico, S.M. Bilenky, Neutrino oscillations, masses and mixing. Phys. Part. Nucl. **35**, 297 (2004). arXiv:hep-ph/0306239
182. M.C. Gonzalez-Garcia, Y. Nir, Neutrino masses and mixing: evidence and implications. Rev. Mod. Phys.**75**, 345 (2003). arXiv:hep-ph/0202058
183. C. Giunti, M. Laveder, Neutrino mixing, in *Developments in Quantum Physics—2004*, ed. by F. Columbus, V. Krasnoholovets (Nova Science Publishers, Hauppauge, 2003), pp. 197–254. arXiv:hep-ph/0310238
184. V. Barger, D. Marfatia, K. Whisnant, Progress in the physics of massive neutrinos. Int. J. Mod. Phys. E **12**, 569 (2003). arXiv:hep-ph/0308123
185. S.M. Bilenky, C. Giunti, W. Grimus, Phenomenology of neutrino oscillations. Prog. Part. Nucl. Phys. **43**, 1 (1999). arXiv:hep-ph/9812360
186. P. Fisher, B. Kayser, K.S. McFarland, Neutrino mass and oscillation. Ann. Rev. Nucl. Part. Sci. **49**, 481 (1999). arXiv:hep-ph/9906244
187. S.M. Bilenky, S.T. Petcov, Massive neutrinos and neutrino oscillations. Rev. Mod. Phys. **59**, 671 (1987)
188. S.M. Bilenky, B. Pontecorvo, Lepton mixing and neutrino oscillations. Phys. Rep. **41**, 22 (1978)

Neutrinoless Double β-Decay

189. J.D. Vergados, H. Ejiri, F. Simkovic, Neutrinoless double beta decay and neutrino mass. Int. J. Mod. Phys. E **25**, 1630007 (2016). arXiv:1612.02924
190. S. Dell'Oro, S. Marcocci, M. Viel, F. Vissani, Neutrinoless double beta decay: 2015 review. Adv. High Energy Phys. **2016**, 2162659 (2016). arXiv:1601.07512
191. J.J. Gomez-Cadenas, J. Martin-Albo, Phenomenology of neutrinoless double beta decay. PoS **GSSI14**, 004 (2015). arXiv:1502.00581; Gran Sasso Summer Institute 2014 Hands-On Experimental Underground Physics at LNGS (GSSI14), September 2014
192. H. Pas, W. Rodejohann, Neutrinoless double beta decay. New J. Phys. **17**, 115010 (2015). arXiv:1507.00170
193. S.M. Bilenky, C. Giunti, Neutrinoless double-beta decay: a probe of physics beyond the standard model. Int. J. Mod. Phys. A **30**, 0001 (2015). arXiv:1411.4791
194. F.T. Avignone III, S.R. Elliott, J. Engel, Double beta decay, majorana neutrinos, and neutrino mass. Rev. Mod. Phys. **80**, 481 (2008). arXiv:0708.1033
195. M. Doi, T. Kotani, E. Takasugi, Double beta decay and majorana neutrino. Prog. Theor. Phys. Suppl. **83**, 1 (1985)

Solar Neutrinos

196. M. Maltoni, A.Yu. Smirnov, Solar neutrinos and neutrino physics. Eur. Phys. J. A **52**, 87 (2016). arXiv:1507.05287
197. F. Vissani, Solar neutrino physics on the beginning of 2017. arXiv:1706.05435
198. V. Antonelli, L. Miramonti, C. Pena-Garay, A. Serenelli, Solar neutrinos. Adv. High Energy Phys. **2013**, 351926 (2013). arXiv:1208.1356
199. M. Blennow, A. Yu. Smirnov, Neutrino propagation in matter. Adv. High Energy Phys. **2013**, 972485 (2013). arXiv:1306.2903
200. T.K. Kuo, J. Pantaleone, Neutrino oscillations in matter. Rev. Mod. Phys. **61**, 937 (1989)

Neutrino in Cosmology

201. A.B. Balantekin, G.M. Fuller, Neutrinos in cosmology and astrophysics. Prog. Part. Nucl. Phys. **71**, 162 (2013). arXiv:1303.3874
202. G. Hinshaw et al., Five year Wilkinson Microwave Anisotropy Probe (WMAP) observations: data processing, sky maps, basic results. Astrophys. J. Suppl. **180**, 225 (2009). arXiv:0803.0732
203. W.C. Haxton, *Neutrino Astrophysics* (2008). arXiv:0808.0735
204. J. Lesgourgues, S. Pastor, Massive neutrinos and cosmology. Phys. Rep. **429**, 307 (2006). arXiv:astro-ph/0603494

Leptogenesis

205. C.S. Fong, E. Nardi, A. Riotto, Leptogenesis in the universe. Adv. High Energy Phys. **2012**, 158303 (2012). arXiv:1301.3062 [hep-ph]
206. W. Buchmuller, Leptogenesis: theory and neutrino masses. Nucl. Phys. Proc. Suppl. **235–236**, 329 (2013). arXiv:1210.7758 [hep-ph]; XXV International Conference on Neutrino Physics, Kyoto
207. P. Di Bari, Developments in leptogenesis. Nucl. Phys. B, Proc. Suppl. **229–232**, 305 (2012). arXiv:1102.3409 [hep-ph]; Neutrino 2010
208. P. Di Bari, An introduction to leptogenesis and neutrino properties. Contemp. Phys. **53**, 315 (2012). arXiv:1206.3168 [hep-ph]
209. S. Davidson, E. Nardi, Y. Nir, Leptogenesis. Phys. Rep. **466**, 105 (2008). arXiv:0802.2962 [hep-ph]
210. W. Buchmuller, P. Di Bari, M. Plumacher, Some aspects of thermal leptogenesis. New J. Phys. **6**, 105 (2004). hep-ph/0406014

Models of Neutrino Masses and Mixing

211. G. Altarelli, Status of neutrino mass and mixing. Int. J. Mod. Phys. A **29**, 1444002 (2014). arXiv:1404.3859; International Conference on Flavor Physics and Mass Generation, Singapore, February 2014

212. S.F. King, A. Merle, S. Morisi, Y. Shimizu, M. Tanimoto, Neutrino mass and mixing: from theory to experiment. New J. Phys. **16**, 045018 (2014). arXiv:1402.4271

Books

Neutrino Physics

213. V. Barger, D. Marfatia, K. Whisnant, *The Physics of Neutrinos* (Princeton University Press, Princeton, 2012). ISBN 978-0691128535

214. Z.-Z. Xing, S. Zhou, *Neutrinos in Particle Physics, Astronomy and Cosmology* (Zhejiang University Press, Hangzhou, 2011). ISBN 978-7308080248

215. S. Bilenky, *Introduction to the Physics of Massive and Mixed Neutrinos*. Lecture Notes in Physics, vol. 817 (Springer, Berlin, 2010). ISBN 978-3-642-14042-6

216. F. Close, *Neutrino* (Oxford University Press, Oxford, 2010)

217. C. Giunti, C.W. Kim, *Fundamentals of Neutrino Physics and Astrophysics* (Oxford University Press, Oxford, 2007). ISBN 978-0-19-850871-7

218. R.N. Mohapatra, P.B. Pal, *Massive Neutrinos in Physics and Astrophysics*, 3rd edn. Lecture Notes in Physics, vol. 72 (World Scientific, Singapore, 2004)

219. M. Fukugita, T. Yanagida, *Physics of Neutrinos and Applications to Astrophysics* (Springer, Berlin, 2003)

220. K. Zuber, *Neutrino Physics*. Series in High Energy Physics (Institute of Physics Publishing, Bristol, 2003)

221. C.W. Kim, A. Pevsner, *Neutrinos in Physics and Astrophysics*. Contemporary Concepts in Physics, vol. 8 (Harwood Academic Press, Reading, 1993)

222. F. Boehm, P. Vogel, *Physics of Massive Neutrinos* (Cambridge University Press, Cambridge, 1992)

223. K. Winter (ed.), *Neutrino Physics* (Cambridge University Press, Cambridge, 1991)

224. B. Kayser, F. Gibrat-Debu, F. Perrier, *The Physics of Massive Neutrinos*. Lecture Notes in Physics, vol. 25 (World Scientific, Singapore, 1989)

225. J.N. Bahcall, *Neutrino Astrophysics* (Cambridge University Press, Cambridge, 1989)

Weak Interaction. The Standard Model

226. C. Burgess, G. Moore, *The Standard Model: A Primer* (Cambridge University Press, Cambridge, 2007)

227. D.A. Bromley, *Gauge Theory of Weak Interactions* (Springer, Berlin, 2000)

228. S.M. Bilenky, *Introduction to Feynman Diagrams and Electroweak Interactions Physics* (Editions Frontières, Gif-sur-Yvette, 1994)

229. P. Renton, *Electroweak Interactions: An Introduction to the Physics of Quarks and Leptons* (Cambridge University Press, Cambridge, 1990)

230. F. Halzen, M.D. Alan, *Quarks and Leptons: An Introductory Course in Modern Particle Physics* (Wiley, Hoboken, 1984)

231. E.D. Commins, P.H. Bucksbaum, *Weak Interactions of Leptons and Quarks* (Cambridge University Press, Cambridge, 1983)

232. S.M. Bilenky, *Introduction to the Physics of Electroweak Interactions* (Pergamon Press, Oxford, 1982)

233. L.B. Okun, *Leptons and Quarks* (Elsevier, Amsterdam, 1982)

234. R.E. Marshak, Riazuddin, C.P. Ryan, *Theory of Weak Interactions in Particle Physics* (Wiley, Hoboken, 1969)

Cosmology

235. J. Lesgourgues, G. Mangano, G. Miele, S. Pastor, *Neutrino Cosmology* (Cambridge University Press, Cambridge, 2013). ISBN 9781139012874
236. S. Weinberg, *Cosmology* (Oxford University Press, Oxford, 2008)
237. L. Bergstrom, A. Goobar, *Cosmology and Particle Astrophysics* (Springer, Berlin, 2004)
238. D.H. Perkins, *Particle Astrophysics* (Oxford University Press, Oxford, 2004)
239. E.V. Lindner, *First Principles of Cosmology* (Addison-Wesley, Boston, 1997)
240. E.W. Kolb, M.S. Terner, *The Early Universe*. Frontiers in Physics (Addison-Wesley, Boston, 1990)

Index

© Springer International Publishing AG, part of Springer Nature 2018
S. Bilenky, *Introduction to the Physics of Massive and Mixed Neutrinos*,
Lecture Notes in Physics 947, https://doi.org/10.1007/978-3-319-74802-3

Printed in the United States
By Bookmasters